COMPREHENSIVE SERIES IN PHOTOCHEMISTRY AND PHOTOBIOLOGY

Series Editors

Donat P. Häder
Professor of Botany

and

Giulio Jori
Professor of Chemistry

European Society for Photobiology

COMPREHENSIVE SERIES IN PHOTOCHEMISTRY AND PHOTOBIOLOGY

Series Editors: Donat P. Häder and Giulio Jori

Titles in this Series

COMPREHENSIVE SERIES IN PHOTOCHEMISTRY AND PHOTOBIOLOGY – VOLUME 5

From DNA Photolesions to Mutations, Skin Cancer and Cell Death

Editors

Évelyne Sage
UMR2027 CNRS/Institut Curie
Institut Curie, Bat. 110
Centre Universitaire
Orsay, France
F-91405

Régen Drouin
Service of Genetics
Department of Paediatrics
Faculty of Medicine and Health Sciences
University of Sherbrooke
3001, 12th Avenue North
Sherbrooke, Quebec, J1H 5N4
Canada

Mahmoud Rouabhia
Faculté de médecine dentaire
Université Laval, Québec, QC
Canada
G1K 7P4

RSC Publishing

ISBN 0-85404-326-8

A catalogue record for this book is available from the British Library

Published by The Royal Society of Chemistry,
Thomas Graham House, Science Park, Milton Road,
Cambridge CB4 0WF, UK

Registered Charity Number 207890

For further information see our web site at www.rsc.org

Typeset by Alden Bookset, Northampton, UK
Printed by Biddles Ltd, King's Lynn, Norfolk, UK

Dedication

This book is dedicated to Professor Raymond Latarjet (1911–1998), member of the French Académie des Sciences, who was one of the Fathers and driving forces of modern photobiology. He was a co-founder of the Comité International de Photobiologie (previous Comité International de la Lumière from 1928 to 1951). His name is associated with the Institut du Radium (now Institut Curie) in Paris, where he built one of the first UV-radiometers (see the photograph). Who, as a photobiologist, did not use once in his life the Latarjet dosimeter?

This book is also dedicated to Professor Claude Hélène (1938–2003), member of the French Académie des Sciences. He began his scientific career in Photochemistry with his PhD (1966) on "Energy transfer and Photochemical Reactions in Nucleic Acids" and distinguished himself as an efficient promotor of photobiology. One of his major contributions to photochemistry was the description of the cleaving catalytic activity of the tripeptide Lys-Trp-Lys, the "smallest specific endonuclease", through intercalation at an apurinic site (Nature 1981). Surprisingly enough, this peptide was also able to photosensitize the thymine dimer cleavage, thus mimicking the photoreactivation enzyme (Photochem. Photobiol., 1977). He was a member of the Editorial board of Photochemistry and Photobiology from 1974 to 1986.

Preface for the ESP series in photochemical and photobiological sciences

"Its not the substance, it's the dose which makes something poisonous!" When Paracelsius, a German physician of the 14th century made this statement he probably did not think about light as one of the most obvious environmental factors. But his statement applies as well to light. While we need light for example for vitamin D production too much light might cause skin cancer. The dose makes the difference. These diverse findings of light effects have attracted the attention of scientists for centuries. The photosciences represent a dynamic multidisciplinary field which includes such diverse subjects as behavioral responses of single cells, cures for certain types of cancer and the protective potential of tanning lotions. It includes photobiology and photochemistry, photomedicine as well as the technology for light production, filtering and measurement. Light is a common theme in all these areas. In recent decades a more molecular centered approach changed both the depth and the quality of the theoretical as well as the experimental foundation of photosciences.

An example of the relationship between global environment and the biosphere is the recent discovery of ozone depletion and the resulting increase in high energy ultraviolet radiation. The hazardous effects of high energy ultraviolet radiation on all living systems is now well established. This discovery of the result of ozone depletion put photosciences at the center of public interest with the result that, in an unparalleled effort, scientists and politicians worked closely together to come to international agreements to stop the pollution of the atmosphere.

The changed recreational behavior and the correlation with several diseases in which sunlight or artificial light sources play a major role in the causation of clinical conditions (e.g. porphyrias, polymorphic photodermatoses, Xeroderma pigmentosum and skin cancers) have been well documented. As a result, in some countries (e.g. Australia) public services inform people about the potential risk of extended periods of sun exposure every day. The problems are often aggravated by the phototoxic or photoallergic reactions produced by a variety of environmental pollutants, food additives or therapeutic and cosmetic drugs. On the other hand, if properly used, light-stimulated processes can induce important beneficial effects in biological systems, such as the elucidation of several aspects of cell structure and function. Novel developments are centered around photodiagnostic and phototherapeutic modalities for the treatment of cancer, artherosclerosis, several autoimmune diseases, neonatal jaundice and others. In addition, classic research areas such as vision and photosynthesis are still very active. Some of these developments are unique to photobiology, since the peculiar physico-chemical properties of electronically excited biomolecules often lead to the promotion of reactions which are characterized by high levels of selectivity in space and time. Besides the biologically centered areas, technical developments have paved the way for the harnessing of solar energy to produce warm water and electricity or the development of environmentally friendly techniques for addressing problems of large social impact (e.g. the decontamination of polluted waters). While also in use in Western countries, these techniques are of great interest for developing countries.

The European Society for Photobiology (ESP) is an organization for developing and coordinating the very different fields of photosciences in terms of public knowledge and scientific interests. Due to the ever increasing demand for a comprehensive overview of the photosciences the ESP decided to initiate an encyclopedic series, the "Comprehensive Series in Photochemical and Photobiological Sciences". This series is intended to give an in-depth coverage over all the very different fields related to light effects. It will allow investigators, physicians, students, industry and laypersons to obtain an updated record of the state-of-the-art in specific fields, including a ready access to the recent literature. Most importantly, such reviews give a critical evaluation of the directions that the field is taking, outline hotly debated or innovative topics and even suggest a redirection if appropriate. It is our intention to produce the monographs at a sufficiently high rate to generate a timely coverage of both well established and emerging topics. As a rule, the individual volumes are commissioned; however, comments, suggestions or proposals for new subjects are welcome.

Donat-P. Häder and Giulio Jori
Spring 2002

Volume preface

For 5 thousand million years Sun has been providing Earth with heat and light. Who would not love to feel sun beams on skin by a sunny spring day after a long winter! Our leisure society has developed outdoor activities and sunbathing fashion, with, as a consequence, a dramatic increase in skin carcinomas and melanomas in wealthy countries from the north to south hemispheres. The photons of sunlight generate a series of genetic events in skin, leading to cancer. One of the very first event is the formation of DNA photolesions, which, by misfortune, may evolve into somatic mutations, when not eliminated by DNA repair mechanisms. The balance between clonal expansion of a single mutated cell and cell death is a key parameter in skin carcinogenesis. A strong knowledge of the basic phenomena underlying all these events is of major importance. We were a bunch of photobiologists from both sides of Atlantic to wish to gather the best scientists in the field, in order to provide the state of art on the question. We took advantage of the meeting of the *American Society for Photobiology*, held in Québec city (Canada) on July 13–17, 2002 to develop this idea. This is how was « born » this monograph of the *Comprehensive series in Photosciences* based on the contributions of imminent scientists in the field. This book comprises 18 chapters that present the current knowledge on many aspects of the basic processes occurring from the production of DNA photolesions to skin cancer. This book is intended to serve as a source of information on several key topics, which should be useful for graduate students as well as for senior photobiologists interested in a survey of basis in photocarcinogenesis.

The first chapter by Cadet et al. is a survey of recent aspects of cellular DNA photochemistry which could be assessed due to major progress in the development of new, acurate and sensitive methods for DNA damage detection, such as HPLC-MS/MS (liquid chromatography tandem mass spectrometry). The authors also pointed out the lack of information on two types of DNA damage potentially induced by UVA, i.e. DNA-protein crosslinks and those resulting from breakdown products of lipid peroxides. In chapter 2, Davis et al. propose chemical sequencing method for screening new compounds as photosensitizers of DNA damage. It is shown how, from DNA damage profiles photoinduced by various drugs used in therapy, the potency and the mode of photosensitization of these compounds can be assessed. Chapter 3 by Sage et al. puts the emphasis on a role for UVA radiation in solar mutagenesis by investigating UVA-induced DNA damage using immunochemical and enzymatic methods, as well as LC-MS/MS and Ligation-mediated PCR. *Via* these technical developments, Sage et al. helped to demonstrate that cyclobutane pyrimidine dimers, mainly TT dimers, and not 8-oxoguanine or other oxidative DNA damage, are the major class of lesions induced by UVA radiation. Chapter 4 by Pfeifer et al. reviews mutation spectra produced by simulated or natural solar light, underlines the role of cytosine deamination in UV mutagenesis and shows that cyclobutane pyrimidine dimers are the main culprit. While it is well established that UV-induced mutations arise as a consequence of DNA synthesis past a dipyrimidine photoproduct by DNA polymerases, it is only recently that the polymerases has been identified. With the outstanding review by Taylor in chapter 5, we penetrate into the intimacy of replicative and translesional synthesis polymerases when passing a photoproduct.

Follows a series of reports on DNA repair. Among solar UV-induced photo-lesions, 8-oxoguanine is an important one, in particular in the case of UVA as opposed to UVB (second most abundant lesion after exposure of mammalian cells to UVA radiation). Using *Saccharomyces cerevisiae*, Boiteux, in chapter 6, described the interplay of Ogg1 protein with other repair and replication pathways into a model which explains how several repair pathways can act in synergy to prevent the mutagenic effect of 8-oxoguanine in eukaryotic cells. Yeast *Saccharomyces cerevisiae* is a good model organism to identify novel eukaryotic genes implicated in the cellular response to DNA damage. This is illustrated in chapter 7 (by Ramotar et al.) which focus on the *yPTPA1* gene, encoding a putative phosphatase activator, which may play a role in signaling the oxidative stress response. It is proposed that in response to 4-NQO/UVA, yPtpa1 becomes activated, and after several phosphorylation/dephosphorylation of effectors, activates a kinase, which then phosphorylates histones to cause their release from the DNA, thus promoting gene expression or allowing DNA repair. Importantly, *ptpa1* strain is the first reported yeast mutant hypersensitive to UVA. It provides an important tool to study the mechanism of UVA defense in eukaryotic cells. Bulky DNA adducts, like dipyrimidine photoproducts induced by UV radiation, are repaired by nucleotide excision repair (NER). NER pathway comprises transcription-coupled repair (TCR), which rapidly repairs actively transcribing genes, and global genomic repair (GGR), which repairs non-transcribed sequences. Chapter 8 is an extensive review on transcription-coupled repair by Conconi et al., with an emphasis on the repair in RNA polymerase I transcribed gene. Until recently, it was believed that transcription-coupled repair operated exclusively on RNA polymerase II transcribed genes, with the RNA polymerase II transcription factor TFIIH providing a link between transcription and repair. However, when repair in rDNA genes actively transcribed by RNA polymerase I was re-examined in UVC-irradiated yeast, transcription-coupled repair of those genes was observed and likely depended on a Rad26 (human CSB gene)-independent pathway. Whether or not transcription-coupled repair of ribosomal genes occur in mammalian cells is still an open question. The next chapter by Volker and Mullenders present an overview of recent findings regarding nucleotide excision repair (GGR pathway) *in vivo*. It is a fascinating story on how the key players interact, associate at a site of DNA lesion within chromatin and enters into the game to play their role and fulfill accurate repair. As reported by the authors, it turns out that nucleotide excision repair is a highly dynamic process *in vivo*. The way it is organized provides the system speed, flexibility and accuracy. Recent findings led to the notion that XPA is not a damage recognition factor, but rather a 'damage verification protein' involved at a later stage of NER to ascertain that the DNA covered by the NER complex indeed harbors a lesion. Then is raised the question of repair in differentiated cells. Indeed, terminally differentiated cells do not generally repair their bulk DNA as efficiently as their undifferentiated precursors, presumably because mutagenic replication of damaged DNA is no longer an issue. In Chapter 10, Oh and Yeh show that over 24 hours, differentiated keratinocytes in culture repair dipyrimidine photoproducts in their global genome as efficiently as the undifferentiated precursors or normal fibroblasts. It is not clear yet if TCR functions in those keratinocytes. In skin repetitively exposed to solar UV radiation, a differentiating keratinocyte of the epidermis may need to repair not only genes that are necessary for its actual layer-specific state, but also genes that will be necessary for its future roles in the upper layers. Insufficient repair would lead to apoptosis, at the detriment of an efficient epidermal barrier formation.

Whether or not a phenomenon equivalent to Weigle reactivation for bacteriophage and the SOS response in bacteria exist in mammalian cells has been a matter of debate for thirty years. In chapter 11, based on experiments on reactivation of UV-damaged viruses and reporter genes in mammalian cells, Rainbow et al. propose a model for UVinducible DNA repair that includes a p53 dependent transcriptional activation of one or more genes involved in nucleotide excision repair. It is known that the activity of the p53 tumour suppressor increases in response to UVirradiation and that p53 stimulates the repair of UV-induced DNA damage, presumably by regulating the level of expression of genes such as XPC and DDB2. However, as pointed out by McKay et al. in chapter 12, most repair occurs while transcription of these gene is inhibited by blocking DNA damage. Thus, the authors raise the questions: « Why would human cells utilize a DNA damage-inducible transcription factor to regulate the basal levels of expression of DNA repair genes? Why would cells increase the expression of p53 regulated DNA repair genes after repair is complete? » They speculate that p53 could participate in an adaptive response to chronic daily exposure to UV light. This would assure an obvious benefit for skin cells and it is relevant on an evolutionary point of view. Is functional p53 required for GGR or for both GGR and TCR of cyclobutane pyrimidine dimers, the major photoproduct produced by 254 nm UVC and polychromatic UVB (290–320-nm) ? Desnoyers et al. in Chapter 13 reconcile conflicting results and show that it is a question of wavelength of the inducing light! They elegantly demonstrate that functional p53 is required for GGR and TCR following cell irradiation with UVB, but in cells exposed to 254 nm UVC, it is essential for GGR and completely dispensable for TCR. These results highlight the risk to employ the "nonsolar" model mutagen 254-nm UV rather than environmentally-relevant UV sources, when investigating the basis of sunlight-induced mutation and cancer. The next chapter by Cheung and Li discusses on the exact relationship between p53 and NER, i.e. direct involvement in DNA repair, *vs* implication through transcriptional gene regulation. It has recently been shown that $p33^{ING1}$, a tumor suppressor gene, cooperate with p53 in a variety of tumor suppressive functions. The authors now demonstrate that $p33^{ING1}$ cooperates with p53 to enhance NER.

Cells have evolved intricate defense mechanisms to counteract the harmful effects of UV irradiation; DNA repair is in first line. Higher organisms use apoptosis as a second line of defense to eliminate cells that have been irreversibly damaged by solar UV radiation. The best known example of UV-induced apoptosis is the formation of sunburn cells, i.e., apoptotic keratinocytes in the skin overexposed to sunlight. Chapter 15 by Vodenicharov and Shah reviews the major nuclear (*via* DNA damage and p53 activation) and non-nuclear signaling events that initiate apoptosis in UV irradiated cells. Non-nuclear events that initiate apoptosis in the UV damaged cells can be Fas or TNF receptor aggregation at the cell surface and activation in the cytoplasm of MAP kinase family members, such as JNK and p38 MAP kinase. Then comes a new concept in chapter 16 by Zhang and Brash: UV-induced apoptosis has opposing roles in early skin cancer. The authors first review the evidence that elimination of UV-damaged cells by apoptosis is an epidermal cancer prevention mechanism. Then, they describe how they came to the idea that UV-induced apoptosis may also *facilitate* the development of skin cancer. They conclude that: UV mutates the *p53* gene (as well as others such as *PTCH*); a *p53* mutation confers apoptosis-resistance; clonal expansion of apoptosis-resistant cells is driven by a UV-induced physiological event, not a second mutation; and this physiological event may be apoptosis of surrounding normal cells. Chapter 17 by Ehrhart et al. reviews the consequences of acquired activation of signaling pathways in skin tumours, in particular $P14^{ARF}$, $P16^{INK4A}$ and Sonic Hedgehog pathways. Those genes, in addition to p53, SMO, are highly mutated, but at different extent, in squamous and basal cell carcinomas of *xeroderma pigmentosum* patients. It is likely that activation of one or several of those pathways can seriously alter the checkpoints activated by the cellular response to solar UV radiation. Chapter 18 by Yarosh shows how the chaos theory and self-organized criticality describe the DNA damage signal transduction network. The model helps explain the genesis of the signal transduction network and shifts the paradigm of signal transduction from linear pathways to network interactions.

This book was made possible with the precious support of Mélanie Pierre (Institut Curie).

Evelyne Sage, Régen Drouin, Mahmoud Rouabhia

Contributors

Nathalie Bastien
Service of Genetics
Department of Paediatrics
Faculty of Medicine and Health Sciences
University of Sherbrooke
3001, 12th Avenue North
Sherbrooke, Quebec, J1H 5N4
Canada

Cecilia Becerril
Centre for Cancer Therapeutics
Ottawa Regional Cancer Centre
503 Smyth Rd, 3rd floor
Ottawa Ontario K1H 1C4
Canada

Vyacheslav A. Bespalov
Biochemistry and Biophysics
School of Molecular Biosciences
Washington State University
Pullman, WA 99164-4660
USA

Serge Boiteux
CEA, Département des Sciences de la Vie
Département de Radiobiologie et Radiopathologie
UMR217 CNRS-CEA
"Radiobiologie Moléculaire et Cellulaire"
92265-Fontenay aux Roses
France
E-mail: boiteux@dsvidf.cea.fr

Ihor P. Boszko
Department of Biology
McMaster University
Hamilton
Ontario L8S 4K1
Canada

Douglas E. Brash
Departments of Therapeutic Radiology, Dermatology and Genetics
Yale School of Medicine
New Haven, CT 06520
USA
E-mail: douglas.brash@yale.edu

Jean Cadet
Laboratoire "Lésions des Acides Nucléiques"
Service de Chimie Inorganique et Biologique
DRFMC
CEA/Grenoble
F-38054 Grenoble Cedex 9
France
E-mail: jcadet@cea.fr

Colleen Caney
Department of Biology
McMaster University
Hamilton
Ontario L8S 4K1
Canada

K-John Cheung Jr.
Division of Dermatology
Department of Medicine
University of British Columbia
Vancouver
British Columbia V6H 3Z6
Canada

Antonio Conconi
Biochemistry and Biophysics
School of Molecular Biosciences
Washington State University
Pullman, WA 99164-4660
USA

Raphaël M.Culerrier
Instabilité Génétique et Cancer
CNRS UPR 2169
Institut Gustave Roussy-PR2
39 rue Camille Desmoulins
94805 Villejuif
France

Jocelyn David
Maisonneuve-Rosemont Hospital
Centre de Recherche Guy Bernier
5415, Boulevard de l'Assomption
Montreal, Quebec H1T 2M4
Canada

R. Jeremy H. Davies
School of Biology and Biochemistry
Medical Biology Centre
Queen's University,
Belfast BT9 7BL
Northern Ireland UK
E-mail: j.davies@qub.ac.uk

Julie Desnoyers
Faculty of Medicine
University of Montreal
and Guy-Bernier Research Center
Hôpital Maisonneuve-Rosemont
Montréal, Québec H1T 2M4
Canada

Thierry Douki
Laboratoire "Lésions des Acides
Nucléiques"
Service de Chimie Inorganique
et Biologique
CEA/DSM/DRFMC
CEA
F-38054 Grenoble Cedex 9
France

Elliot Drobetsky
Hôpital Maisonneuve-Rosemont
Centre de Recherche Guy Bernier
5415, Boulevard de l'Assomption
Montreal, Quebec, H1T 2M4
Canada
E-mail: elliot.drobetsky@umontreal.ca

Régen Drouin
Service of Genetics,
Departement of Paediatrics
Faculty of Medicine and Health
Sciences
University of Sherbrooke
3001, 12th Avenue North
Sherbrooke, Quebec, J1H 5N4
Canada

Jean-Claude Ehrhart
Instabilité Génétique et Cancer
CNRS UPR 2169
Institut Gustave Roussy-PR2
39 rue Camille Desmoulins
94805 Villejuif
France

Deirdre Fahy
Biochemistry and Biophysics
School of Molecular Biosciences
Washington State University
Pullman, WA 99164-4660
USA

Murray A. Francis
Hospital for Sick Children
8129 Elm Wing
555 University Avenue
Toronto
Ontario M5G 1X8
Canada

Fabien P.Gosselet
Instabilité Génétique et Cancer
CNRS UPR 2169
Institut Gustave Roussy-PR2
39 rue Camille Desmoulins
94805 Villejuif
France

Pál Gróf
Semmelweis University of Medicine
Institute of Biophysics
PO Box 263
H-1444 Budapest
Hungary

Dong-Hyun Lee
Department of Biology
City of Hope
Duarte, CA 91010
USA

Caroline Léger
Faculty of Medicine
University of Montreal and
Guy-Bernier Research Center
Hôpital Maisonneuve-Rosemont
Montréal, Québec H1T 2M4
Canada

Gang Li
Jack Bell Research Center
2660 Oak Street
Vancouver
British Columbia V6H 3Z6
Canada
E-mail: gangli@interchange.ubc.ca

Géraldine Mathonnet
Faculty of Medicine
University of Montreal and
Guy-Bernier Research Center
Hôpital Maisonneuve-Rosemont
Montréal, Québec H1T 2M4
Canada

Bruce C. McKay
Centre for Cancer Therapeutics
Ottawa Regional Cancer Centre
503 Smyth Rd, 3rd floor
Ottawa Ontario K1H 1C4
Canada

Leon H.F. Mullenders
Medical Genetics Center
Department of Toxicogenetics
Leiden University Medical Center
Wassenaarseweg 72
2333 AL, Leiden
The Netherlands
E-mail: l.mullenders@lumc.nl

Dennis H. Oh
Department of Dermatology
University of California, San Francisco
Dermatology Service (190)
San Francisco VA Medical Center
4150 Clement Street
San Francisco, CA 94121
USA
E-mail: dennisoh@itsa.ucsf.edu

Daniel Perdiz
Laboratoire de Santé
Publique-Environnement
Faculté de Pharmacie
Université de Paris XI
F-92296 Chatenay-Malabry
France

Gerd P. Pfeifer
Department of Biology
City of Hope
Duarte, CA 91010
USA
E-mail: gpfeifer@coh.org

Photini Pitsikas
Department of Biology
McMaster University
Hamilton
Ontario L8S 4K1
Canada

Jean-Pierre Pouget
Laboratoire "Lésions des Acides Nucléiques"
Service de Chimie Inorganique et Biologique
DRFMC
CEA/Grenoble
F-38054 Grenoble Cedex 9
France

Andrew J. Rainbow
Department of Biology
McMaster University
Hamilton
Ontario L8S 4K1
Canada
E-mail: rainbow@mcmaster.ca

Dindial Ramotar
Maisonneuve-Rosemont Hospital,
Centre de Recherche, Guy Bernier
5415, Boulevard de l'Assomption
Montreal, Quebec H1T 2M4
Canada
E-mail: dramotar.hmr@ssss.gouv.qc.ca

Jean-Luc Ravanat
Laboratoire "Lésions des Acides Nucléiques"
Service de Chimie Inorganique et Biologique
DRFMC
CEA/Grenoble
F-38054 Grenoble Cedex 9
France

Anne Reynaud-Angelin
UMR 2027 CNRS/Institut Curie
Institut Curie - Bât 110
Centre Universitaire
F- 91405 Orsay Cedex
France

Patrick J. Rochette
Service of Genetics
Department of Paediatrics
Faculty of Medicine and Health Sciences
University of Sherbrooke
3001, 12th Avenue North
Sherbrooke, Quebec, J1H 5N4
Canada

Evelyne Sage
UMR 2027 CNRS/Institut Curie
Institut Curie - Bât 110
Centre Universitaire
F- 91405 Orsay Cedex
France
E-mail: Evelyne.Sage@curie.u-psud.fr

Alain Sarasin
Instabilité Génétique et Cancer
CNRS UPR 2169
Institut Gustave Roussy-PR2
39 rue Camille Desmoulins
94805 Villejuif
France
E-mail: sarasin@igr.fr

Girish M. Shah
Laboratory for Skin Cancer Research
CHUL Research Center (CHUQ)
Faculty of Medicine
Laval University
2705 Laurier Boulevard
Quebec, Québec G1V 4G2
Canada
E-mail: girish.shah@crchul.ulaval.ca

Michael J. Smerdon
Biochemistry and Biophysics
School of Molecular Biosciences
Washington State University
Pullman, WA 99164-4660
USA
E-mail: smerdon@mail.wsu.edu

Jennifer C. Spronck
Centre for Cancer Therapeutics
Ottawa Regional Cancer Centre
503 Smyth Rd, 3rd floor
Ottawa Ontario K1H 1C4
Canada

Sharon M. Starrs
School of Biology and Biochemistry
Medical Biology Centre
Queen's University,
Belfast BT9 7BL
Northern Ireland UK

Clarke S. Stevenson
School of Biology and Biochemistry
Medical Biology Centre
Queen's University,
Belfast BT9 7BL
Northern Ireland UK

John-Stephen Taylor
Department of Chemistry
Campus Box 1134
Washington University
One Brookings Drive
St. Louis, MO 63130
USA
Email: taylor@wustl.edu

Momchil D. Vodenicharov
Laboratory for Skin Cancer Research
CHUL Research Center (CHUQ)
Faculty of Medicine
Laval University
2705 Laurier Boulevard
Quebec, Québec G1V 4G2
Canada

Marcel Volker
Medical Genetics Center
Department of Toxicogenetics
Leiden University Medical Center
Wassenaarseweg 72
2333 AL, Leiden
The Netherlands

Daniel B. Yarosh
AGI Dermatics
205 Buffalo Avenue
Freeport, New York 11520
USA
E-mail: DanYarosh@agiderm.com

Kelvin Yeh
Dermatology Service
University of California, San Francisco
4150 Clement Street
San Francisco, CA 94121
USA

Jung-Hoon Yoon
Department of Biology
City of Hope
Duarte, CA 91010
USA

Young-Hyun You
Lawrence Berkeley
National Laboratory
Berkeley, CA 94720
USA

Wengeng Zhang
Departments of Therapeutic Radiology
and Genetics
Yale School of Medicine
New Haven, CT 06520
USA

Contents

Chapter 1

UVB and UVA induced formation of photoproducts within cellular DNA

Jean Cadet, Thierry Douki, Jean-Pierre Pouget and Jean-Luc Ravanat

Table of contents

Abstract

Emphasis is placed in this short survey on recent aspects of the photochemistry of cellular DNA that involve both the effects of UVB and UVA radiations. Direct excitation of the purine and pyrimidine bases of DNA is known to mostly generate dimeric pyrimidine photoproducts in oxygen independent photoreactions. Interestingly, the twelve possible dimeric photoproducts at the four main bipyrimidine sites can now be singled out as dinucleoside monophosphates. This is achieved using a specific and sensitive assay that associates high performance liquid chromatography to tandem mass spectrometry (HPLC-MS/MS) operating in the electrospray ionization (ESI) mode. Thus, it was found that UVB irradiation of human monocyte cells gives rise predominantly to *cis-syn* cyclobutadithymine, thymine-cytosine pyrimidine(6-4) pyrimidone adduct and related cyclobutyl dimer. In contrast the dimeric photoproducts at (di)cytosine sites are generated in very low yields although characteristic tandem mutations of UV-B irradiation are observed at the latter CC sequences. Further, cytosine photohydrate and Dewar valence isomers of the (6-4) photoproducts are at the best minor UV-B photoproducts. Relevant information on UVA-sensitized oxidative damage to cellular DNA was gained from measurements using chromatographic methods and the modified comet assay. Thus, it was shown that 8-oxo-7,8-dihydro-2′-deoxyguanosine is the predominant DNA oxidation product, as mostly, the result of singlet oxygen oxidation. In addition, oxidized pyrimidine bases and DNA strand breaks whose formation involves •OH radical, are produced in much lower yields. Work is in progress to assess the UVA-induced formation of other markers of oxidative stress. These include on the one hand DNA-protein crosslinks and on the other hand DNA adducts with reactive aldehydes that arise from the breakdown of initially generated lipid peroxides.

1.1 Introduction

Solar UV radiation appears to be the main etiological factor responsible for the induction of skin cancers in human population. It is well established that UVB and UVA radiations act mostly on cellular DNA via direct and photosensitized reactions respectively [for earlier reviews, see 1,2]. Precise assessment of the final products of these photoreactions has been hampered for years by the lack of accurate and quantitative methods of measurement. This particularly concerns the individual determination of dimeric pyrimidine photoproducts including *cis-syn* cyclobutadipyrimidines (CPDs), pyrimidine (6-4) pyrimidone photoadducts (6-4PPs) and related valence Dewar isomers (DewarPPs) for which only limited information was available until recently. However, relevant data on the distribution and repair of the three latter classes of photoproducts within the DNA of plasmids, isolated cells and tissues were gained mostly from serological approaches [3–8]. These include ELISA, RIA and immuno-dot-blot measurements together with immunostaining detection through the availability of monoclonal and polyclonal antibodies [8–12]. We may also mention the recent development of an immunological method aimed at measuring CPDs and 6-4PPs in the DNA of isolated cells in association with the comet assay [13]. Another suitable method that is receiving major attention is the ligation-mediated polymerase chain reaction (LM-PCR) [14–16]. This allows the mapping within DNA at the nucleotide resolution of dimeric pyrimidine photoproducts and particularly of CPDs, the latter lesions being revealed through the nicking activity of T4 endonuclease V. Thus, it was found that methylation of cytosine residues in 5′-CCG and 5′-TCG sequences leads to a 10-fold increase in the UVB formation of CPDs [17]. Interestingly the latter lesions were found to constitute major p53 mutation hot spots in mouse skin tumors [18]. It was also shown that accumulation of dimeric DNA photoproducts takes place at the same locations of the *p53* gene in both human skin and epidermal cells of Hupki (human p53 knock-in) mice [19]. Another striking information inferred from LC-PCR analysis is the predominant implication of CPDs in a vast majority of UVB-induced mutations in mammalian cells [20]. LM-PCR measurement of CPDs in the DNA of the basal layer of engineered human skin led to the conclusion that upper layers of epidermis protected against the genotoxic effect of UVC but not from those of solar UVB radiation [21]. Evidence was also provided, still using LM-PCR detection of CPDs, that human cells either genetically or functionally compromised for p53 function, are defective in both global and transcription-coupled nucleotide excision repair (TCNER) [22]. Interestingly, cells functionally compromised for retinoblastoma tumor suppressor protein function are only defective in TCNER [21]. Preferential repair of CPDs was found to occur in the promoter and quiescent initiation domain of the *CDC2* gene in both quiescent and proliferating human fibroblasts [23].

There is an increasing attention devoted to the assessment of the molecular effects of UVA radiation on DNA in relation with the increased formation of reactive oxygen species (ROS) that for the bulk is mediated by endogenous photosensitizers. This interest is explained, at least, partly by the likely association of UVA with skin cancer risk [24,25] and particularly with skin melanoma incidence whose observation in heavily pigmented hybrids of *Xiphophorus* fish [26] requires, however, further support before to be considered as a suitable model for human. In that respect, UVA does not appear to be a specific mutagen in contrast to UVB that induces a characteristic mutation fingerprint at bipyrimidine sites and more precisely at TC and CC sequences [27]. Thus, the incidence of p53 mutations in UVA-induced skin tumors in hairless mice is very low without any specificity [27]. The relatively high incidence of A:T→T:A point mutations observed on the *LacZ* gene upon exposure of human cells to UVA radiation cannot be correlated with the formation of any known DNA lesions [28,29]. In contrast, the UVA-mediated increase in the frequency of T→G transversions in the *aprt* locus of Chinese hamster ovary cells that was not observed in the nucleotide excision repair-deficient cells may be accounted for by the damaging effects of ROS on DNA [30,31]. It may be pointed out that indirect evidence for the exaltation of the formation of ROS upon UVA irradiation of human skin fibroblasts was provided by the observed induction of heme oxygenase [32] and the release of free iron from ferritin [33]. More direct proofs for the occurrence of oxidation reactions within human and CHO cells upon exposure to UVA radiation was the observed increase in the level of 8-oxo-7,8-dihydroguanine, an ubiquitous biomarker of oxidative processes, in both nuclear DNA and RNA [34–38]. In addition, relevant information on several classes of oxidative damage induced by UVA and visible radiation to cellular DNA was gained from application of the alkaline elution technique that involves the use ot two DNA repair enzymes, namely formamidopyrimiidne DNA *N*-glycosylase (Fpg) and endonuclease III (endo III). Thus it was shown that the formation of Fpg-sensitive sites, likely to mostly involve 8-oxoGua, are predominant with respect to DNA strand breaks and endo III-sensitive sites (mostly oxidized pyrimidine bases) [39–41].

Emphasis is placed in this short survey on recent aspects on the formation of UVB and UVA-mediated damage to cellular DNA that mostly involved dimeric pyrimidine photoproducts and oxidative lesions. The bulk of the measurement of DNA photoproducts was achieved using the recently available HPLC-tandem mass spectrometry technique and the modified comet assay.

1.2 Distribution of UVB radiation-induced dimeric pyrimidine photoproducts within cellular DNA

The advent of the HPLC in the mid 70's together with the availability of new stationary phases including octadecylsilyl silica gel (ODS) packing material has provided a strong impetus to the development of sensitive and highly resolutive analytical method aimed at monitoring the formation of tiny amounts of lesions within cellular DNA. Interestingly, it was shown as early applications that the *cis-syn* isomer of cyclobutadithymine (*c,s* Thy<>Thy) was efficiently separated from the DNA hydrolysis products [42,43]. It should be reminded that the only one available sensitive detection approach at that time was the radioactive measurement of the content of HPLC fractions; however, this was not achieved on line due to the lack of suitable detector. One of the main advantages provided by the HPLC separation on the ODS columns is that thymine is eluted less rapidly than the targeted *c,s* Thy<>Thy, avoiding any contamination of the fractions containing the latter minor photoproduct due to the trailing of [^{3}H]-thymine [42,43]. However, the measurement of *cis-syn* cyclobutane dimer involving cytosine and [^{3}H]-thymine that is released as Ura<>Thy was difficult due to a co-eluting radioactive contaminant, the likely 5-hydroxy-5-methylhydantoin that arises from self-radiolysis process [44]. Interestingly, the assay despite some limitations has received several relevant applications including the assessment of repair kinetic of *c,s* Thy<>Thy in UVC-irradiated of normal and xeroderma pigmentosum fibroblast cells [43]. Subsequently, a suitable HPLC method that does not require radioactive pre-labeling of DNA has become available for monitoring the formation of fluorescent pyrimidine (6-4) pyrimidone photoproducts (6-4PPs).[45] This has required the use of HF-pyridine at room temperature as a mild hydrolytic reagent to obtain a quantitative release of relatively unstable 6-4PPs as nucleobase derivatives. The detection of the latter photoproducts that exhibit a fluorescence emission spectra with maxima around 380 nm upon excitation in the UVB range was achieved at the output of the HPLC column using a fluorescence detector. The distribution of the 6-4PPs including either two thymine or one thymine and a cytosine was assessed in

UVC irradiated DNA. However, the relatively low sensitivity of the steady-state fluorescence detection technique has prevented the application of the assay for monitoring the formation of 6-4PPs within nuclear DNA of UVC or UVB irradiated cells.

Interestingly, most of the limitations that have been encountered for measuring the dimeric pyrimidine photoproducts (vide supra) were overcome using the accurate HPLC-MS/MS method operating in the electrospray ionization (ESI) mode. This method that recently became available has already found interesting applications for the measurement of DNA damage including several oxidized pyrimidine and purine nucleosides and nucleobases [46,47]. The overall strategy involves enzymatic digestion of DNA [48,49] from UVB irradiated cells by a cocktail of several enzymes including 3′-and 5′-exonucleases after the quantitative conversion of Cyt<>Cyt and 5′-end cytosine 6-4PPs into the corresponding uracil derivatives through deamination. Therefore, in one HPLC analytical run it is possible to accurately measure at the output of the chromatographic column the twelve possible bipyrimidine adducts at TT, TC, CT and CC sequences both in isolated and cellular DNA upon exposure to low doses of UVC and UVB radiations [50,51]. Interestingly, the tandem mass spectrometric measurement which is achieved in the sensitive multiple reaction monitoring (MRM) mode provides also a specific way of distinguishing CPD from 6-4PP for a given bipyrimidine site due to the occurrence of a different fragmentation pattern [52]. Similar trends in the photoproduct distribution are observed in isolated and cellular DNA upon either UVC or UVB irradiation. As a first remark it may be noted that the formation of the Dewar valence isomers of 6-4-PPs is barely detectable upon exposure of cellular DNA to biologically relevant doses of UVB radiation. Under the latter conditions, only the Dewar isomer at CC sites (Figure 1) was found to be produced, however, in a very low yield. The three main UVB-induced dimeric

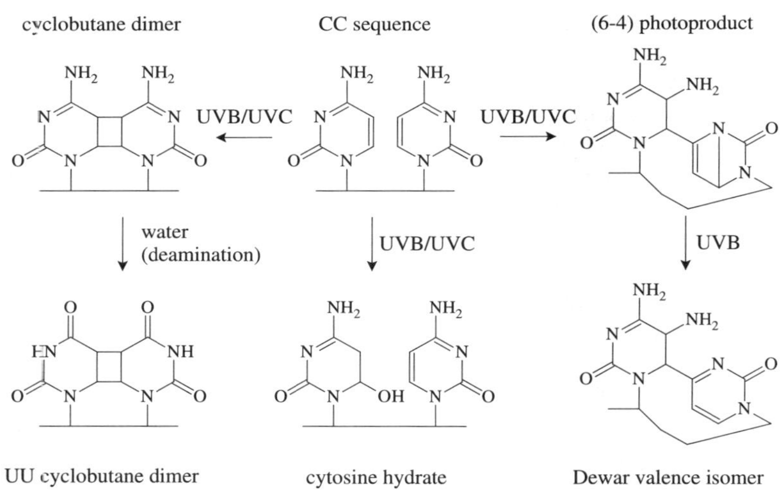

Figure 1. Chemical structure of the main UVB-induced monomeric and dimeric cytosine photoproducts. An example of secondary deamination reaction that may affect 5,6-dihydrocytosine residues is provided for the cyclobutane dimer.

photoproducts appear to be generated in the following decreasing order of importance: *c,s* Thy <> Thy > 6-4PP at TC sequence > Thy <> Cyt (Table 1). In contrast, the CT sites and to a lower extent, the CC sequences exhibit a low photoreactivity as inferred from the low yield formation of both CPDs and 6-4-PPs. It should be reminded that the CC photoproducts although generated with a low efficiency exhibit a high mutagenic potential that leads to the observation of the characteristic UV-induced tandem mutation CC→TT. However, further work is required to definitively establish the nature of the highly mutagenic UVB-induced photolesion(s). As a final remark, it may be stressed that the comparison of the HPLC-MS/MS measurements and LM-PCR analysis of dimeric pyrimidine photoproducts in the DNA of UVC irradiated cells [53] shows that application of the latter method leads to a strong underestimation of the yield of 6-4PPs at TT and TC sites. This is likely due to the low efficiency of the piperidine-mediated conversion of the 6-4PPs into strand cleavage at the 3′-side since only the related valence Dewar isomers appear to be strongly alkali-labile.

1.3 UVC and UVB-induced formation of monomeric base photoproducts

Is was recently confirmed that UVB is able to generate 8-oxoGua in DNA of mouse keratinocytes [54], mouse epidermis [55] and. Chines hamster ovary cells [56]. However, as shown in a comparative study on the UVB-induced-formation of several classes of photodamage to cellular DNA, the contribution of 8-oxoGua is rather low. Thus, the yield of 8-oxoGua measured as the corresponding 2′-deoxyribonucleoside by HPLC-electrochemical detection in the DNA of UVB irradiated CHO cells was 2.1 lesions per 10^6 bases and kJ.m^{-2} [56]. This is about two orders of magnitude lower than the level of CPDs that was assessed by immunodetection. Further work is required to better delineate the mechanisms of 8-oxoGua formation that may involve a Fenton type chemistry, hole migration from initially photo-ionized pyrimidine and adenine bases or singlet oxygen oxidation. In that respect, we may anticipate a notable contribution of $^{\bullet}$OH radical or related reactive oxygen through the Fenton reaction since UVB irradiation of cellular DNA was found to lead to similar yields of strand breaks and Fpg-sensitive sites when detected using the alkaline elution technique [39,40].

Table 1. Distribution of bipyrimidine photoproducts* within DNA of human THP1 monocytes upon exposure to UVB radiation expressed in number of lesions per kJ.m^2 and 10^4 bases (dose range 0–2.6 kJ.m^2)

	TT	TC	CT	CC
CPD	3.147±0.07	1.286±0.047	0.577±0.51	0.279±0.067
6–4PP	0.245±0.007	1.4000±0.034	<0.01	0.062±0.028
DewarPP	<0.01	<0.01	<0.01	<0.03

*From Douki and Cadet [51].

Table 2. Oxidative damage to DNA of human THP1 monocytes upon exposure to UVA and ionizing radiations expressed in number of lesions per either $kJ.m^2$ (dose range 0–2.6 $kJ. m^2$) or per Gy (dose range 0–40 Gy)[e]

Lesions (per10^9 bases)	Control	UVA radiation (per $kJ.m^2$)	Gamma rays (per Gy)
8-OxodGuo[a]		11	0.8
FapyGua[b]		27	not determined
DNA strand breaks[c]	265	130	0.9
Fpg-sensitive sites[d]	190	18	1.9
Endo III-sensitive sites[d]	195	53	0.3

[a]HPLC-ECD.
[b]HPLC/GC-MS.
[c]Comet assay (single strand breaks, double strand breaks and alkali-labile sites).
[d]Modified comet assay.
[e]From Pouget et al. [62].

Another putative UVB DNA photodamage that has received a lot of attention in early model studies is 6-hydroxy-5,6-dihydrocytosine the so-called "cytosine photohydrate" that arises from hydration of singlet excited state cytosine [for a comprehensive review, see 1]. A relevant piece of information on the formation of cytosine photohydrate in both isolated and cellular DNA was recently gained from the application of a suitable HPLC-MS/MS assay [57]. This allows the measurement of 2′-deoxycytidine photohydrates as the 6*R* and 6*S* diastereomers of 6-hydroxy-5,6-dihydro-2′-deoxyuridine upon DNA enzymatic hydrolysis and quantitative deamination. Thus, it was found that UVC-induced formation of cytosine hydrate in isolated DNA is a minor photochemical event with a yield of formation which is about 2 orders of magnitude lower than that of CPDs. The formation ratio CPDs/cytosine hydrate was found to be even lower by a factor of 10 in cellular DNA as the likely result of lower accessibility of water molecules for hydration of the cytosine moieties in compacted cellular DNA. These data that, at the best, suggest a minor contribution of cytosine hydrate to the overall biological effect of far-UV radiation are in agreement with a previous estimation of endonuclease III-sensitive sites that were to be 2 orders of magnitude lower than the level of CPDs [58].

1.4 Damage induced by UVA radiation to cellular DNA

Several lines of evidence that underline the major role played by endogenous photosensitizers in promoting oxidative reactions to cellular DNA upon activation by UVA radiation are now available [for a recent review, see 59] are available. Another indirect support for the UVA-induced generation of reactive oxygen species is provided by the observation of the enhancement of the cytotoxic and DNA damaging effects of this component of solar radiation upon addition of L-arginine to human keratinocyte HaCaT cell cultures [60]. A reasonable explanation involves the implication of peroxynitrite as a damaging species as the result of the reaction of L-arginine–stimulated formation of $NO^{\bullet}$ with photogenerated $O_2^{\bullet -}$. As a more direct proof of the occurrence of oxidation reactions, it was recently shown that UVA irradiation of cellular DNA gives rise to 8-oxodGuo as assessed by HPLC-ECD measurements [56,61]. Insights into the mechanism of 8-oxodGuo formation were gained from a study that has involved the measurement of three main classes of UVA-induced oxidative damage to DNA in human monocytes using a modified comet assay [62]. These included frank DNA strand breaks together with alkali-labile lesions on the one hand and additional nicks provided by incubation with Fpg and endo III enzymes respectively on the other hand. Interestingly it was found in agreement with previous measurements achieved using the alkaline elution technique that the level of Fpg-sensitive sites was much higher than that of either strand breaks or lesions recognized by endo III [39,40]. Interestingly, the distribution pattern of the oxidative lesions is different from that induced by exposure to gamma rays (Table 2). Under the latter conditions where $^{\bullet}OH$ radical is the predominant reactive species, the yield of endo- and Fpg-sensitive sites is similar, each of them being about three times lower than that of DNA strand breaks. It should be added that recent investigations using a suitable derivatized naphthalene endoperoxide as a chemical source of singlet oxygen ($^1O_2$) [63] have shown that the latter ROS reacts in a highly specific way with the guanine moiety of both isolated and cellular DNA to produce exclusively 8-oxoGua [64,65]. It was found that $^1O_2$ is not able to act as a one-electron oxidant as inferred from the lack of formation of 2,6-diamino-4-hydroxy-5-formamidopyrimiidine (FapyGua) which is also generated by the reaction of $^{\bullet}OH$ radical with guanine. It should be added that attempts to detect FapyGua in the DNA of UVA-irradiated cells were unsuccessful [62], suggesting either the lack or the low involvement of $^{\bullet}OH$ radical and/or one-electron oxidation process in the formation of 8-oxoGua In fact the predominance of the latter compound over DNA strand breaks and endoIII-sensitive sites, mostly oxidized pyrimidine bases, may be rationalized in terms of predominant participation of $^1O_2$ (85%) together with a low contribution of $^{\bullet}OH$ radical (15%). It is expected that the qualitative and quantitative formation of the different classes of oxidatively generated damage to DNA is likely to vary with the nature of the cells since they are expected to contain different types of photosensitizers. This may explain the absence of detection of 8-oxoGua in the DNA of UVA-irradiated human epidermoid carcinoma cells [66].

It clearly appears that 8-oxoGua, the main oxidative lesion identified so far in the DNA of UVA-irradiated cells is at the best generated in very low amounts. In fact exposure of monocyte cells to a dose of UVA radiation up to 50 $kJ.m^{-2}$ is required to double the level of steady-state level of 8-oxoGua [62] that arises mostly from oxidative metabolism in non-irradiated cells. This strongly suggests that the contribution of 8-oxoGua to the overall biological deleterious effects of UVA radiation is expected to be very low. In that respect, the role of CPDs that have been shown to be formed in much higher yield than 8-oxoGua (ratio 19) has to be further investigated. Interestingly, it was recently found that Thy<>Thy is the predominant UVA-induced dimeric pyrimidine photoproduct at the exclusion of other CPDs and 6-4PPs [67,68], suggesting the occurrence of an energy transfer process for its formation.

1.5 Conclusions

Major progress has been recently made in the assessment of the main UVB and UVA-induced damage to cellular DNA through accurate HPLC and biochemical measurements. Thus, the complete pattern of the dimeric pyrimidine photoproducts is now available for the main bipyrimidine sequences. It is confirmed that there is a strong primary sequence effect on the formation of both CPDs and 6-4PPs. However, information is still lacking on the UVB-induction of 5-methylcytosine (5-MeCyt) containing dimeric pyrimidine photoproducts in both isolated and cellular DNA. Indirect measurement has suggested that the presence of a 5-MeCyt residue in a bipyrimidine sequence would prevent the formation of related 6-4PP photoproducts in cellular DNA [69]. However, this contrasts with the fact that both CPDs and 6-4PPs were found to be efficiently produced in 5-MeCyt containing dinucleoside monophosphates [70]. Further applications of the HPLC-MS/MS assay are expected for assessing the kinetic of repair of individual bi-pyrimidine photoproducts in various cell lines as already shown for the thymine dimeric lesions within the DNA of *Arabidopsis thaliana* [71]. A major analytical development is expected for the latter assay with the association of either micro-HPLC or capillary electrophoresis to the tandem mass spectrometry detection that may result in a significant increase in sensitivity. Other types of oxidative DNA damage have to be considered as potential UVA-induced lesions. Relevant candidates are represented by adducts that arise from the reaction of aminobases with aldehydes such as malonaldehyde and 4-hydroxy-2-nonenal, breakdown products of lipid peroxides for which a suitable HPLC-MS/MS is available [72]. A second class of photosensitized DNA photodamage for which there is still a paucity of structural and mechanistic information deals with DNA-protein cross-links.

Acknowledgements

The authors acknowledge the financial support from the Centre National d'Etudes Spatiales.

References

1. J. Cadet, P. Vigny (1990). The photochemistry of nucleic acids, In: H. Morrison (Ed.) *Bioorganic Photochemistry: Photochemistry and the Nucleic Acids* (pp. 1–272). Wiley and Sons, New York, NY.
2. J. Cadet, M. Berger, T. Douki, B. Morin, S. Raoul, J.-L. Ravanat, S. Spinelli (1997). Effects of UV and visible radiations on DNA - Final base damage. *Biol. Chem.*, **378**, 1275–1286.
3. E. Otoshi, T. Yagi, T. Mori, T. Matsunaga, O. Nikaido, S.-T. Kim, K. Hitomi, M. Ikenaga, T. Todo (2000). Respective roles of cyclobutane dimmers, (6-4)photoproducts, and minor photoproducts in ultraviolet mutagenesis of repair-deficient xeroderma pigmentosum A cells. *Cancer Res.*, **60**, 1729–1735.

4. D. Perdiz, P. Grof, M. Mezzina, O. Nikkaido, E. Moustacchi, E. Sage (2000). Distribution and repair of bipyrimidine photoproducts in solar UV-irradiated mammalian cells. *J. Biol. Chem.*, **275**, 26732–26742.
5. H. Steige, L. Roza, A.A. Vink, T. Ruzicka, S. Grether-Beck, J. Krutmann (2000). Enzyme plus light therapy to repair DNA damage in ultraviolet-B-irradiated human skin. *Proc. Natl. Acad. Sci. USA*, **97**, 1790–1795.
6. S. Horiki, H. Miyauchi-Hashimoto, K. Tanaka, O. Nikaiddo, T. Horio (2000). Protective effects of sunscreening agents on photocarcinogenesis, photoaging, and DNA damage in XPA gene knockout mice. *Arch. Dermatol. Res.*, **292**, 511–518.
7. D.B. Yarosh, S. Boumakis, S. Brown, M.T. Canning, J.W., Galvin, D.M., Both, E. Kraus, A. O'Connor, D.A. Brown (2002). Measurement of UVB-induced DNA damage and its consequences in models of immunosuppression. *Methods*, **28**, 55–62.
8. D.L. Mitchell, R. Greinert, F.R. de Gruijl, K.L.H. Guikers, E.W. Breitbart, M. Byrom, M.M. Gallmeier, M.G. Lowerey, B. Volkmer (1999). Effects of chronic low-dose ultraviolet B radiation on DNA damage and repair in mouse skin. *Cancer Res.*, **59**, 2875–2884.
9. T. Mori, M. Nakane, T. Hattori, T. Matsunaga, M. Ihara, O. Nikaido (1991). Simultaneous establishment of monoclonal antibodies specific for either dimmer or (6-4)photoproduct from the same mouse immunized with ultraviolet-irradiated DNA. *Photochem. Photobiol.*, **54**, 225–232.
10. T. Mizuno, T. Matsunaga, M. Ihara, O. Nikaido (1991). Establishment of a monoclonal antibody recognizing cyclobutane-thymine dimer in DNA: a comparative study with 64M1 antibody specific for 6-4photoproducts. *Mutat Res.*, **254**, 175–184.
11. T. Matsunaga, Y. Hatakeyama, M. Ohta, T. Mori, O. Nikaido (1993). Establishment and characterization of a monoclonal antibody recognizing the Dewar isomers of (6-4) photoproducts. *Photochem. Photobiol.*, **57**, 934–940.
12. L. Roza, K.J.M. Van der Wulp, S.J. MacFarlane, P.H.M. Lohman, R.A. Baan (1988). Detection of cyclobutane thymine dimers in DNA of human cells with monoclonal antibodies raised against a thymine-dimer containing tetranucleotide. *Photochem. Photobiol.*, **48**, 627–634.
13. S. Sauvaigo, C. Serres, N. Signorini, N. Emmonet, M.-J. Richard, J. Cadet (1998). Use of the single cell gel electrophoresis assay for the immunofluorescent detection of a specific DNA damage. *Anal. Biochem.*, **259**, 1–7.
14. S. Tornaletti, D. Rozek, G.P. Pfeifer (1993). The distribution of UV photoproducts along the human p53 gene in relation to mutations and skin cancer. *Oncogene*, **8**, 2051–2057.
15. S. Tommasi, M.F. Denissenko, G.P. Pfeifer (1997). Sunlight induces pyrimidine dimers preferentially at 5-methylcytosine bases. *Cancer Res.* **57**, 4727–4730.
16. R. Drouin, H. Rodriguez, G.P. Holmquist, S.A. Akman (1996). In: G.P. Pfeifer (Ed.), *Technologies for DNA Damage and Mutations* (pp. 37–43). Plenum Press, New York, NY.
17. Y.H. You, G.P. Pfeifer (2001). Similarities in sunlight-induced mutational spectra of CpG-methylated transgenes and the p53 gene in skin cancer point to an important role of 5-methylcytosine residues in solar UV mutagenesis. *J. Mol. Biol.*, **305**, 389–399.
18. Y.H. You, P.E. Szabo, G.P. Pfeifer (2000). Cyclobutane pyrimidine dimers form preferentially at the major p53 mutational hotspot in UVB-induced mouse skin tumors. *Carcinogenesis*, **21**, 2113–2117.
19. J.-L. Luo, W.-M. Tong, J.-H. Yoon, M. Hergenhahn, R. Koomagi, Q. Yang, D. Galendo, G.P. Pfeifer, Z.-Q. Wang, M. Hollstein (2001). UV-induced DNA damage and mutations in Hupki (human p53 knock-in) mice recapitulate p53 hotspot alterations in sun-exposed human skin. *Cancer Res.*, **61**, 8158–8162.
20. Y.-H. You, D.-H. Lee, J.-H. Yoon, S. Nakajima, A. Yasui, G.P. Pfeifer (2001). Cyclobutane pyrimidine dimers are responsible for the vast majority of mutations induced by UVB irradiation in mammalian cells. *J. Biol. Chem.*, **276**, 44688–44694.

21. J.-P. Therrien, M. Rouabhia, E.A. Drobetsky, R. Drouin (1999). The multilayered organization of engineered human skin does not influence the formation of sunlight-induced cyclobutane pyrimidine dimers in cellular DNA. *Cancer Res.*, **59**, 285–289.
22. J.-P. Therrien, R. Drouin, C. Baril, E.A. Drobetsky (1999). Human cells compromised for p53 function exhibit effective global and transcription-coupled nucleotide excision repair, whereas cells compromised for pRb function are defective only in global repair. *Proc. Natl. Acad. Sci. USA*, **96**, 15038–15043.
23. S. Tommasi, A.B. Oxyzoglou, G.P. Pfeifer (2000). Cell cycle-independent removal of UV-induced pyrimidine dimers from the promoter and the transcription initiation domain of the human CDC2 gene. *Nucleic Acids Res.*, **28**, 3991–3998.
24. R.M. Tyrrell, S.M. Keyse (1990). The interaction of UVA radiation with cultured cells. *J. Photochem. Photobiol.B: Biol.*, **4**, 349–361.
25. A. de Laat, J.C. van der Leun, F.R. de Gruijl (1997). Carcinogenesis induced by UVA (365 nm) radiation: the dose-time dependence of tumor promotion in hairless mice. *Carcinogenesis*, **18**, 1013–1020.
26. R.B. Setlow, E. Grist, K. Thompson, A.D. Woodhead (1993). Wavelengths effective in induction of malignant melanoma. *Proc. Natl. Acad. Sci. USA*, **90**, 6666–6670.
27. H.J. van Kranen, A. de Lat, J. van der Leun, P.W. Wester, A. de Vries, R.J.W. Berg, C.F. van Kreijl, F.R. de Gruijl (1997). Low incidence of *p53* mutations in UVA (365 nm)-induced skin tumors in hairless mice. *Cancer Res.*, **57**, 1238–1240.
28. A. Stary, C. Robert, A. Sarasin (1997). Deleterious effects of ultraviolet radiation in human cells. *Mutat. Res.*, **383**, 1–8.
29. E. Sage (1993). Distribution and repair of photolesions in DNA: genetic consequences and the role of sequence context. *Photochem. Photobiol.*, **57**, 163–174.
30. E.A. Drobetsky, J. Turcotte, A. Chateauneuf (1995). A role for ultraviolet A in solar mutagenesis. *Proc. Natl. Acad. Sci. USA*, **92**, 2350–2354.
31. E. Sage, B. Lamolet, E. Brulay, E. Moustacchi, A. Chateauneuf, E.A. Drobetsky (1996), Mutagenic specificity of solar UV light in nucleotide excision repair-deficient rodent cells. *Proc. Natl. Acad. Sci. USA*, **93**, 176–180.
32. G.F. Vile, R.M. Tyrrell (1993). Oxidative stress resulting from ultraviolet A irradiation of human skin fibroblasts leads to a heme oxygenase-dependent increase in ferritin. *J. Biol. Chem.*, **268**, 14678–14681.
33. C. Pourzand, R.D. Watkin, J.E. Brown, R.M. Tyrrell (1999). Ultraviolet A radiation induces immediate release of iron in human primary skin fibroblasts: the role of ferritin. *Proc. Natl. Acad. Sci. USA*, **96**, 6751–6756.
34. A. Fisher-Nilsen, S. Loft, K.G. Jensen (1993). Effect of ascorbate and 5-aminosalisylic acid on light-induced 8-hydroxydeoxyguanosine formation in V79 Chinese hamster cells. *Carcinogenesis*, **14**, 2431–2433.
35. J.E. Rosen, A.K. Prahalad, G.M. Williams (1996). 8-Oxodeoxyguanosine formation in the DNA of cultured cells after exposure to H_2O_2 alone or with UVB or UVA irradiation. *Photochem. Photobiol.* **64**, 117–122.
36. W.G. Warmer, R.R. Weiss (1997). *In vitro* photooxidation of nucleic acids by ultraviolet radiation. *Photochem. Photobiol.*, **65**, 560–563.
37. E. Kvam, R.M. Tyrrell (1997). Induction of oxidative DNA base damage in human skin cells by UV and near visible radiation. *Carcinogenesis* **18**, 2379–2384.
38. X. Zhang, B.S. Rosenstein, Y. Wang, M. Lebwohl, D.L. Mitchell, H. Wei (1997). Induction of 8-oxo-7,8-dihydro-2′-deoxyguanosine by ultraviolet radiation in calf thymus DNA and HeLa cells. *Photochem. Photobiol.*, **65**, 119–124.
39. C. Kielbassa, L. Roza, B. Epe (1997). Wavelength dependence of oxidative DNA damage induced by UV and visible light. *Carcinogenesis*, **18**, 811–816.

40. M. Pfaum, C. Kielbassa, M. Garmyn, B. Epe (1998). Oxidative DNA damage induced by visible light in mammalian cells: extent, inhibition by antioxidants and genotoxic effects. *Mutat. Res.*, **408**, 137–146.
41. C. Kielbassa, B. Epe (2000). DNA damage induced by ultraviolet and visible light and its wavelength dependence. *Methods Enzymol.*, **319**, 436–445.
42. H.J. Niggli, P.A. Cerutti (1983). Cyclobutane-type pyrimidine formation and excision in human skin fibroblasts after irradiation with 313 nm ultraviolet light. *Biochemistry*, **22**, 1390–1395.
43. J. Cadet, N.E. Gentner, B. Rozga, M.C. Paterson (1983). Rapid quantitation of ultraviolet-induced thymine-containing dimers in human cell DNA by reversed-phase high-performance liquid chromatography. *J. Chromatogr.*, **280**, 99–108.
44. J. Cadet, M. Berger (1985). Radiation-induced decomposition of the purine bases within DNA and related model compounds. *Int. J. Radiat. Biol.*, **47**, 127–143.
45. T. Douki, L. Voituriez, J. Cadet (1995). Measurement of pyrimidine (6-4) pyrimidone photoproducts in DNA by a mild acidic hydrolysis-HPLC fluorescence assay. *Chem. Res. Toxicol.* **8**, 244–253.
46. S. Frelon, T. Douki, J.-L. Ravanat, J.-P. Pouget, C. Tornabene, J. Cadet (2000). High-performance liquid chromatography-tandem mass spectrometry measurement of radiation-induced basedamage to isolated and cellular DNA. *Chem. Res. Toxicol.*, **13**, 1002–1010.
47. S. Frelon, T. Douki, J. Cadet (2002). Radical oxidation of the adenine moiety of nucleoside and DNA. 2-Hydroxy-2′-deoxyadenosine is a minor decomposition product. *Free Radic. Res.*, **36**, 499–508.
48. M. Liuzzi, M. Weinfeld, M.C. Paterson (1989). Enzymatic analysis of isomeric trithymidilates containing ultraviolet light-induced cyclobutane pyrimidine dimers 1. Nuclease P1 hydrolysis of the intradimer phosphodiester linkage. *J. Biol. Chem.*, **264**, 6355–6363.
49. T. Douki, T. Zalizniak, J. Cadet (1997). Far-UV-induced dimeric photoproducts in short oligonucleotides: sequence effects. *Photochem. Photobiol.*, **66**, 171–179.
50. T. Douki, M. Court, S. Sauvaigo, F. Odin, J. Cadet (2000). Formation of the main UV-induced thymine dimeric lesions within isolated and cellular DNA as measured by high performance liquid chromatography-tandem mass spectrometry. *J. Biol. Chem.*, **275**, 16678–11685.
51. T. Douki, J. Cadet (2001). Individual determination of the yield of the main UV-induced dimeric pyrimidine photoproducts in DNA suggests a high mutagenicity of CC photolesions. *Biochemistry*, **40**, 2495–2501.
52. T. Douki, M. Court, J. Cadet (2000). Electrospray-mass spectrometry characterization and measurement of far-UV-induced thymine photoproducts. *J. Photochem. Photobiol. B: Biol.*, **54**, 145–154.
53. J.-H. Yoon, C.-S. Lee, T.R. O'Connor, A. Uasui, G.P. Pfeiffer (2000). The DNA damage spectrum produced by simulated sunlight. *J. Mol. Biol.*, **299**, 681–693.
54. M.S. Stewart, G.S. Cameron, B.C. Pence (1996). Antioxidant nutrients protect against UVB-induced oxidative damage to DNA of mouse keratinocytes in culture. *J. Invest. Dermatol.*, **106**, 1086–1089.
55. H. Wei, Q. Cai, M. Lebwohl (1998). Tamoxifen reduces endogenous and UV light-induced oxidative damage to DNA, lipid and protein in vitro and in vivo. *Carcinogenesis*, **19**, 1013–1018.
56. T. Douki, D. Perdiz, P. Grof, Z. Kuluncsics, E. Moustacchi, J. Cadet, E. Sage (1999). Oxidation of guanine in cellular DNA by solar UV radiation: biological role. *Photochem. Photobiol.*, **70**, 184–190.

57. T. Douki, G. Vadesne-Bauer, J. Cadet (2002). Formation of 2′-deoxyurdine hydrates upon exposure of nucleotides to gamma radiation and UVC-irradiation of isolated and cellular DNA. *Photochem. Photobiol. Sci.*, **1**, 565–569.
58. J. Jen, D.L. Mitchell, R.P. Cunningham, C.A. Smith, J.-S. Taylor, J.E. Cleaver (1997). Ultraviolet irradiation produces novel endonucleases III-sensitive cytosine photoproducts at dipyrimidine sites. *Photochem. Photobiol.*, **65**, 323–329.
59. J. Cadet, T. Douki, J.-P. Pouget, J.-L. Ravanat, S. Sauvaigo (2001). Effects of UV and visible radiations on cellular DNA. *Curr. Probl. Dermatol.*, **29**, 62–73.
60. C. Didier, N. Emonet-Piccardi, J.-C. Béani, J. Cadet, M.-J. Richard (1999), L-arginine increases UVA cytotoxicity in irradiated human keratinocyte cell line: potential role of nitric oxide. *FASEB J.*, **13**, 1817–1824.
61. P. Duez, M. Hanocq, J. Dubois (2001). Photodynamic DNA damage mediated by δ-aminolevulinic acid-induced porphyrin, Carcinogenesis, **22**, 771–778.
62. J.-P. Pouget, T. Douki, M.J. Richard, J. Cadet (2000). DNA damage induced in cells by gamma and UVA radiation as measured by HPLC/GC-MS and HPLC-EC and Comet assay. *Chem. Res. Toxicol.*, **13**, 541–549.
63. G.R. Martinez, J.-L; Ravanat, M.H.G. Medeiros, J. Cadet, P. Di Mascio (2000). Synthesis of a naphthalene endoperoxide as a source of ^{18}O-labeled singlet oxygen for mechanistic studies *J. Am. Chem. Soc* **122**, 10212–10213.
64. J.-L. Ravanat, C. Saint-Pierre, P. Di Mascio, G.R. Martinez, M.H.G. Medeiros, J. Cadet (2001). Damage to isolated DNA mediated by singlet oxygen. *Helv. Chim. Acta*, **84**, 3702–3709.
65. J.-L. Ravanat, P. Di Mascio, G.R. Martinez, J. Cadet (2000). Singlet oxygen induces oxidation of cellular DNA. *J. Biol. Chem.*, **275**, 40601–40604.
66. E. Sage, D. Perdiz, P. Grof, A. Reynaud-Angelin, T. Douki, J. Cadet, P. Rochette, N. Bastien, R. Drouin, DNA damage induced by UVA irradiation: role in solar mutagenesis. In: R. Drouin, E. Sage, M. Rouabhia (Eds) From DNA Photolesions to Mutations, Skin Cancer and Cell Death, Royal Society of Chemistry, Cambridge UK, this volume pp 33–47.
67. T. Douki, A. Reynaud-Angelin, J. Cadet, E. Sage (2003). Bipyrimidine photoproducts rather than oxidative lesions are the main DNA damage involved in the genotoxic effect of solar UVA radiation. *Biochemistry* **42**: 9221–9226.
68. P.J. Rochette, J.P. Therrien, R. Drouin, D. Perdiz, N. Bastien, E.A. Drobetsky, E. Sage (2003). UVA-induced cyclobutane pyrimidine dimers form predominantly at thymine-thymine dipyrimidines and correlate with the mutation spectrum in rodent cells. *Nucleic Acids Res.* **31**, 2786–2794.
69. B.W. Glickman, R.M. Schuaper, W.A. Haseltine, R.L. Dunn, D.E. Brush (1986). The C-G (6-4) UV photoproducts is mutagenic in *Escherichia coli. Proc. Natl. Acad. Sci.*, USA, **83**, 6945-6949.
70. T. Douki, J. Cadet (1994). Formation of cyclobutane dimers and (6-4) photoproducts upon far-UV photolysis of 5-methylcytosine-containg dinucleoside monophosphates. *Biochemistry*, **33**, 11942–1950.
71. A-L. Dany, T. Douki, C. Triantaphylides, J. Cadet (2001). Repair of the main UV-induced thymine dimeric lesions within Arabidopsis thaliana DNA: evidence for the major involvement of photoreactivation pathways. *J. Photochem. Photobiol. B: Biol.*, **65**, 127–135.
72. S. Frelon, T. Douki, A. Favier, J. Cadet (2002). Comparative study of base damage induced by gamma radiation and Fenton reaction in isolated DNA. *J. Chem. Soc. Perkin Trans. I* 2866–2870.

Chapter 2

Chemical sequencing profiles of photosensitized DNA damage

R. Jeremy H. Davies, Sharon M. Starrs and Clarke S. Stevenson

Table of contents

Abstract

Gel sequencing protocols utilising end-labeled single and double stranded DNA molecules provide a versatile and informative methodology for probing the mechanism of action of water-soluble photosensitizers. Direct strand breaks can be readily distinguished from alkali-labile sites, and from photoproducts that are recognised by lesion-specific DNA glycosylase enzymes. In the case of photooxidation, where guanine bases are selectively targeted, characteristic patterns of alkali-labile damage are associated with type I (one-electron transfer) and type II (singlet oxygen) reactions. Mechanistic inferences drawn from the profiles of DNA damage can be consolidated by observing the effects of quenchers and scavengers, by spectroscopic measurements, and by diagnostic product analysis. This article reviews applications of the chemical sequencing approach to a selection of UV-absorbing photosensitizers with emphasis on small organic molecules that do not bind tightly to DNA. These include benzophenone-like drugs and quinolone antibiotics that are associated with cutaneous phototoxicity, 2-phenylbenzimidazoles, and certain quinones. The examples of suprofen, nalidixic acid and lomefloxacin are discussed in some detail.

2.1 Introduction

The ability of certain light-absorbing molecules to act as photosensitizers of DNA damage has wide ramifications in biology and medicine [1,2]. Most notably, these include the roles of endogenous and exogenous sensitizers in promoting photomutagenesis and photocarcinogenesis, and the clinical promise of various modalities of photodynamic therapy [3]. To achieve a detailed understanding of the mechanism of action of individual photosensitizers it is generally advisable to employ a combination of complementary experimental approaches. These extend from time-resolved photophysical measurements, through chemical studies of product formation and characterization, to biochemical assays of DNA damage at the cellular level. In this context, gel sequencing protocols with end-labeled DNA molecules that can map damaged sites at nucleotide resolution have proved very informative [4]. This brief review focuses on how DNA chemical sequencing methodology [5] can be exploited to identify potential photosensitizers and gain insight into the mechanisms responsible for DNA modification. The coverage is largely restricted to examples of low molecular weight bioactive compounds that absorb at UVA or UVB wavelengths, and exhibit no significant binding affinity for DNA. It specifically excludes mention of the elegant applications to DNA complexes with organometallic ligands and tethered sensitizers that have been so effective in exploring long-range hole transport through the double helix [6,7]. It also excludes discussion of the powerful DNA sequencing approaches for detecting photolesions that are based on polymerase-catalysed primer extension, such as ligation-mediated PCR [8,9].

2.2 Mechanistic considerations

A large number of photosensitizers are now known that will damage DNA when they are excited at UV and/or visible wavelengths. Most commonly this occurs *via* photooxidation, but triplet energy transfer and photoadduct formation are also important possibilities.

In the case of photooxidation, two main mechanistic pathways can be distinguished that are conveniently denoted type I and type II [10]. In type I, the reaction involves electron or hydrogen transfer between DNA and the excited state sensitizer. In type II, highly reactive singlet oxygen, $^1O_2$ ($^1\Delta_g$), is generated by triplet energy transfer from the sensitizer to ground state ($^3\Delta_g$) molecular oxygen. Owing to their relatively low ionisation potential, guanine nucleobases are the most susceptible targets in DNA for photomodification by either type I or type II attack. The resulting lesions (*vide infra*) can generally be detected as alkali-labile strand cleavage sites when the DNA is subsequently heated with piperidine. However, distinctly different (and diagnostically useful) patterns of damage are associated with the type I and type II pathways. In line with theoretical predictions [11], type I electron transfer from native DNA to an excited state sensitizer predominantly damages guanines that are situated 5′ to another guanine or (to a lesser degree) adenine. In contrast, type II modification by singlet oxygen tends to damage all guanines equally irrespective of their sequence context [12].

Photosensitizer radical anions produced by electron abstraction from DNA may be quenched by molecular oxygen to yield the superoxide anion. Although this species is unreactive towards DNA, it readily generates hydrogen peroxide which, in presence of transition metal ions, can decompose to hydroxyl radicals through Fenton-type reactions. As well as adding to nucleobase moieties, hydroxyl radicals will abstract hydrogen atoms from deoxyribose residues and thus induce random cleavage of the polynucleotide backbone. In some instances, excited state photosensitizers are sufficiently reactive to abstract hydrogen atoms directly from deoxyribose and similarly cause overt strand breaks [2].

Ketonic and other photosensitizers sometimes function by transfer of triplet energy to DNA and thereby promote the formation of cyclobutane pyrimidine dimers [10]. Owing to its low lying triplet state, dimerization of thymine is favoured in this process. Close association between a sensitizer and DNA may lead to the formation of covalent photoadducts that can be isolated following nuclease digestion [13]. The photo-crosslinking of duplex DNA by psoralen derivatives is good example of this phenomenon [10].

It is frequently the case that photosensitizers modify DNA by two or more competing mechanisms whose relative efficiencies are likely to be strongly influenced by the experimental conditions, the availability of oxygen, and the conformation of the DNA.

2.3 Experimental procedure

The most straightforward system for detecting photosensitized strand breaks in DNA is the supercoiled-plasmid nicking assay. Because of its great sensitivity and operational simplicity it is still in widespread use. However, it is restricted to studies of double stranded DNA, gives no information regarding sequence specificity, and is not suitable for the detection of alkali-labile lesions. It is possible to overcome these limitations by using the following gel sequencing approach which is generally applicable to investigating the properties of water-soluble photosensitizers.

The target molecules are single or double stranded synthetic oligonucleotides (typically up to 50 bases in length) that have been labeled [14] with ^{32}P at the 3′ or 5′-end of one of the strands. Restriction endonuclease fragments from natural DNAs can also be used for this purpose but synthetic oligonucleotides have the advantage that they can be designed to incorporate defined sequence contexts or structural motifs such as hairpin loops or mismatched bases. A solution containing the DNA and photosensitizer at suitable (sub-millimolar) concentrations in neutral buffer is exposed to a metered fluence of UV radiation. Following exposure, the DNA is recovered by ethanol precipitation and analysed by standard chemical sequencing methods [5,15] for the presence of direct or latent strand cleavage sites. In regard to the latter, heating with 1 M piperidine (90°C, 30 min) is routinely used to reveal alkali-labile base lesions which are a prominent feature of photooxidative DNA damage. More discrimination is possible by inducing strand cleavage with DNA glycosylase enzymes such as the *E. coli* Fpg protein and endonuclease III that act respectively on oxidised purines and pyrimidines. T4 endonuclease V is especially useful for recognising cyclobutane pyrimidine photodimers.

To determine sites of base damage, the cleavage fragments are fractionated by electrophoresis and visualised by autoradiography or phosphorimaging. Their mobility is then compared to that of the bands in reference base-specific chemical sequencing ladders. Unless the cleavage fragments contain unmodified nucleobases and possess phosphorylated termini, their positions may not coincide exactly with bands in the reference lanes. On this point, it should be noted that the bands in A+G reference ladders produced by acidic depurination of short oligonucleotides bearing a 3′-dideoxyadenosine label display anomalous electrophoretic mobility [16].

To obtain gel sequencing profiles that accurately reflect the probability of damage at each nucleotide residue, it is important that the experimental conditions used for photosensitization produce on average approximately one 'hit' per labeled DNA molecule. With doubly damaged molecules, only fragments derived from the lesion nearer to the labeled end will be detected. Consequently, in evaluating potential photosensitizers it is necessary to conduct experiments over a range of photosensitizer concentrations and irradiation fluence. In assessing the photoreactivity of new compounds it is advisable to include a positive control for an efficient photosensitizer, such as riboflavin, for comparative purposes.

Whenever possible, it is desirable to investigate the activity of photosensitizers on both single and double stranded DNA as the pattern of induced damage is often sensitive to polynucleotide conformation. Single stranded oligonucleotides should be designed so that they are devoid of self-complementary sequence elements that could form stable segments of base-paired secondary structure. The corresponding duplex is readily prepared by annealing with an oligonucleotide of complementary sequence. As illustrated below, experiments that test the effects of scavengers and quenchers of reactive oxygen species can provide definitive information about the molecular mechanisms underlying photosensitization.

2.4 Photosensitizers and profiles of DNA damage

The utility of chemical sequencing techniques in probing the effects of photosensitizers on DNA was convincingly demonstrated in several early studies. Although its theoretical basis had yet to be established, the type I photooxidation signature (i.e. alkali-labile sites at 5′-guanines in GG doublets) was observed to result from the photosensitization of duplex DNA with 3-carbethoxypsoralen [17] and with riboflavin [18]. On the other hand, photosensitization with the singlet oxygen generator hematoporphyrin [19] (which only modifies single stranded DNA) showed the type II pattern, where piperidine-sensitive sites are created uniformly at all guanines in the sequence. As will be evident from the examples that follow, these distinctive profiles have since been encountered with many other photosensitizers. The systems selected for discussion mainly encompass UV-absorbing photosensitizers that are not bound tightly by DNA and they include some drugs that are associated with phototoxic side effects.

2.4.1 Suprofen and other benzophenone-type drugs

Suprofen (Figure 1) belongs to the aryl propionate family of non-steroidal anti-inflammatory drugs, several members of which (including ketoprofen and naproxen) have a tendency to sensitize phototoxic and/or photoallergic responses in skin [20]. In common with ketoprofen and the hypolipidemic agent fenofibrate, suprofen can

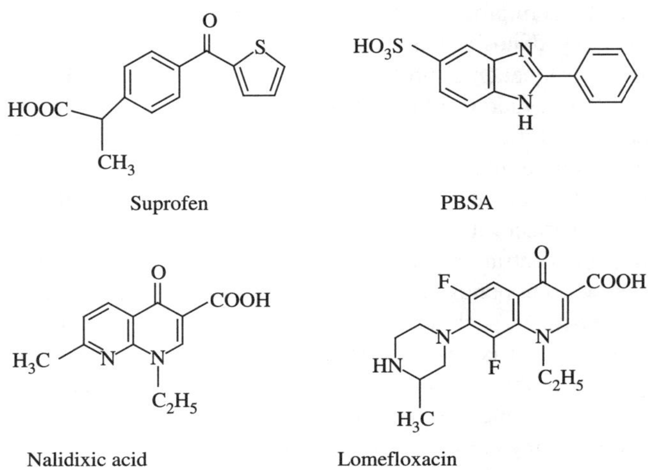

Figure 1. Chemical structures of the DNA photosensitizers suprofen, 2-phenylbenzimidazole 5-sulfonic acid (PBSA), nalidixic acid, and lomefloxacin.

be looked upon as a derivative of the ketonic photosensitizer benzophenone [21]. It absorbs maximally at 300 nm and when irradiated at UVB wavelengths efficiently photosensitizes DNA damage [22]. It therefore serves as an appropriate example with which to illustrate experimental aspects of the gel sequencing method and its application in practice.

Photosensitization with suprofen does not cause detectable levels of direct strand breakage in target DNA molecules but, as shown in Figure 2, it induced piperidine-labile lesions at guanine residues in both the single and double stranded forms of a 3′-end labeled oligonucleotide. Corresponding cleavage patterns for a 5′-end labeled oligomer can be found in reference [22]. The extent of modification was observed to increase with both the concentration of suprofen and the UVB fluence. Although photochemical modification is essentially confined to the sites of guanine residues, the damage profile differs strikingly in the single and double stranded conformations.

The pattern for single stranded DNA (see Figure 2, lanes 4–7), showing even cleavage at all the guanines, is as expected for type II attack by singlet

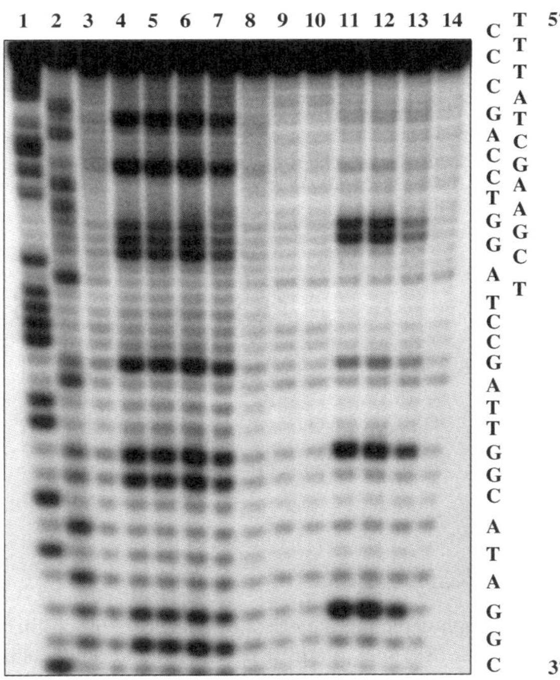

Figure 2. Gel autoradiograph showing the induction of alkali-labile DNA damage, photosensitized by suprofen (SP), in a single stranded (ss) and double stranded (ds) 3′-end labeled 44mer oligonucleotide as a function of SP concentration and UV fluence. Lane 1: C+T; lane 2: A+G; lane 3: UVB (10 kJ m^{-2}) + ss; lanes 4, 5, and 6: UVB (10 kJ m^{-2}) + ss + 50, 100 and 200 μM SP respectively; lanes 7 and 8: 200 μM SP + ss + 5 kJ m^{-2} and 0 kJ m^{-2} UVB respectively; lanes 9 and 11–14: as for lanes 3 and 5–8, except containing ds DNA. All samples were heated with piperidine.

oxygen – which suprofen is known to generate when irradiated in aqueous solution [23]. The results from quenching experiments are consistent with this interpretation. As is evident from Figure 3 (lanes 7–10), the damage is strongly suppressed by sodium azide and diazabicyclooctane (DABCO) but not by the hydroxyl radical scavenger mannitol or by superoxide dismutase (SOD). The involvement of

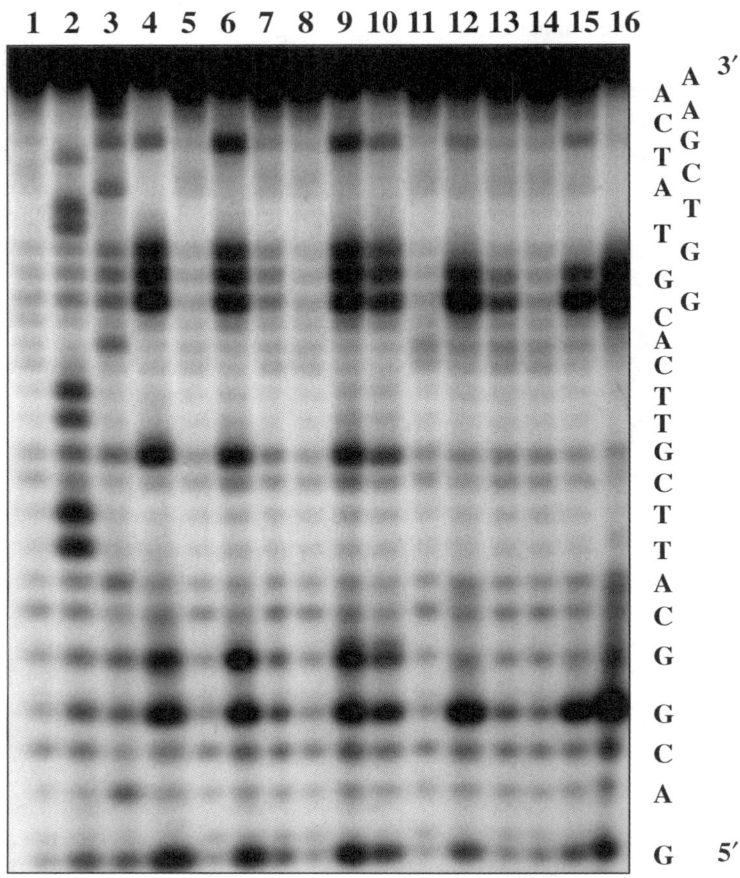

Figure 3. Gel autoradiograph depicting the effect of quenchers on profiles of DNA damage photosensitized by suprofen (SP) in a single stranded (ss) and double stranded (ds) 5′-end labeled 34mer oligonucleotide. In ss samples, the SP concentration was 10 μM and the UVB fluence 10 kJ m^{-2}; in ds samples, the SP concentration was 75 μM and the UVB fluence 20 kJ m^2. Quenched samples contained sodium azide, DABCO, or mannitol at a concentration of 10 mM, or 200 U of superoxide dismutase (SOD). Lane 1: piperidine-treated 34mer; lane 2: T; lane 3: A+G; lane 4: G; lane 5: UVB + ss; lane 6: UVB + ss + SP; lanes 7–10: UVB + ss DNA + SP + sodium azide, DABCO, mannitol and SOD respectively; lane 11: UVB + ds; lane 12: UVB + ds + SP; lanes 13–16: as for lanes 7–10 except with ds DNA. All samples were heated with piperidine.

singlet oxygen is also supported by the increased level of damage that is observed when photosensitization is carried out in D_2O solution [22].

With duplex DNA (see Figure 2, lanes 11–13), the alkali-labile sites map almost exclusively to guanines that are 5′ to another guanine or, much less prominently, to an adenine. Isolated guanine residues are unreactive. As first recognised by Saito and coworkers [11,24], this cleavage pattern characteristically signifies one-electron transfer from DNA nucleobases to an excited state photosensitizer. Positive electronic holes produced in this manner tend to migrate through the DNA base stack until they become trapped at the 5′-guanines in GG and GA doublets because of their especially low ionisation potentials. Subsequent reaction of the guanine radical cations with water and oxygen generates alkali-labile lesions such as triamino-oxazolone.

As shown in Figure 3 (lanes 13, 14), the photosensitization of duplex DNA is inhibited by the presence of sodium azide or DABCO. This may be due to their quenching excited state suprofen molecules as well as singlet oxygen. Although mannitol (lane 15) had no effect, superoxide dismutase (lane 16) appeared to enhance the photodamage. There is precedent for this behaviour which has been attributed to removal of a protective effect arising from the neutralization of ionised nucleobases by electron donation from superoxide anion [4].

To account for the contrasting photoreactivity of single and double stranded DNA, we assume that it reflects competition between a type I process involving electron abstraction that prevails in duplex DNA and type II attack by singlet oxygen which is dominant in single stranded DNA. It is well established that singlet oxygen generated by unbound sensitizers modifies denatured DNA much more efficiently than native DNA [19,25]. Conversely, type I electron transfer accompanied by positive hole migration is strongly favoured by the regular array of stacked bases within the double helix [11].

Although not absolutely definitive, corroboration of the existence of a type I and/or type II photooxidation mechanism can be obtained by analysing the products that are formed by photosensitization of the model compound 2′-deoxyguanosine. It is assumed that formation of the deoxyribosides of 2,5-diamino-imidazol-4-one (dIz) and 2,2,4-triamino-oxazolone (dZ) is diagnostic of electron abstraction to generate the guanine radical cation [26], while reaction with singlet oxgen favours production of the deoxyriboside of spiroiminodihydantoin (originally identified as 4,8-dihydro-4-hydroxy-8-oxo-2′-deoxyguanosine) [27]. When deoxyguanosine is irradiated in presence of suprofen, all three of these marker compounds can be isolated by HPLC as predicted for a mixed type I/type II mechanism.

Naproxen is a weaker photosensitizer of DNA damage than suprofen. Following piperidine treatment, there is an elevated background of random chain breaks that is suggestive of free radical attack. Nonetheless, prefential targeting of GG and GA doublets is clearly apparent in duplex DNA [22].

On UVA irradiation, ketoprofen, fenofibric acid (the main metabolite of fenofibrate) and benzophenone all resemble suprofen in producing type II sequencing profiles with single stranded DNA and type I profiles with double stranded. In a thorough study, Lhiaubet et al. [28] showed that at high fluence all of the guanines in duplex DNA were progressively modified in the order GG > GA > GC > GT in agreement with calculated ionisation potentials. A minor pathway involving free radical damage to cytosines was also identified. Significantly, all three compounds were found to photosensitize the formation of cyclobutane thymine dimers in native DNA in addition to oxidative damage.

2.4.2 Pterins and phenylbenzimidazoles

Pterin (2-amino-4-hydroxypteridine) and several of its derivatives are capable of photosensitizing DNA damage on exposure to UVA radiation. To characterize their mode of action, Ito and Kawanishi [29] targeted ^{32}P-end-labeled restriction fragments of a human protooncogene that contained around 300 base pairs of DNA. Piperidine treatment of the photosensitized samples of native DNA caused the highly selective cleavage at 5′-guanines in GG doublets that is typical of a type I mechanism. In support of this, the feasibility of pterin radical anions being formed by photoinitiated electron abstraction from dGMP was demonstrated through spin destruction experiments monitored by electron spin resonance (ESR) spectroscopy. The reactivity of photoexcited pterins with dGMP broadly matched their efficiency in inducing DNA damage, i.e. pterin $\sim$ 6-carboxypterin $>$ biopterin $\sim$ neopterin $\gg$ folic acid. With denatured single stranded DNA, the level of pterin-photosensitized damage was decreased and, while most guanines were affected, the intensity of cleavage varied considerably.

In gel sequencing experiments with oligonucleotides, we have shown that the common sunscreen ingredient 2-phenylbenzimidazole 5-sulfonic acid (PBSA, Figure 1) and its parent 2-phenylbenzimidazole (PBI) are efficient photosensitizers of oxidative DNA damage when they are irradiated with a UVB source or with natural sunlight [30]. There is no indication of direct strand breakage in either single or double stranded DNA but guanine-specific alkali-labile lesions are generated in the same manner as for suprofen. Thus, there is a very clear dichotomy in the profiles for single stranded DNA, where all the guanines are damaged to a similar extent, and duplex DNA where only the 5′-guanines in GG and GA doublets are reactive. The results of quenching studies are consistent with the conclusion that attack by singlet oxygen is the major pathway of photooxidation in the single stranded conformation, while one-electron transfer is predominant with double stranded DNA. Spectroscopic measurements have established that the photophysical properties of PBSA and PBI are pH-dependent and that both compounds are capable of generating a variety of free radicals and active oxygen species on photoexcitation [31]. However, the gel sequencing data suggest that free radicals do not contribute significantly to the DNA damage produced under physiological conditions.

2.4.3 Quinolone antibiotics

Much interest has centred on the quinolone class of antibacterial drugs on account of their strong association with clinical photosensitivity reactions [32]. Nalidixic acid (Figure 1) is the progenitor of this group which includes a variety of fluorinated analogues. Of these, lomefloxacin (Figure 1) has attracted special attention as it headed the list of adverse drug reports compiled by the FDA in the mid-1990s [33] and has been shown to photoinitiate tumour formation in mouse skin [32]. When examined by the gel sequencing method, the photoreactivity of nalidixic acid has been observed to differ markedly from that of the fluoroquinolones lomefloxacin, fleroxacin and ciprofloxacin all of which generate similar patterns of DNA damage [34].

As illustrated in Figure 4, photosensitization by nalidixic acid leads to the familiar signatures of guanine-specific photodamage implying preferential type II modification of single stranded DNA and a type I mechanism in the double helix. At the relatively high UVB fluence employed, some formation of thymine-cytosine (6-4) photoadducts occurs which accounts for the piperidine-induced cleavage seen at cytosine positions in the 3′-end-labeled duplex oligonucleotide. A major role for singlet oxygen in the reaction with single stranded DNA is compatible with the known ability of nalidixic acid to photosensitize its production in aqueous media,

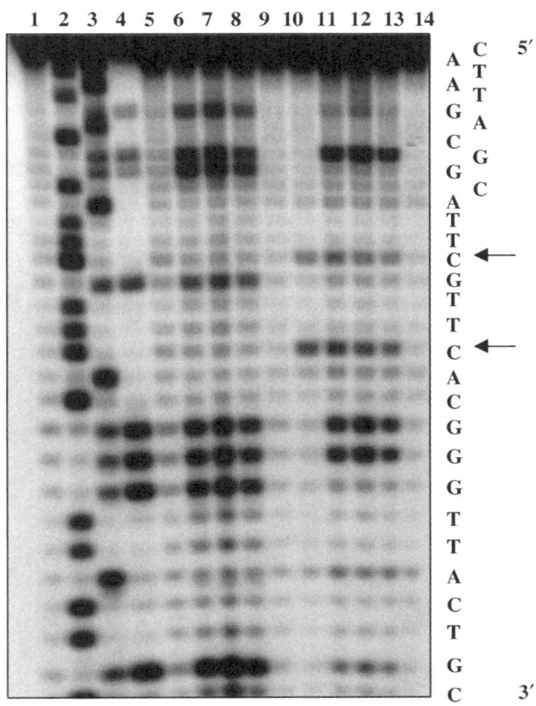

Figure 4. Gel autoradiograph comparing profiles of nalidixic acid (NA)-photosensitized damage in a single stranded (ss) and double stranded (ds) 3′-end labeled 34mer. Lane 1: piperidine-treated 34mer; lane 2: C+T; lane 3: A+G; lane 4: G; lane 5: 20 kJ m^{-2} UVB + ss; lanes 6 and 7: 20 kJ m^{-2} UVB + ss + 90 and 180 μM NA; lanes 8 and 9: NA (180 μM) + ss + 10, and 0 kJ m^{-2} UVB; lanes 10–14: as for lanes 5–9 only with ds DNA. All samples were heated with piperidine. Arrows indicate the positions of (6-4) photoadducts in ds DNA.

damage enhancement in D_2O solution, and the effects of quenchers [35]. In regard to a type I pathway, Hiraku and Kawanishi [34] have demonstrated by the ESR spin destruction technique that photoexcited nalidixic acid can oxidize dGMP by one-electron transfer. Their gel sequencing data, obtained with large (>200 bp) restriction fragments, are in good agreement with our own although the intensity of the cleavage bands, as determined by laser densitometry, varies considerably at different points in the DNA molecules. A possible complication with denatured DNA of this length is that it may contain some secondary structure arising from base pairing between short sections of complementary sequence.

We have found that when lomefloxacin is irradiated with DNA at pH 7.4 there is little evidence of direct strand breakage but the main effect is to introduce alkali-labile sites in a non-specific manner at all nucleotides in the sequence. The resulting cleavage pattern, which equates to a full sequence ladder, is shown for single stranded DNA in Figure 5 (lanes 5–7). Duplex DNA is affected in the same way but it is a lot more resistant to photomodification. The damage is clearly suppressed

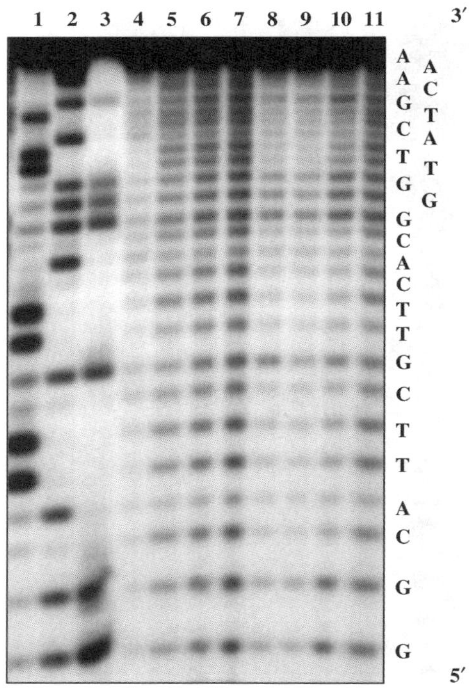

Figure 5. Gel autoradiograph showing piperidine-induced cleavage photosensitized by 400 μM lomefloxacin (LFX) in a single stranded oligonucleotide at pH 7.4, and its dependence on UVB fluence (lanes 5–7). Lanes 8–11 represent the effects of quenchers (20 mM) and SOD (200 U) on LFX-photosensitized DNA damage. Lane 1: T; lane 2: A+G; lane 3: G; lane 4: 25 kJ m^{-2} UVB; lanes 5–7: 12.5, 25, and 50 kJ m^{-2} UVB + LFX; lanes 8–11: sodium azide, DABCO, mannitol and SOD + 25 kJ m^{-2} UVB + LFX. All samples were heated with piperidine.

by sodium azide and DABCO, suggesting possible involvement of singlet oxygen, but mannitol and superoxide dismutase have little or no effect (Figure 5). We surmise that this behaviour can be explained by predominant attack by the carbene-like species that is formed in high yield by photodefluorination of lomefloxacin [36]. Carbenes are very reactive electrophiles and are known to be capable of producing alkali-labile lesions in DNA without any base specificity [37]. It is noteworthy that if the pH is shifted from physiological to pH 4 there is much greater selectivity for all the guanine residues [35]. This may be due to a predicted reduction in the stability of the carbenoid species as lomefloxacin is protonated allowing increased competition from singlet oxygen.

In a separate study of lomefloxacin photosensitization, Hiraku and Kawanishi [34] concluded that singlet oxygen is the primary causative agent of DNA damage. This was based on the selective modification of guanines in single stranded and (less obviously) double stranded DNA. Although excitation with UVA instead of UVB was used, the reason for the discrepancy with our own findings is unclear. However, the photoproduct analyses reported by Sauvaigo et al. [38] indicate that lomefloxacin and other fluoroquinolones efficiently sensitize thymine dimerization in DNA *via* triplet energy transfer so that this process predominates over photo-oxidation. Furthermore, they found lomefloxacin to behave more like a type I than a type II photosensitizer.

Lomefloxacin thus provides an apt illustration of the complexity of mechanisms that may be operative in photosensitized reactions involving DNA.

2.4.4 *Quinones*

Studies on a series of cationic anthraquinone derivatives have demonstrated that their mode of binding to duplex DNA is a crucial factor in determining how they damage DNA upon photoexcitation [4,39]. Thus, intercalative binding favours electron transfer leading to alkali-labile sites at the 5′-guanines in GG steps. By contrast, irradiation of an anthraquinone that binds in the minor groove causes random direct chain cleavage suggestive of hydrogen abstraction from deoxyribose [40]. Thirdly, when an anionic anthraquinone that does not associate with DNA is irradiated in solutions containing chloride ion, it induces very efficient sequence-neutral DNA cleavage, possibly *via* generation of chlorine atoms [41]. Photosensitization with a tetracationic anthraquinone derivative has been found to selectively target bases in the single stranded loop region of hairpin structures [42].

Menadione (2-methyl-1,4- naphthoquinone) behaves as a type I photosensitizer in its action on double stranded DNA [43] whereas UVA irradiation with 3-*p*-toluidinyl-1,5- azulenequinone is reported to create piperidine-sensitive sites at all guanine residues [44].

2.4.5 *Other examples*

A lysine-naphthalimide conjugate that absorbs strongly at UVA wavelengths is a very effective sensitizer of guanine-specific photodamage, giving a type I profile in duplex DNA but a type II pattern with single strands [45]. D_2O was observed to enhance the piperidine-induced cleavage in the latter case but it had no effect on the cleavage at GG steps in duplex molecules. The introduction of nitro groups into the naphthalimide nucleus alters the photoreactivity so that thymine bases are also modified. Photosensitization with the singlet oxygen generator perinaphthenone [30] produces alkali-labile lesions at all guanines in both single and double stranded DNA, though some weak cleavage at other bases is apparent with the latter. UVA excitation with 2-amino-3-cyano-quinoxaline di-*N*-oxide sensitizes guanine-specific DNA damage under aerobic and anaerobic conditions [46]. Despite ambiguous evidence from quenching experiments concerning the involvement of singlet oxygen, electron transfer is inferred to be the predominant mechanism. In an interesting recent application, Oikawa et al. [47] have found that photosensitization with riboflavin generates Fpg cleavage sites (attributed to 8-oxo-guanine formation) at the central guanines of GGG triplets in oligonucleotides containing telomeric DNA repeats. This supports the contention that telomeric DNA sequences are especially susceptible to oxidative damage.

2.5 Concluding remarks

The gel sequencing approach described here offers a convenient and practical method for screening new compounds as photosensitizers of DNA damage and for assessing their potency. It must be stressed that as it is an 'in vitro' system the results may not extend to chromosomal DNA in vivo. For this to happen, an essential pre-requisite is that the photosensitizer is able to cross the cell membrane and gain access to the nucleus. In favourable cases, this may be established by using the single cell gel electrophoresis (comet) assay. Irrespective of the threat that a photosensitizer poses to intracellular DNA, its action on isolated DNA may give clues as to how it will modify other cellular components. For example, if a photosensitizer shows evidence of type II attack by singlet oxygen it may well have the capacity to cause membrane damage through lipid oxidation.

Ideally, mechanistic conclusions derived from gel sequencing experiments should be confirmed by other means including spectroscopic techniques and product analyses. While the type I and type II mechanisms are readily distinguished on the basis of damage profiles in duplex DNA, the situation is more complicated with single stranded DNA where all guanines are liable to modification by both pathways. Hence the importance of corroborating the involvement of singlet oxygen by examining the effects of quenchers and D_2O on the reaction. An interesting aspect of the type I mechanism concerns the extent to which the damage observed at GG doublets occurs by direct electron donation from these sites to the excited sensitizer as opposed to being caused by hole migration from adjacent bases. As electron transfer is favoured by close association of the reactants, it is curious that small anionic molecules like PBSA and suprofen (Figure 1) which have no measurable binding affinity for native DNA [22,30] nonetheless modify it in type I fashion. It should be possible to gain some insight into this process by incorporating recognised 'hole traps' such as 8-oxo-guanine [48] and 7-deazaguanine [49] at strategic positions in double helical oligonucleotides and monitoring how this affects the cleavage pattern. Owing to their particularly low ionisation potentials, oxidative damage should be funnelled to the hole traps with concomitant protection of the GG and GA doublets that are usually targeted.

In principle, any photosensitizer exhibiting differential activity towards single and double stranded DNA has potential as a probe of DNA secondary structure. PBSA [30] and a tetracationic anthraquinone [42] have, for example, been shown to selectively modify bases in the loop region of hairpin structures. There is considerable scope for more development of this application in regard to other structural motifs and, especially, to major types of secondary structure such as 'Z'-form, triplex and quadruplex DNA. There are also major opportunities to extend studies of photosensitized damage at the sequence level to synthetic and functional RNA molecules. That significant differences may be encountered is illustrated by the case of riboflavin which, though acting as a typical type I photosensitizer towards DNA, cleaves RNA molecules very specifically at G.U base pairs [50].

References

1. E. Kohen, R. Santus, J.G. Hirschberg (1995). *Photobiology*, Academic Press, New York.
2. I.E. Kochevar, D.A. Dunn (1990) Photosensitized reactions of DNA: cleavage and addition. In: H. Morrison (Ed.), *Bioorganic Photochemistry*, (Vol. 1, pp. 273–315). John Wiley & Sons, New York.
3. T.J. Dougherty, C.J. Gomer, B.W. Henderson, G. Jori, D. Kessel, M. Korbelik, J. Moan, Q. Peng (1998). Photodynamic therapy. *J. Nat. Cancer Inst.*, **90**, 889–905.
4. B. Armitage (1998). Photocleavage of nucleic acids. *Chem. Rev.*, **98**, 1171–1200.
5. A.H. Maxam, W. Gilbert (1980). Sequencing end-labeled DNA with base-specific chemical cleavages. *Methods Enzymol.*, **65**, 499–560.
6. R.E. Holmlin, P.J. Dandliker, J.K. Barton (1997). Charge transfer through the DNA base stack. *Angew. Chem. Int. Ed. Engl.*, **36**, 2714–2730.
7. G.B. Schuster (2000). Long-range charge transfer in DNA: transient structural distortions control the distance dependence. *Acc. Chem. Res.* **33**, 253–260.
8. H. Morrison, H. Harmon (2000). 'Hot spots' associated with the photoinduced binding of *cis*-dichloro bis(1,10 phenanthroline)rhodium(III) chloride to HIV-1 and c-raf DNA. *Photochem. Photobiol.*, **72**, 731–738.
9. G.P. Pfeifer, R. Drouin, G.P. Holmquist (1993). Detection of DNA adducts at the DNA sequence level by ligation-mediated PCR. *Mutation Res.*, **288**, 39–46.
10. J. Cadet, P. Vigny (1990). The photochemistry of nucleic acids. In: H. Morrison (Ed.), *Bioorganic Photochemistry*, (Vol. 1, pp. 1–272) John Wiley & Sons, New York.
11. H. Sugiyama, I. Saito (1996). Theoretical studies of GG-specific photocleavage of DNA via electron transfer: significant lowering of ionisation potential and 5′-localization of HOMO of stacked GG bases in B-form DNA. *J. Am. Chem. Soc.*, **118**, 7063–7068.
12. J. Cadet, M. Berger, T. Douki, J.-L. Ravanat (1997). Oxidative damage to DNA: formation, measurement, and biological significance. *Rev. Physiol. Biochem. Pharmacol.*, **131**, 1–87.
13. L. Jacquet, R.J.H. Davies, A. Kirsch-De Mesmaeker, J.M. Kelly (1997). Photoaddition of $Ru(tap)_2(bpy)^{2+}$ to DNA: a new mode of covalent attachment of metal complexes to duplex DNA. *J. Am. Chem. Soc.*, **119**, 11763–11768.
14. J. Sambrook, E.F. Fritsch, T. Maniatis (1989). *Molecular Cloning – A Laboratory Manual*, 2nd ed., Cold Spring Harbor Laboratory Press, Cold Spring Harbor.

15. W.A. Haseltine, C.P. Lindan, A.D. D'Andrea, L. Johnsrud (1980). The use of DNA fragments of defined sequence for the study of DNA damage and repair. *Methods Enzymol.*, **65**, 235–248.
16. S. Starrs, R.J.H. Davies (1999). Prevention of anomalous band mobility in A+G sequencing ladders of 3′-end labelled DNA. *BioTechniques*, **27**, 36–46.
17. E. Sage, T. Le Doan, V. Boyer, D.E. Helland, L. Kittler, C. Helene, E. Moustacchi (1989). Oxidative DNA damage photo-induced by 3-carbethoxypsoralen and other furocoumarins. *J. Mol. Biol.*, **209**, 297–314.
18. K. Ito, S. Inoue, K. Yamamoto, S. Kawanishi (1993). 8-Hydroxydeoxyguanosine formation at the 5′ site of 5′-GG-3′ sequences in double-stranded DNA by UV radiation with riboflavin. *J. Biol. Chem.*, **268**, 13221–13227.
19. S. Kawanishi, S. Inoue, S. Sano, H. Aiba (1995) Photodynamic guanine modification by hematoporphyrin is specific for single-stranded DNA with singlet oxygen as a mediator. *J. Biol. Chem.*, **261**, 6090–6095.
20. F. Boscá, M.L. Marín, M.A. Miranda (2001). Photoreactivity of the nonsteroidal anti-inflammatory 2-arylpropionic acids with photosensitizing side effects. *Photochem. Photobiol.*, **74**, 637–655.
21. F. Boscá, M.A. Miranda (1998). Photosensitizing drugs containing the benzophenone chromophore. *J. Photochem. Photobiol. B: Biol.*, **43**, 1–26.
22. S.M. Starrs, R.J.H. Davies (2000). Sequence specificity of alkali-labile DNA damage photosensitized by suprofen. *Photochem. Photobiol.*, **72**, 291–297.
23. S. Sortino, G. De Guidi, G. Marconi, S. Monti (1998). Triplet photochemistry of suprofen in aqueous environment and in the β-cyclodextrin inclusion complex. *Photochem. Photobiol.*, **67**, 603–611.
24. I. Saito, M. Takayama, H. Sugiyama, K. Nakatani, A. Tsuchida, M. Yamamoto (1995). Photoinduced DNA cleavage via electron transfer: demonstration that guanine residues located 5′ to guanine are the most electron-donating sites. *J. Am. Chem. Soc.*, **117**, 6406–6407.
25. M.E. Hogan, T.F. Rooney, R.H. Austin (1987). Evidence for kinks in DNA folding in the nucleosome. *Nature*, **328**, 554–557.
26. T. Douki, J. Cadet (1999). Modification of DNA bases by photosensitized one-electron oxidation. *Int. J. Radiat. Biol.*, **75**, 571–581.
27. J.-L. Ravanat, J. Cadet (1995). Reaction of singlet oxygen with 2′-deoxyguanosine and DNA. Isolation and characterization of the main oxidation products. *Chem. Res. Toxicol.*, **8**, 379–388.
28. V. Lhiaubet, N. Paillous, N. Chouini-Lalanne (2001). Comparison of DNA damage photoinduced by ketoprofen, fenofibric acid and benzophenone via electron and energy transfer. *Photochem. Photobiol.*, **74**, 670–678.
29. K. Ito, S. Kawanishi (1997). Photoinduced hydroxylation of deoxyguanosine in DNA by pterins: sequence specificity and mechanism. *Biochemistry*, **36**, 1774–1781.
30. C. Stevenson, R.J.H. Davies (1999). Photosensitization of guanine-specific DNA damage by 2-phenylbenzimidazole and the sunscreen agent 2-phenylbenzimidazole-5-sulfonic acid. *Chem. Res. Toxicol.*, **12**, 38–45.
31. J.J. Inbaraj, P. Bilski, C.F. Chignell (2002). Photophysical and photochemical studies of 2-phenylbenzimidazole and UVB sunscreen 2-phenylbenzimidazole-5-sulfonic acid. *Photochem. Photobiol.*, **75**, 107–116.
32. J. Ferguson (1995). Fluoroquinolone photosensitization: a review of clinical and laboratory studies. *Photochem. Photobiol.*, **62**, 954–958.
33. R.E. Osterberg, A. Szarfman (1996). Assessment of risk for photocarcinogenesis: regulatory reviewer viewpoint. *Photochem. Photobiol.*, **63**, 362–365.

34. Y. Hiraku, S. Kawanishi (2000). Distinct mechanisms of guanine-specific DNA photodamage induced by nalidixic acid and fluoroquinolone antibacterials. *Arch. Biochem. Biophys.*, **382**, 211–218.
35. S.M. Starrs (2000). Ph.D. Thesis, Queen's University, Belfast.
36. E. Fasani, A. Profumo, A. Albini (1998). Structure and medium-dependent photodecomposition of fluoroquinolone antibiotics. *Photochem. Photobiol.*, **68**, 666–674.
37. P.E. Nielsen, C. Jeppesen, M. Egholm, O. Buchardt (1988). Adenosine-guanosine preferential photocleavage of DNA by azido-benzoyl- and diazocyclopentadienylcarbonyloxy derivatives of 9-aminoacridine. *Nucleic Acids Res.*, **16**, 3877–3888.
38. S. Sauvaigo, T. Douki, F. Odin, S. Caillat, J.-L. Ravanat, J. Cadet (2001). Analysis of fluoroquinolone-mediated photosensitization of 2′-deoxyguanosine, calf thymus and cellular DNA: determination of type-I, type-II and triplet-triplet energy transfer mechanism contribution. *Photochem. Photobiol.*, **73**, 230–237.
39. D.T. Breslin, C. Yu, D. Ly, G.B. Schuster (1997). Structural modification changes the DNA binding mode of cation-substituted anthraquinone photonucleases: association by intercalation or minor groove binding determines the DNA cleavage efficiency. *Biochemistry*, **36**, 10463–10473.
40. D.T. Breslin, J.E. Coury, J.R. Anderson, L. McFail-Isom, Y. Kan, L.D. Williams, L.A. Bottomley, G.B. Schuster (1997). Anthraquinone photonuclease structure determines its mode of binding to DNA and the cleavage chemistry observed. *J. Am. Chem. Soc.*, **119**, 5043–5044.
41. B. Armitage, G.B. Schuster (1997). Anthraquinone photonucleases: a surprising role for chloride in the sequence-neutral cleavage of DNA and the footprinting of minor groove-bound ligands. *Photochem. Photobiol.*, **66**, 164–170.
42. P.T. Henderson, B. Armitage, G.B. Schuster (1998). Selective photocleavage of DNA by anthraquinone derivatives: targeting the single-strand region of hairpin structures. *Biochemistry*, **37**, 2991–3000.
43. K. Ito, S. Kawanishi (2000). Sequence specificity of ultraviolet A-induced DNA damage in the presence of photosensitizer. *Methods Enzymol.*, **319**, 417–427.
44. T. Nozoe, C.C. Lin, S.-C. Tsay, S.-F. Yu, L.C. Lin, P.W. Yang, J.R. Hwu (1997). Single-strand cleavage of DNA with site-specificity by photolysis of azulenequinones. *Bioorg. Med. Chem. Lett.*, **7**, 975–978.
45. I. Saito, M. Takayama, S. Kawanishi (1995). Photoactivatable DNA-cleaving amino acids: highly sequence-selective DNA photocleavage by novel L-lysine derivatives. *J. Am. Chem. Soc.*, **117**, 5590–5591.
46. T. Fuchs, K.S. Gates, J.-T. Hwang, M.M. Greenberg (1999). Photosensitization of guanine-specific DNA damage by a cyano-substituted quinoxaline di-*N*-oxide. *Chem. Res. Toxicol.*, **12**, 1190–1194.
47. S. Oikawa, S. Tada-Oikawa, S. Kawanishi (2001). Site-specific DNA damage at the GGG sequence by UVA involves acceleration of telomere shortening. *Biochemistry*, **40**, 4763–4768.
48. S. Steenken, S.V. Jovanovic, M. Bietti, K. Bernhard (2000). The trap depth (in DNA) of 8-oxo-7,8-dihydro-2′-deoxyguanosine as derived from electron-transfer equilibria in aqueous solution. *J. Am. Chem. Soc* , **122**, 2373–2374.
49. K. Nakatani, C. Dohno, I. Saito (2000). Modulation of DNA-mediated hole-transport efficiency by changing superexchange electronic interaction. *J. Am. Chem. Soc.*, **122**, 5893–5894.
50. P. Burgstaller, M. Famulok (1997). Flavin-dependent photocleavage of RNA at G.U base pairs. *J. Am. Chem. Soc.*, **119**, 1137–1138.

Chapter 3

DNA damage induced by UVA radiation: role in solar mutagenesis

Evelyne Sage, Daniel Perdiz, Pál Gróf, Anne Reynaud-Angelin, Thierry Douki, Jean Cadet, Patrick J. Rochette, Nathalie Bastien and Régen Drouin

Table of contents

Abstract

Exposure to solar ultraviolet radiation is a major risk factor in the development of skin cancer. Bipyrimidine photoproducts, i.e. cyclobutane pyrimidine dimers (CPDs) and (6-4) photoproducts, caused in DNA by UVB radiation are presumed to constitute a primary event in the initiation of skin tumours. There has been little attention on DNA damage induced by the UVA counterpart and by sunlight. However, consecutively to the increased human exposure to UVA, the deleterious effect of such radiation has recently appeared as a source of concern in public health. We highlight here the main findings that have emerged regarding the formation of bipyrimidine photoproducts and oxidative damage by UVA radiation and simulated sunlight (SSL) in mammalian cells, and their correlation with mutations. In particular, we report the formation of CPDs by UVA radiation, predominantly at TT sites, and the induction of 8-oxo-7,8-dihydroguanine (8-oxoGua), though with a lower efficiency, whereas other oxidized base damage and strand breaks are rather infrequently formed. Taken together, the present data support the idea that bipyrimidine photoproducts (CPDs) rather than photooxidative damage may be responsible for the genotoxic effect of UVA in mammalian cells, a notion which is confirmed by the mutational specificity of UVA. In addition, the induction of CPD at biologically relevant doses of UVA radiation lead us to suggest that UVA radiation may be involved in solar mutagenesis.

3.1 Introduction

Clinical, epidemiological and molecular evidence demonstrated that DNA damage and the subsequent mutations induced by the ultraviolet (UV) component of sunlight are critical events in the incidence of skin tumours [1]. Mutations generated in gatekeeper genes are in fact primary factors in the initiation of multistage photocarcinogenesis. UVB ($\lambda > 295$ nm) and UVA (320–340 nm), the environmentally relevant UV radiation, constitute approximately 0.4 and 5.5% of the terrestrial solar energy, respectively. The two predominant DNA lesions caused by direct absorption of UVB photons, i.e. *cis-syn* cyclobutane pyrimidine dimers (CPDs) and pyrimidine (6-4) pyrimidone ((6-4) photoproducts), are assumed to be responsible for the mutations observed in skin tumours. The UVA component of solar radiation is extremely poorly absorbed by DNA, and rather excites other endogenous chromophores, generating reactive oxygen species, some of which possibly involved in the outcome of photocarcinogenesis [1,2]. UVA radiation also damages DNA, either through the generation of singlet oxygen or by a type-1 photosensitization reaction, resulting in photooxidation of DNA bases, mainly guanine [3,4]. The widespread use of efficient UVB-absorbing sunscreens which is often associated with greatly prolonged periods of sunbathing, has resulted in an increase of human exposure to UVA. The deleterious effect of UVA has, as a consequence, recently emerged as a source of concern for public health. In order to better understand the relative contribution of the different UV components in solar mutagenesis, and in particular the role of UVA, we investigated the formation of bipyrimidine photoproducts and oxidative DNA damage in Chinese hamster ovary (CHO) cells upon exposure to either UVB, UVA or simulated sunlight (SSL, $\lambda > 300$ nm), as well as to the non-solar model 254-nm UVC. The damage distribution which greatly varies depending on the wavelength, was further compared to the mutational specificity of these different types of radiation which was previously reported [5–7].

3.2 Induction of bipyrimidine photoproducts by solar UV radiation

We quantitatively analyzed the induction of each of the three bipyrimidine photoproducts, i.e. CPDs, (6-4) photoproducts and their Dewar valence isomers (Dewar photoproducts) in CHO cells exposed to UVC, UVB, UVA or simulated sunlight (SSL), using a calibrated immunodotblot technique [8]. The monoclonal antibodies employed, TDM-2, 64M-2 and DEM-1, exhibited a higher affinity for the photolesions at TT and TC sites than at CT and CC sites [9]. The high sensitivity of the method allowed the quantitative assessement of the three photolesions in the same irradiated DNA sample, at physiologically relevant doses, i.e. those giving low toxicity in cultured cells and corresponding to normal human environment.

CPDs were readily produced by UVC, UVB and SSL, and less expectedly by UVA. In fact, the first observation on CPDs induction by radiation of wavelengths >320 nm is not recent [10]. It was observed in bacteria, mammalian cells and human skin [10–12]; it was, however, an object of controversy [13]. Recently, this was reassessed in various laboratories including ours, using more sensitive methods [3,8,14–16]. CPDs were detected at similar yields ($2\text{–}3 \cdot 10^{-7}$ /kbp /Jm^{-2}) in plasmid DNA or DNA from cells irradiated with UVA [8,15]. As compared to UVC, we found that CPDs were produced 10^2 and 10^5 fold less efficiently by UVB and UVA, respectively [8,15]. This observed rate of CPDs formation by UVA could be compatible with direct absorption of the shortest UVA photons (320–360 nm) by DNA [17]. However, since the induction of CPDs by UVA has been a matter of debate for over 20 years, we calculated the relative spectral effectivenesses, which correspond to the relative efficiency of the different regions of UV spectrum at producing a damage (CPDs) using the action spectrum for this lesion (from [3]) and the spectral distribution of our UV sources (from [15]). It showed that the majority of CPDs formed by our UVA irradiation device was induced by photons above 345 nm, and that no more than 12% of the CPDs were produced by the rare UVB photons present in the emission spectrum of our lamp [8,16,18]. This demonstrated that, indeed, CPDs are formed by UVA radiation, at doses compatible with human exposure. In the meantime, this raised the question on the mechanism of CPD formation at such wavelengths. In contrast, no more than 11% of CPDs produced by SSL arose from photons in the UVA range, that comprised 8% by UVA2 wavelengths (320–345 nm) [8,16,18].

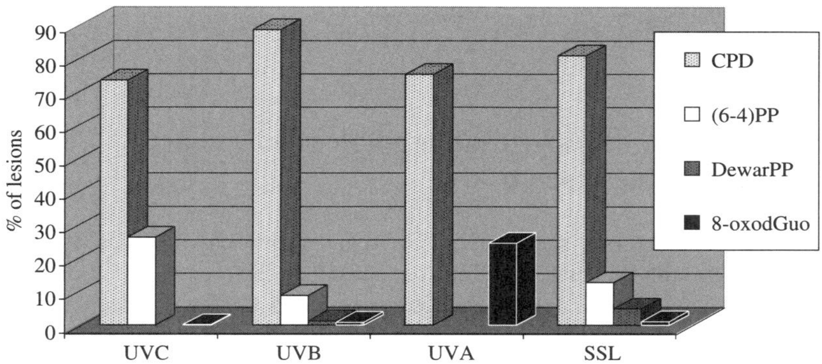

Figure 1. The relative induction of CPDs, (6-4), Dewar photoproducts and 8-oxodGuo varies depending on the type of UV radiation. [From Douki et al. [16] and Perdiz et al. 8].

Under our irradiation conditions, (6-4) photoproducts were readily produced by UVC, UVB and SSL, but could not be detected following UVA exposure. In addition, Dewar photoproducts were not observed after UVC and UVA irradiation, at the biologically relevant doses used [8].

We show in Figure 1 and Table 1 that the relative formation of bipyrimidine photoproducts greatly varied depending on the type of radiation. CPDs were the quantitatively most important photolesions produced by all types of radiation. The ratio of the Dewar to the (6-4) photoproducts was 1 to 8 for UVB irradiation, but 1 to 3 for SSL, as revealed by calibrated immunodotblots. In line with this last finding it was reported that in human cells exposed to natural sunlight (6-4) photoproducts were produced at a low level and did not increase with duration of exposure, in opposite to Dewar photoproducts whose induction increased largely as a function of time exposure [19]. Together, these observations demonstrate that Dewar photoproducts are a quantitatively important class of DNA photodamage produced by natural and simulated sunlight. It also suggests that UVA wavelengths are mainly responsible for

Table 1. Distribution of bipyrimidine photoproducts and 8-oxodGuo within DNA of CHO cells exposed to solar UV radiation*

	UVC	UVB	UVA	SSL
(6-4)/CPD	1:4	1:8		1:6
Dewar/CPD		1:70		1:16
Dewar/(6-4)		1:8		1:3
CPD/8oxodGuo	≈800	≈100	≈3	≈40
Dewar/8oxodGuo		≈1		≈5

*From Douki et al. [16] and Perdiz et al. [8]

the photoisomerisation of the UVB-induced (6-4) photoproducts into Dewar photoproducts

3.3 Distribution of bipyrimidine photoproducts at the nucleotide resolution

There were some reports mentioning that the distribution of CPDs at the four possible bipyrimidine sequences differs, depending on the type of radiation [20,21].

A recently developed and highly sensitive HPLC-MS/MS technique [22] was used to simultaneously and individually quantify CPDs, (6-4) and Dewar photoproducts at TT, TC, CT and CC sites, in CHO cells exposed to UVB, UVA and SSL [23]. Following UVB, most of the photolesions were observed at TT and TC sites, with a ratio of CPDs to (6-4) photoproducts of 3, whereas CPDs at CC sites were rare and Dewar photoproducts not detected at the low doses employed. In contrast, upon UVA irradiation, only T-containing CPDs were detected, with a large majority of CPDs at TT sites (Figure 2). As described above, (6-4) and Dewar photoproducts were not observed. In the case of SSL, the proportion of CPDs at TT sites was higher than for UVB, probably reflecting the UVA contribution. More remarkable was the high yield of formation of the Dewar photoproducts at the expense of the corresponding (6-4) photoproducts (Figure 2). The (6-4) photoproducts at TC sites were more frequent than at TT sites, but less readily converted into their Dewar isomers. These observations were confirmed by *in vitro* photoconversion experiments on UVC-irradiated DNA further exposed to UVA radiation [23]. We conclude that the UVA photons comprised in SSL (or natural sunlight), as opposed to UVB, are not shielded by the DNA bases and are absorbed

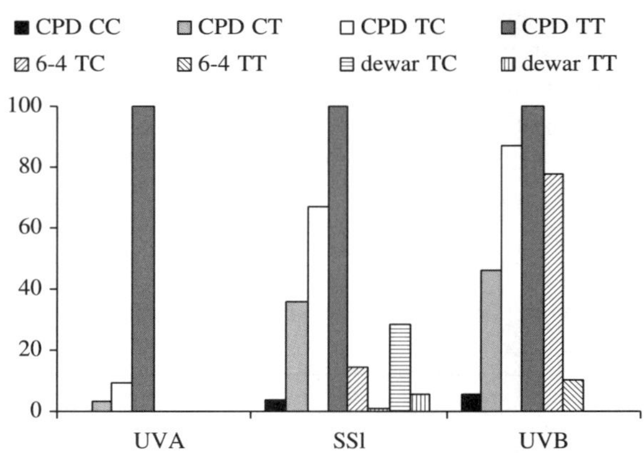

Figure 2. Distribution of CPDs, (6-4) and Dewar photoproducts at the four possible bipyrimidine sites within DNA of CHO cells exposed to either UVA, UVB or SSL, as revealed by HPLC-MS/MS. The yields of formation were obtained by linear regression of the amount of lesions with respect to the dose, and a normalization on the induction of CPDs at TT bipyrimidine (100%) was applied.

by (6-4) photoproducts which exhibit maximal absorption at 315 nm and 325 nm for TC and TT derivatives, respectively. Thus, the (6-4) photoproducts initially produced by UVB are further converted into their Dewar isomers upon excitation by UVA photons.

Using ligation-mediated PCR (LMPCR), we determined the frequency distribution of CPDs at the sequence level along the exon 2 of the adenylphosphoribosyl transferase *(aprt)* gene in CHO cells exposed to UVB (295–320 nm), UVA (340–400 nm) or simulated sunlight (SSL), for comparison with the corresponding mutation spectra previously generated at this locus [5–7]. This powerful technique allows to map CPDs at each bipyrimidine site, and the quantification method associated with, provides the relative quantity of CPDs formed at each of these sites [24]. Forty-four bipyrimidine sites (5 TT, 19 CC, 9 CT and 11 CT sites) on the non-transcribed strand and 21 (3 TT, 5 CC, 7 CT and 6 TC sites) on the transcribed strand of *aprt* gene were analyzed. Results [25] showed that almost all bipyrimidine sites were targets for CPD formation by the three types of radiation. However, the band patterns obtained upon UVB and SSL exposures were rather similar and drastically differed from that produced following UVA irradiation (Figure 3). Indeed, in contrast to what was observed for UVB and SSL, after UVA irradiation a large majority of CPDs was formed at TT bipyrimidine sites, and those produced at CC sites were poorly represented. These data confirmed the preferential induction of CPDs at TT sites by UVA, as observed on total genomic DNA using HPLC-MS/MS.

The principal induction of CPDs at T-containing bipyrimidine sites, and mostly at TT sites is a reminiscence of what was observed by photosensitization reaction of DNA by aromatic ketones [26,27]. We suggest that following cells irradiation with UVA, most of CPDs are formed through triplet energy transfer to thymine which has the lowest triplet state energy with regard to other bases. The UVA-absorbing chromophore tighly bound to DNA remains to be identified.

3.4 Detection of oxidative DNA damage induced by solar UV radiation

Until the 90's, the main reported lesions in UVA-irradiated cells were single strand breaks, possibly produced *via* hydroxyl radicals [28]. More recently, we and others measured the induction of 8-oxo-7,8-dihydro-2'-deoxyguanosine (8-oxodGuo) by UVA radiation in human and CHO cells, using either an enzymatic assay or HPLC associated with electrochemical detection [3,14,16,29,30]. As expected, 8-oxodGuo was readily formed after UVA irradiation, whereas it was weakly produced after UVB and barely detected following exposure to biologically relevant doses of UVC, as shown in Figure 1. Meanwhile, after UVA irradiation, 8-oxodGuo was less efficiently produced than CPDs, and the ratio of CPDs to 8-oxodGuo was approximately 3 (Table 1). Such a ratio was reproducibly observed using different UVA sources and various assays for damage measurements (Douki and Sage, unpublished).

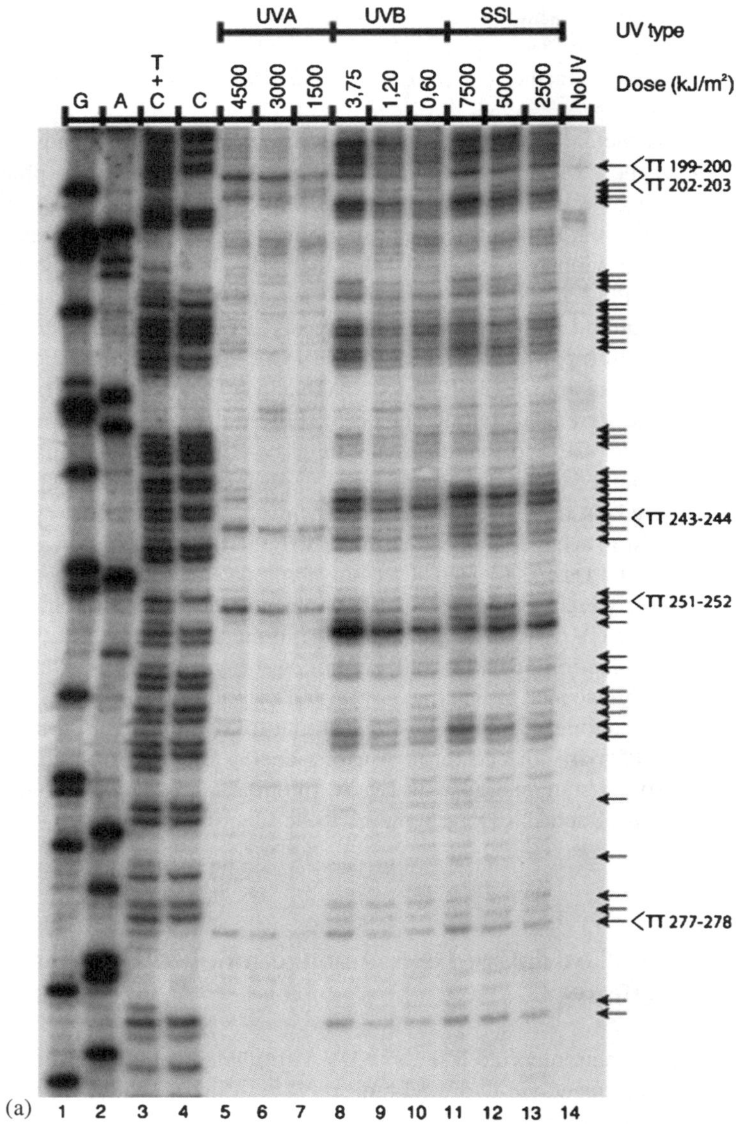

Figure 3. Distribution of CPDs induced by UVA, UVB and SSL, along the non-transcribed (a) and transcribed (b) strands of the *aprt* exon 2 gene of CHO cells. Cells were either unirradiated (No UV) or irradiated as indicated and CPD formation was analyzed by LMPCR using the T4 endonuclease. The first four lanes from the left of the autoradiogram exhibit LMPCR of DNA treated in standard Maxam-Gilbert cleavage reactions. Arrows indicate bipyrimidine sites quantified with phosphorimager and represent adducted positions. Following UVB and SSL irradiation, CPDs were distributed among all possible dimeric sites, TT, TC, CT and CC, with various frequencies depending on the sequence context. In (a) the two strong bands that show up upon UVA exposure (well visible at the dose of 4500 kJm^{-2}) are located at TT positions. Much weaker bands appear at the other bipyrimidine sites for the highest doses. Figure 3a is from [25].

(b) 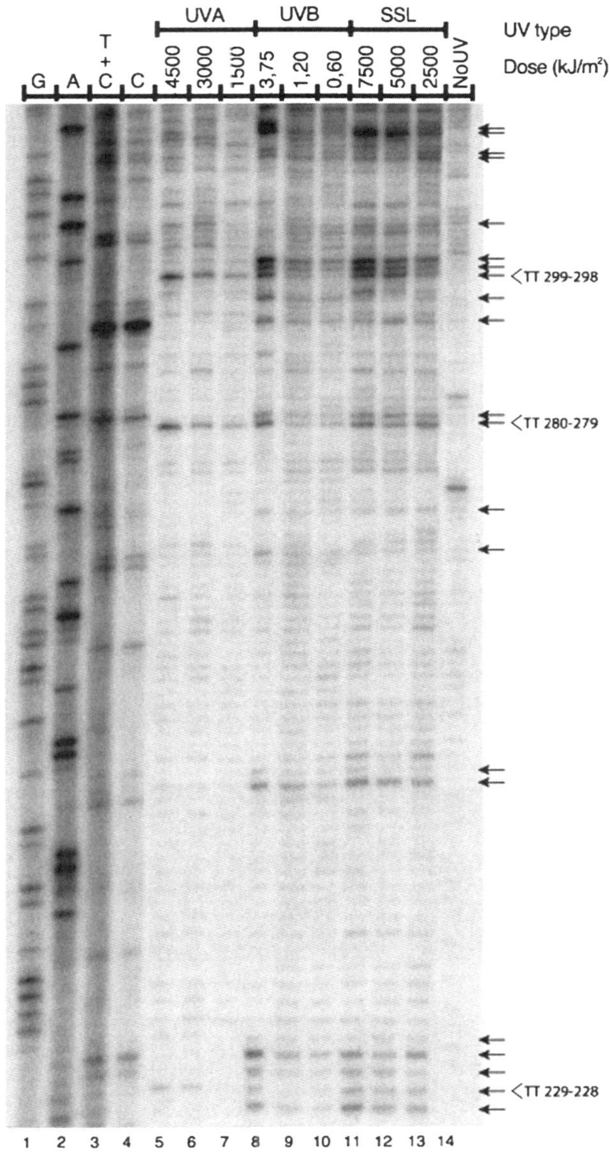

Figure 3. (Continued)

We explored the possibilility that UVA could also produce other oxidative damage, in particular at pyrimidine bases. CHO cells were exposed to UVA, UVB or SSL, and DNA was extracted and incubated in the presence or in the absence of DNA repair enzymes, prior to alkaline agarose gel electrophoresis. Fpg and Nth proteins from *Escherichia coli* revealed oxidation products of purine (mainly 8-oxodGuo) and pyrimidine respectively, whereas denV protein from T4 phage served to detect CPDs. The induction of four classes of lesions, i.e. single strand breaks, oxidized pyrimidines, oxidized purines and CPDs, could be observed in a single assay. Figure 4 shows that Nth-sensitive sites were not detected after UVB or SSL irradiation, but appeared only at high doses of UVA. Fpg-sensitive sites were not observed after UVB, barely detected after SSL, but clearly revealed in a dose-dependent manner after UVA. In contrast, CPDs were the major class of photoproducts formed by UVB, SSL, and by UVA radiation as well. The yields of formation of these various classes of damage could be assessed from electrophoretic profiles. After UVA irradiation, the comparative rates of formation were approximatively 1: 0.3: 0.1: 0.1 for CPDs, Fpg-sensitive sites, Nth-sensitive sites and strand breaks plus alkaline sensitive sites, respectively [23]. The very low frequency of oxidized pyrimidine (Nth-sensitive sites) was in accord with previously reported data obtained by alkaline elution and comet assay [3,30]. It was also in line with the lack of detection of various oxidized pyrimidine bases,

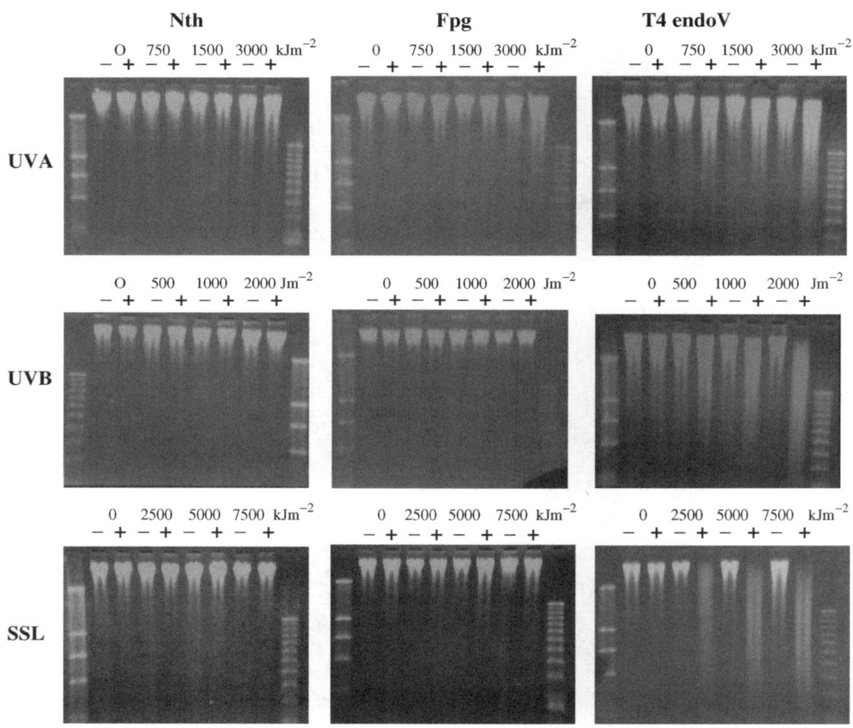

Figure 4. Gel electrophoresis analysis of DNA photolesions induced in CHO cells exposed to UVA, UVB or Simulated Sunlight (SSL). After irradiation of CHO cells by UVA, UVB or SSL, at the doses indicated, DNA was immediately extracted and incubated in the presence (+) or in the absence (−) of bacterial Nth or Fpg protein, or T4 endonuclease V protein, which cleaved DNA at oxidized pyrimidines, oxidized purines and CPD respectively. Cleavage products are resolved by 0.6% alkaline agarose gel electrophoresis. The size markers in the first and last lanes of each gel are the λ DNA-HindIII digest marker and the 1kb DNA ladder.

including thymine glycol, by HPLC-MS/MS, whereas 8-oxodGuo was unambiguously observed (Douki and Sage, unpublished). The low yield of production of oxidized pyrimidine and single strand breaks was in favor of a minor contribution of hydroxyl radicals in UVA-induced damage. 8-oxodGuo would be rather produced by singlet oxygen or *via* a type I process [30].

Altogether, this last series of results confirmed our previous data [16] and led us to conclude that although UVA radiation produces oxidative DNA damage (mainly 8-oxodGuo), CPDs are the most frequent photolesions formed.

3.5 DNA damage and mutation induced by solar UV radiation

The mutational specificities of UVC, UVB, UVA and SSL were characterized at the *aprt* locus in CHO cells proficient and deficient in nucleotide excision repair [5–7]. Mutation spectra obtained for UVC and UVB were characterized by a majority of GC to AT transition at bipyrimidine sites, suggesting bipyrimidine photoproducts as premutagenic lesions [31]. This class of base substitution represented about one-third of SSL-induced mutation. Moreover, a correlation was observed between CPDs damage hotspots along exon 2 of the *aprt* gene revealed by LMPCR and mutational hotspots, for UVC, UVB, SSL [25]. However, at most of these sites (6-4) photoproducts could also be formed. As those are quickly removed by nucleotide excision repair, they should not constitute a major premutagenic lesion in repair-proficient cells, as opposed to CPDs [32,33]. After SSL irradiation, a significant decrease in GC to AT base substitution and increase in tandem CC to TT mutation were observed [5]. Since a large fraction of (6-4) photoproducts formed are photoisomerized into Dewar photoproducts upon SSL exposure (see above), we wonder if Dewar photoproducts, a poorly repaired class of damage [8,34], could not be responsible for some of the tandem CC to TT mutations.

Upon UVA irradiation, GC to AT transitions represented 22% and 65% of the overall base changes in repair proficient and deficient cells, respectively, and are recovered at bipyrimidine sites [6,7]. In addition, the induction of CPDs at TC and CC sites revealed by LMPCR correlated with GC to AT transitions and tandem CC to TT changes at mutational hotspots [25]. Since in this case neither (6-4) nor Dewar photoproducts were formed, this strongly designated CPDs as a culprit. Although CC to TT tandem mutations have been recovered on plasmid treated *in vitro* with reactive oxygen species and replicated in bacteria [35], the implication of oxidized pyrimidine bases causing those classes of mutation by UVA is unprobable, since oxidative damage at pyrimidine sites was rarely observed after UVA (see above). Even though an unfrequent but highly mutagenic oxidative product cannot be totally excluded at these positions, it is likely that C-containing-CPDs are the premutagenic lesions responsible for the GC to AT base changes and the few tandem CC to TT mutations observed after UVA.

In repair proficient CHO cells, the most prevalent UVA-induced mutational class was AT to CG transversion [6]. Such events were also frequently recovered in SSL-induced mutation spectrum, as opposed to what observed after UVC or UVB. This could be attributed to the UVA component of SSL. In human cells, GC to AT transitions still predominated the UVA-induced mutation spectrum in the *lac*Z' gene carried on plasmid, although an increase in mutations at AT base pairs was observed in comparison with UVB [36]. Most notably, the highly frequent CPDs formed by UVA at TT sites along the exon2 of the *aprt* gene in CHO cells corresponded well with the frequently mutated sites where AT to CG changes occurred [25]. This observation was in favor of an involvement of CPDs at TT sites in the generation of mutations at AT base pairs following UVA exposure.

8-oxodGuo mainly induces GC to TA transversions. This class of mutation was poorly represented after UVC or UVB irradiation, in accord with the low induction of this oxidized base. Surprisingly, even though 8-oxodGuo was the second most frequent lesion formed by UVA, GC to TA mutations were not significantly recovered following UVA and SSL exposure [6]. This indicated that this 8-oxodGuo is well repaired and poorly mutagenic in mammalian cells. It is in contrast to what observed in *E. coli* where near-UV radiation caused an increase in GC to TA transversions which depended on the presence of Fpg protein involved in the repair of 8-oxodGuo [37]. In yeast also, 8-oxodGuo seemed to play a role in mutagenesis by sunlight [38], SSL and UVA radiation [39, Kozmin and Sage, unpublished].

Altogether, these observations strongly suggest that CPDs are the major premutagenic photolesions in UVA-induced mutagenesis in mammalian cells, as opposed to oxidative damage.

3.6 Conclusion: the dark side of UVA

1- In contrast to what initially speculated, oxidative DNA damage is not frequently produced by UVA radiation. Single strand breaks and oxidation products at pyrimidines are rather infrequent, whereas 8-oxodGuo is the main oxidative lesion readily detected. However, GC to TA transversion, the hallmark of 8-oxodGuo, is poorly recovered in mammalian cells exposed to UVA radiation. This indicates that 8-oxodGuo, or other oxidative damage, plays a minor role in UVA-induced mutagenesis in mammalian cells.

2- CPDs represent the major class of damage produced by UVA in mammalian cells. The frequency of C-containing CPDs correlates well with the occurrence of GC to AT transitions at TC, CC and CT sites. The most prevalent type of CPDs is formed at TT sites, likely *via* photosensitization (the photosensitizer(s) has to be identified). CPDs at TT sites seem to correlate with the frequently recovered mutations at AT base pairs. Taken together, these results suggest that CPDs are the major premutagenic damage in UVA-induced mutagenesis in mammalian cells.

3- (6-4) and Dewar photoproducts are not formed after UVA irradiation at biologically relevant doses.

4- However, UVA photons comprised in SSL are responsible for the photoisomerisation of (6-4) photoproducts, a well repaired photolesion produced by UVB, into the persistent and very mutagenic Dewar photoproducts.

5- The contribution of UVA wavelengths appears in the mutation spectrum produced by simulated sunlight.

It is worth mentioning that such conclusions were drawn from experiments performed in rodent cultured cells exposed to rigorously filtered UVA radiation [18], at doses which correspond to sunlight exposures of approximately 1 hr 30 min to 5 hrs at zenith in summer at Paris latitude, according to [15]. Some of these points deserve to be reinvestigated in human keratinocytes and melanocytes. However, our data indicate that UVA radiation may be involved in solar mutagenesis, and clearly point out CPDs as the major premutagenic lesions, whereas oxidative damage, in particular 8-oxodGuo, would play a minor role, in mammalian cells.

Acknowledgements

This work was supported by grants from la Ligue Nationale Française contre le Cancer, Institut Curie (Genotoxicology program), Centre National de la Recherche Scientifique to ES, National Cancer Institute of Canada (NCIC) (with funds from the Canadian Cancer Society and the Terry Fox Run), to RD. DP is the recipient of a PhD fellowship from the Ministère de l'Éducation Nationale de l'Enseignement Supérieur et de la Recherche and from the Fondation pour la Recherche Médicale. PJR holds student fellowships from the Canadian Institute for Health Research (CIHR). RD is a scholar of the Fonds de la Recherche en Santé du Québec (FRSQ).

References

1. H.S. Black, F.R. deGruijl, P.D. Forbes, J.E. Cleaver, H.N. Ananthaswamy, E.C. deFabo, S.E. Ullrich, R.M. Tyrrell (1997). Photocarcinogenesis: an overview. *J. Photochem. Photobiol. B: Biol.*, **40**, 29–47.
2. R.M. Tyrrell, S.M. Keyse (1990). New trends in photobiology. The interaction of UVA radiation with cultured cells. *J. Photochem. Photobiol. B: Biol.*, **4**, 349–361.
3. C. Kielbassa, L. Roza, B. Epe (1997). Wavelength dependence of oxidative DNA damage induced by UV and visible light. *Carcinogenesis*, **18**, 811–816.
4. J.L. Ravanat, T. Douki, J. Cadet (2001). Direct and indirect effects of UV radiation on DNA and its components. *J. Photochem. Photobiol. B: Biol.*, **63**, 88–102.
5. E.A. Drobetsky, E. Moustacchi, B.W. Glickman, E. Sage (1994). The mutational specificity of simulated sunlight at the aprt locus in rodent cells. *Carcinogenesis*, **15**, 1577–1583.
6. E.A. Drobetsky, J. Turcotte, A. Chateauneuf (1995). A role for ultraviolet A in solar mutagenesis. *Proc. Natl. Acad. Sci. USA*, **92**, 2350–2354.
7. E. Sage, B. Lamolet, E. Brulay, E. Moustacchi, A. Chateauneuf, E.A. Drobetsky (1996). Mutagenic specificity of solar UV light in nucleotide excision repair-deficient rodent cells. *Proc. Natl. Acad. Sci. USA*, **93**, 176–180.
8. D. Perdiz, P. Gróf, M. Mezzina, O. Nikaido, E. Moustacchi, E. Sage (2000). Distribution and repair of bipyrimidine photoproducts in solar UV- irradiated mammalian cells. Possible role of Dewar photoproducts in solar mutagenesis. *J. Biol. Chem.*, **275**, 26732–26742.
9. T. Matsunaga, Y. Hatakeyama, M. Ohta, T. Mori, O. Nikaido (1993). Establishment and characterization of a monoclonal antibody recognizing the Dewar isomers of (6-4) photoproducts. *Photochem. Photobiol.*, **57**, 934–940.

10. R.M. Tyrrell (1973). Induction of pyrimidine dimers in bacterial DNA by 365 nm radiation. *Photochem. Photobiol.*, **17**, 69–73.
11. I.C. Enninga, R.T.L. Groenendijk, A.R. Filon, A.A. van Zeeland, J.W.I.M. Simons (1986). The wavelength dependence of UV-induced pyrimidine dimer formation, cell killing and mutation induction in human diploid skin fibroblast. *Carcinogenesis*, **7**, 1829–1836.
12. S.E. Freeman, H. Hacham, R.W. Gange, D.J. Maytum, J.C. Sutherland, B.M. Sutherland (1989). Wavelength dependence of pyrimidine dimer formation in DNA of human skin irradiated *in situ* with ultraviolet light. *Proc. Natl. Acad. Sci. USA*, **86**, 5605–5609.
13. H. Hacham, S.E. Freeman, R.W. Gange, D.J. Maytum, J.C. Sutherland, B.M. Sutherland (1990). Does exposure of human skin *in situ* to 385 or 405 nm UV induce pyrimidine dimers in DNA? *Photochem. Photobiol.*, **52**, 893–896.
14. X. Zhang, B.S. Rosenstein, Y. Wang, M. Lebwohl, D.L. Mitchell, H. Wei (1997). Induction of 8-oxo-7,8-dihydro-2'-deoxyguanosine by ultraviolet radiation in calf thymus DNA and HeLa cells. *Photochem. Photobiol.*, **65**, 119–124.
15. Z. Kuluncsics, D. Perdiz, E. Brulay, B. Muel, E. Sage (1999). Wavelength dependence of ultraviolet-induced DNA damage distribution: involvement of direct or indirect mechanisms and possible artifacts. *J. Photochem. Photobiol. B: Biol.*, **49**, 71–80.
16. T. Douki, D. Perdiz, P. Grof, Z. Kuluncsics, E. Moustacchi, J. Cadet, E. Sage (1999). Oxidation of guanine in cellular DNA by solar UV radiation: biological role. *Photochem. Photobiol.*, **70**, 184–190.
17. J.C. Sutherland, K.P. Griffin (1981). Absorption spectrum of DNA for wavelengths greater than 300 nm. *Radiat. Res.*, **86**, 399–409.
18. P. Gróf, G. Rontó, E. Sage (2002). A computional study of physical and biological characterization of common UV sources and filters, and their relevance for substituting sunlight. *J. Photochem. Photobiol. B: Biol.*, **68**, 53–59.
19. P.H. Clingen, C.F. Arlett, L. Roza, T. Mori, O. Nikaido, M.H.L. Green (1995). Induction of cyclobutane pyrimidine dimers, pyrimidine(6-4)pyrimidone photoproducts, and Dewar valence isomers by natural sunlight in normal human mononuclear cells. *Cancer Res.*, **55**, 2245–2248.
20. D.L. Mitchell, J. Jen, J.E. Cleaver (1992). Sequence specificity of cyclobutane pyrimidine dimers in DNA treated with solar (ultraviolet B) radiation. *Nucleic Acids Res.*, 20, 225–229.
21. E. Sage (1999). DNA damage and mutations induced by solar UV radiation, In: C. Baumstark-Khan, S. Kozubek, G. Horneck (Eds.), *Fundamentals for the Assessement of Risks from Environmental Radiation* (pp. 115–126). Kluwer Academic Publishers, Netherlands.
22. T. Douki, J. Cadet (2001). Individual determination of the yield of the main UV-induced dimeric pyrimidine photoproducts in DNA suggests a high mutagenicity of CC photolesions. *Biochemistry*, **40**, 2495–2501.
23. T. Douki, A. Reynaud-Angelin, J. Cadet, E. Sage (2003). Bipyrimidine photoproducts rather than oxidative lesions are the main type of DNA damage involved in the genotoxic effect of solar UVA radiation. *Biochemistry*, **42**, 9221–9226.
24. R. Drouin, J.P. Therrien, M. Angers, S. Ouellet (2001). *In vivo* DNA analysis. *Methods Mol. Biol.*, **148**, 175–219.
25. P.J. Rochette, J.P. Therrien, R. Drouin, D. Perdiz, N. Bastien, E.A. Drobetsky, E. Sage (2003). UVA-induced cyclobutane pyrimidine dimers form predominantly at thymine-thymine dipyrimidines and correlate with the mutation spectrum in rodent cells. *Nucleic Acids Res.*, **31**, 2786–2794.
26. A.A. Lamola (1968). Excited state precursors of thymine photodimers. *Photochem. Photobiol.*, **7**, 619–632.

27. M. Charlier, C. Helene (1972). Photochemical reactions of aromatic ketones with nucleic acids and their components. 3. Chain breakage and thymine dimerization in benzophenone photosensitized DNA. *Photochem. Photobiol.*, **15**, 527–536.
28. M.J. Peak, J.G. Peak (1990). Hydroxyl radical quenching agents protect against DNA breakage caused by both 365-nm UVA and by gamma radiation. *Photochem. Photobiol.*, **51**, 649–652.
29. E. Kvam, R.M. Tyrrell (1997). Induction of oxidative DNA base damage in human skin cells by UV and near visible radiation. *Carcinogenesis*, **18**, 2379–2384.
30. J.P. Pouget, T. Douki, M.-J. Richard, J. Cadet (2000). DNA damage induced in cells by gamma and UVA radiation as measured by HPLC/GC-MS and HPLC-EC and Comet assay. *Chem. Res. Toxicol.*, **13**, 541–549.
31. D.E. Brash (1997). Sunlight and the onset of skin cancer. *Trends Genet.*, **13**, 410–414.
32. G.P. Pfeifer (1997). Formation and processing of UV photoproducts: effects of DNA sequence and chromatin environment. *Photochem. Photobiol.*, **65**, 270–283.
33. Y.H. You, D.H. Lee, J.H. Yoon, S. Nakajima, A. Yasui, G.P. Pfeifer (2001). Cyclobutane pyrimidine dimers are responsible for the vast majority of mutations induced by UVB irradiation in mammalian cells. *J. Biol. Chem.*, **276**, 44688–44694.
34. B.S. Rosenstein, D.L. Mitchell (1991). The repair of DNA damages induced in normal human skin fibroblasts exposed to simulated sunlight. *Radiat. Res.*, **126**, 338–342.
35. T.M. Reid, L.A. Loeb (1993). Tandem double CC– >TT mutations are produced by reactive oxygen species. *Proc. Natl. Acad. Sci. USA*, **90**, 3904–3907.
36. C. Robert, B. Muel, A. Benoit, L. Dubertret, A. Sarasin, A. Stary (1996). Cell survival and shuttle vector mutagenesis induced by ultraviolet A and ultraviolet B radiation in a human cell line. *J. Invest. Dermatol.*, **106**, 721–728.
37. C.M. Palmer, D.M. Serafini, H.E. Schellhorn (1997). Near ultraviolet radiation (UVA and UVB) causes a formamidopyrimidine glycosylase-dependent increase in G to T transversions. *Photochem. Photobiol.*, **65**, 543–549.
38. B.A. Kunz, J.D. Armstrong (1998). Differences in the mutational specificities of sunlight and UVB radiation suggest a role for transversion-inducing DNA damage in solar photocarcinogenesis. *Mutation Res.*, **422**, 77–83.
39. S.G. Kozmin, Y.I. Pavlov, T.A. Kunkel, E. Sage (2003). Roles of *Saccharomyces cerevisiae* DNA polymerases Polη and Polζ in response to irradiation by simulated sunlight. *Nucleic Acids Res.*, **31**, 4541–4552.

Chapter 4

Mutations induced by UV and sunlight

Gerd P. Pfeifer, Dong-Hyun Lee, Jung-Hoon Yoon and Young-Hyun You

Table of contents

Abstract

A hallmark of UV mutagenesis is the high frequency of transition mutations at dipyrimidine sequences containing cytosine. In human skin cancers, more than 30% of all mutations in the p53 gene are transitions at dipyrimidines within the sequence context CpG, i.e. 5′-TCG and 5′-CCG, and these are localized at several mutational hotspots. Since CpGs are methylated along the p53 gene in human keratinocytes, these mutations may be derived from solar UV-induced pyrimidine dimers forming at sequences that contain 5-methylcytosine. Previously, it had been shown that CPDs form preferentially at dipyrimidines containing 5-methylcytosine when UVB or sunlight was used for irradiation. In order to define the contribution of 5-methylcytosine to sunlight-induced mutations, the CpG-methylated *lacI* and *cII* transgenes in mouse fibroblasts were used as mutational targets. After UVC irradiation, 6 to 9% of the base substitutions were at dipyrimidines containing 5-methylcytosine and only few mutations were transitions within this sequence context. However, 24 to 32% of the solar light-induced mutations were at dipyrimidines that contain 5-methylcytosine and most of these mutations were transitions. Sunlight-induced mutational hotspots at methylated CpGs correlated with sequences that form high levels of CPDs after irradiation of cells with sunlight but not with UVC. The data make a strong case that CPDs forming preferentially at dipyrimidines with 5-methylcytosine are responsible for a considerable fraction of the mutations induced by sunlight in mammalian cells. Using photoproduct-specific photolyases transfected into mouse cells containing mutational reporter genes, we demonstrated that CPDs (rather than 6-4 photoproducts or other lesions) are responsible for the vast majority of UVB-induced mutations in mammalian cells.

4.1 Mapping of UV damage at the nucleotide level by ligation-mediated PCR

Much information on UV photoproduct formation and repair in mammalian cells has been obtained by the mapping of UV damage at nucleotide resolution using the ligation-mediated PCR technique. Cyclobutane pyrimidine dimers (CPDs) and (6-4) photoproducts are the most prominent DNA damage products formed after UV irradiation [1–3]. The ligation-mediated PCR (LMPCR) amplification technique was developed to analyze the distribution and repair of UV-photoproducts along specific human genes at the DNA sequence level [4,5]. Specific LMPCR methods for the detection of (6-4) photoproducts [4] and CPDs [5] can be used to distinguish these lesions. LMPCR provides a sufficient level of sensitivity even when low UV doses (10–20 J/m^2 of 254 nm UV) are used, and the repair of CPDs can be measured reliably at individual nucleotide positions [6–9]. These earlier studies have shown, for example, that the repair of CPDs is DNA sequence dependent [6,7] and that the UV photoproduct frequency can be strongly modulated by the chromatin environment [5,10]. Using these techniques it has become possible to analyze the effects of chromatin structure, nucleotide sequence composition and transcriptional status of a gene on DNA repair at a high level of resolution.

4.2 The DNA damage spectrum produced by simulated sunlight

DNA photoproducts are responsible for the mutagenic effects of ultraviolet and solar irradiation. Experimental systems for determining the levels and sequence dependence of photoproduct formation in DNA have often used high doses of short wave (254 nm UVC) irradiation. The issue of wave-length dependence of photoproduct formation was re-assessed by using DNA sequencing technologies, such as end-labelling and ligation-mediated PCR, and different doses of UVC as well as more physiologically relevant doses of solar irradiation emitted from a 1000 W Oriel solar UV simulator. Using UVC, the (6-4)photoproducts formed at 5′-TpC and 5′-CpC sequences but only very low levels were seen at all other dipyrimidines including 5′-TpT. Contrary to expectation, however, the (6-4) photoproducts formed at almost undetectable levels under conditions of irradiation for up to 5 h with the solar UV simulator. The same solar light treatment produced high levels of CPDs at various dipyrimidine sites. In addition, DNA glycosylases, which recognize oxidized and ring-opened bases, such as the *E. coli* Fpg and Nth proteins, did not produce significant cleavage of simulated sunlight-irradiated DNA. From this data, it would appear that CPDs are at least 20 to 40-times more frequent than any other DNA photoproduct when DNA or cells are irradiated with simulated sunlight [11]. This data raises some questions about the relative mutagenic potential of UVA in the range of 340–400 nm in which very few CPDs should be formed.

4.3 Similarities in sunlight-induced mutational spectra of CpG-methylated transgenes and the P53 gene in skin cancer indicate an important role of 5-methylcytosine in solar UV mutagenesis

A characteristic signature of UV mutagenesis is a high frequency of transition mutations at dipyrimidine sequences containing cytosine. The p53 tumor suppressor gene is commonly mutated in human basal and squamous cell carcinomas [12]. In human nonmelanoma skin cancers, more than 30% of all mutations in the p53 gene are transitions at dipyrimidines within the sequence context CpG, i.e. 5′-TCG and 5′-CCG, found at several mutational hotspots (codons 196, 213, 245, 248, and 282). Since CpGs are methylated along the p53 gene in all tissues in vivo including keratinocytes, these mutations may be derived from solar UV-induced cyclobutane pyrimidine dimers forming at sequences that contain 5-methylcytosine bases. Previously, it had been demonstrated that CPDs form preferentially at dipyrimidines containing 5-methylcytosine when UVB [13] or natural sunlight [14] were used for irradiation. Mutational hotspots in the p53 gene in skin cancers are related to a combination of high CPD frequencies induced by the UVB component of sunlight and slow repair rates at most methylated 5′-TCG and 5′-CCG sites [6,13,14].

In order to define the contribution of 5-methylcytosine to sunlight-induced mutations, two CpG-methylated transgenes in fibroblasts of the BigBlue® mouse were used as mutational targets [15,16]. In total, more than 500 mutations induced by UVC or the solar UV simulator were sequenced, along with mutations from untreated cells. Only a few of the mutations in untreated cells were transitions at dipyrimidines, but more than 90% of the UVC- and solar irradiation-induced mutations were targeted to such sites, the majority being transition mutations. After UVC irradiation, 6% of the base substitutions in the *lacI* gene were at dipyrimidines containing 5-methylcytosine and only 2% of all mutations were transitions within this sequence context. However, 24% of the solar light-induced mutations were at dipyrimidines that contain 5-methylcytosine and most of them scored as transitions. Several sunlight-induced mutational hotspots at methylated CpGs correlated with sequences that form the highest levels of CPDs after irradiation with sunlight but not with UVC [15].

Similar results were obtained with the *cII* transgene [16]. This mutational target gene is also methylated at most CpG sequences. The mutational spectra produced by irradiation with 254 nm UVC radiation and simulated sunlight, respectively, differed most dramatically by the much more common involvement of dipyrimidines containing 5-methylcytosine in the solar UV mutation spectrum (32% versus 9% of all mutations). A distinct mutational hotspot induced by simulated sunlight occurred at a sequence 5′-TmCG (nucleotide position 196) and was associated with high levels of CPD photoproduct formation. We then carried out a comparison of sunlight-induced mutational spectra of the *cII* and *lacI* transgenes with the *p53* gene in skin tumors. We showed that 5-methylcytosine is involved in 25 to 40% of all mutations in all three systems. The combined data make a strong case that CPDs forming preferentially at dipyrimidines with 5-methylcytosine are responsible for a considerable fraction of the mutations induced by sunlight in human skin cells.

4.4 The role of cytosine deamination in UV mutagenesis

An additional factor that may contribute strongly to UVB mutagenesis is deamination of cytosines or 5-methylcytosines within pyrimidine dimers [17]. Such a mechanism is currently under investigation in several laboratories. There have been a few studies on the deamination of cytosines in CPD-containing dinucleotides [18–20] all of which show an increase of deamination by several orders of magnitude when a cytosine is part of a CPD and the 5,6 double bond is saturated. Since such deamination takes place in vivo [21], it will produce pyrimidine dimers containing either uracil (from deamination of cytosine) or thymine (from deamination of 5-methylcytosine). It can be assumed that these deaminated dimers are bypassed correctly by DNA polymerase eta, which has been shown to correctly bypass thymine-thymine dimers with incorporation of adenine opposite

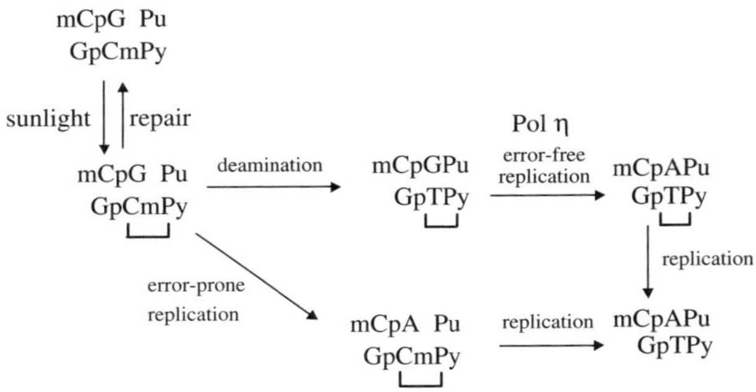

Figure 1. Schematic model for UV mutagenesis at a CPD within the sequence context 5′PymCG. This sequence is particularly prone to CPD formation due to the presence of the 5-methylcytosine base. Bypass of the CPD by a replicative DNA polymerase could produce a C to T mutation. Deamination may occur within the CPD at an elevated rate. If the deaminated CPD is bypassed mostly error-free by DNA polymerase eta, a C to T transition will occur.

the lesion [22,23]. It is not known how damage-tolerant DNA polymerases, such as polymerase eta, bypass CPDs containing non-deaminated cytosines or 5-methylcytosines. Our expectation is that they incorporate guanines in a mostly error-free pathway and a mutation occurs only after deamination. There is genetic evidence from studies in yeast showing that DNA polymerase eta bypasses CPDs containing cytosine in an error-free manner [24]. Thus, deamination could be a significant event leading to the induction of C to T mutations at dipyrimidine sequences in UVB irradiated cells. Figure 1 summarizes two possible pathways involved in UVB mutagenesis.

4.5 CPDs are responsible for a large fraction of mutations induced by UVB in mammalian cells

As mentioned, the most prevalent DNA lesions induced by UVB are the cyclobutane pyrimidine dimers and the (6-4) photoproducts. It has been a long-standing question in the field as to which of the two photoproducts is responsible for mutations in mammalian cells. In order to resolve this issue, we have introduced photoproduct-specific DNA photolyases into a mouse fibroblast cell line carrying the transgenic mutation reporter genes *lacI* and *cII*. To repair CPDs, the CPD photolyase gene of *Potorous tridactylis* was introduced and for repairing (6-4)PPs, the (6-4) photolyase gene from *Arabidopsis thaliana* was introduced into the mouse embryonic fibroblast cell line and stable clones expressing active photolyases were derived.

Exposure of the photolyase-expressing cell lines to 360 nm photoreactivating light resulted in almost complete repair of either CPDs or (6-4) photoproducts within less than three hours. The mutations produced by the remaining, non-repaired photoproducts were scored. The mutant frequency in the *cII* gene after photoreactivation by CPD photolyase was reduced from 127×10^{-5} to 34×10^{-5} (background 8–10 $\times 10^{-5}$). Photoreactivation with (6-4) photolyase did not lower the mutant frequency to a significant extent. In the *lacI* gene the mutant frequency after photoreactivation repair of CPDs was reduced from 148×10^{-5} to 28×10^{-5} (background 6–10 $\times 10^{-5}$). Mutation spectra obtained with and without photoreactivation by CPD photolyase indicated that the remaining mutations produced by UVB were derived from some background mutations, unrepaired CPDs, and other DNA photopoducts including perhaps a small contribution from (6-4) photoproducts. We concluded that CPDs are responsible for at least 80% of the UVB-induced mutations in this mammalian cell model system [25].

References

1. D.L. Mitchell, R.S. Nairn (1989). The biology of the (6-4) photoproduct. *Photochem. Photobiol.*, **49**, 805–819.
2. E. Sage (1993). Distribution and repair of photolesions in DNA: genetic consequences and the role of sequence context. *Photochem. Photobiol.*, **57**, 163–174.
3. G.P. Pfeifer (1997). Formation and processing of UV photoproducts: effects of DNA sequence and chromatin environment. *Photochem. Photobiol.*, **65**, 270–283.
4. G.P. Pfeifer, R. Drouin, A.D. Riggs, G.P. Holmquist (1991). In vivo mapping of a DNA adduct at nucleotide resolution: detection of pyrimidine (6-4) pyrimidone photoproducts by ligation-mediated polymerase chain reaction. *Proc. Natl. Acad. Sci. U.S.A.*, **88**, 1374–1378.
5. G.P. Pfeifer, R. Drouin, A.D. Riggs, G.P. Holmquist (1992). Binding of transcription factors creates hot spots for UV photoproducts in vivo. *Mol. Cell. Biol.*, **12**, 1798–1804.
6. S. Tornaletti, G.P. Pfeifer (1994). Slow repair of pyrimidine dimers at p53 mutation hotspots in skin cancer. *Science*, **263**, 1436–1438.
7. S. Gao, R. Drouin, G.P. Holmquist (1994). DNA repair rates mapped along the human PGK-1 gene at nucleotide resolution. *Science*, **263**, 1438–1440.
8. Y. Tu, S. Tornaletti, G.P. Pfeifer (1996). DNA repair domains within a human gene: selective repair of sequences near the transcription initiation site. *EMBO J.*, **15**, 675–683.
9. J.P. Therrien, R. Drouin, C. Baril, E.A. Drobetsky (1999). Human cells compromised for p53 function exhibit defective global and transcription-coupled nucleotide excision repair, whereas cells compromised for pRb function are defective only in global repair. *Proc. Natl. Acad. Sci. U.S.A.*, **96**, 15038–15043.
10. S. Tornaletti, G.P. Pfeifer (1995). UV-light as a footprinting agent: modulation of UV-induced DNA damage by transcription factors bound at the promoters of three human genes. *J. Mol. Biol.*, **249**, 714–728.
11. J.-H. Yoon, C.-S. Lee, T. O'Connor, A. Yasui, G.P. Pfeifer (2000). The DNA damage spectrum produced by simulated sunlight. *J. Mol. Biol.*, **299**, 681–693.
12. D.E. Brash, J.A. Rudolph, J.A. Simon, A. Lin, G.J. McKenna, H.P. Baden, A.J. Halperin, J. Pontén (1991). A role for sunlight in skin cancer: UV-induced p53 mutations in squamous cell carcinoma. *Proc. Natl. Acad. Sci. U.S.A.*, **88**, 10124–10128.
13. R. Drouin, J.-P. Therrien (1997). UVB-induced cyclobutane pyrimidine dimer frequency correlates with skin cancer mutational hotspots in *p53*. *Photochem. Photobiol.*, **66**, 719–726.

14. S. Tommasi, M.F. Denissenko, G.P. Pfeifer (1997). Sunlight induces pyrimidine dimers preferentially at 5-methylcytosine bases. *Cancer Res.*, **57**, 4727–4730.
15. Y.-H. You, C. Li, G.P. Pfeifer (1999). Involvement of 5-methylcytosine in sunlight-induced mutagenesis. *J. Mol. Biol.*, **293**, 493–503.
16. Y.H. You, G.P. Pfeifer (2001). Similarities in sunlight-induced mutational spectra of CpG-methylated transgenes and the p53 gene in skin cancer point to an important role of 5-methylcytosine residues in solar UV mutagenesis. *J. Mol. Biol.*, **305**, 389–399.
17. N. Jiang, J.-S. Taylor (1993). In vivo evidence that UV-induced C-T mutations at dipyrimidine sites could result from the replicative bypass of cis-syn cyclobutane dimers or their deamination products. *Biochemistry*, **32**, 472–481.
18. F.T. Liu, N.C. Yang (1978). Photochemistry of cytosine derivatives. 1. Photochemistry of thymidylyl-(3′ leads to 5′)-deoxycytidine. *Biochemistry*, **17**, 4865–4876.
19. T. Douki, J. Cadet (1992). Far-UV photochemistry and photosensitization of 2′-deoxycytidylyl-(3′-5′)-thymidine: isolation and characterization of the main photoproducts. *J. Photochem. Photobiol. B*, **15**, 199–213.
20. D.G. Lemaire, B.P. Ruzsicska (1993). Kinetic analysis of the deamination reactions of cyclobutane dimers of thymidylyl-3′,5′-2′-deoxycytidine and 2′-deoxycytidylyl-3′,5′-thymidine. *Biochemistry*, **32**, 2525–2533.
21. Y. Tu, R. Dammann, G.P. Pfeifer (1998). Sequence and time-dependent deamination of cytosine bases in UVB- induced cyclobutane pyrimidine dimers in vivo. *J. Mol. Biol.*, **284**, 297–311.
22. R.E. Johnson, S. Prakash, L. Prakash (1999). Efficient bypass of a thymine-thymine dimer by yeast DNA polymerase, Poleta. *Science*, **283**, 1001–1004.
23. C. Masutani, R. Kusumoto, A. Yamada, N. Dohmae, M. Yokoi, M. Yuasa, M. Araki, S. Iwai, K. Takio, F. Hanaoka (1999). The XPV (xeroderma pigmentosum variant) gene encodes human DNA polymerase eta. *Nature*, **399**, 700–704.
24. S.L. Yu, R.E. Johnson, S. Prakash, L. Prakash (2001). Requirement of DNA polymerase eta for error-free bypass of UV-induced CC and TC photoproducts. *Mol. Cell. Biol.*, **21**, 185–188.
25. Y.-H. You, D.-H. Lee, J.-H. Yoon, S. Nakajima, A. Yasui, G.P. Pfeifer (2001). Cyclobutane pyrimidine dimers are responsible for the vast majority of mutations induced by UVB irradiation in mammalian cells. *J. Biol. Chem.*, **276**, 44688–44694.

Chapter 5

Mechanisms and mutagenic consequences of photoproduct bypass by replicative and DNA damage bypass polymerases

John-Stephen Taylor

Table of contents

Abstract

The mechanism by which UV light causes mutations has been the subject of intense study for over 40 years. Though it is now well appreciated that UV mutations principally arise as a consequence of DNA synthesis past dipyrimidine photoproducts by polymerases, it has only been in the past few years that the polymerases involved have been identified. Initially it was thought that dipyrimidine photoproducts were absolute blocks to DNA synthesis, and most replicative DNA polymerases are indeed severely attenuated by these photoproducts. Recently, some members of the newly discovered umuC/DinB/RAD30 or Y family of polymerases have been found to be able to insert or extend opposite dipyrimidine photoproducts or even bypass them. Insight into why some polymerases are able to bypass DNA photoproducts while others are not, and the basis for nucleotide insertion selectivity is becoming available through detailed physical and structural studies of photoproducts along with kinetic studies and crystal structures of ternary complexes of DNA polymerases. Now that the mutagenic consequences of the bypass of a number of DNA photoproducts by these polymerases is known, the mutagenic consequences of the bypass of other photoproducts can be predicted, and can be used to explain the origin of many, but not all, UV-induced mutations.

5.1 Introduction

UV light principally causes C→T mutations at dipyrimidine sites which have been linked to skin cancer formation [1–8]. The origin of these mutations and the mechanisms by which they arise have been the subject of much speculation and study since evidence was obtained for cis-syn thymine dimer formation in DNA in 1961 [9,10]. Since then, a number of other photoproducts of dipyrimidine sites have been discovered [11–13], such as the (6-4) photoproduct [14] and its Dewar valence isomer [15], and the spore photoproduct [16]. Initially, it was thought that DNA photoproducts were blocks to DNA polymerases [17], but eventually it became apparent that organisms had evolved methods for synthesizing past DNA photoproducts. The polymerases involved in synthesizing past DNA photoproducts and causing mutations then became the focus of intense study.

One of the first bypass systems to be intensively studied was the one involved in SOS mutagenesis [18–21]. Initially, it was thought that the SOS inducible UmuC and UmuD proteins somehow assisted pol III in bypassing UV damage. It was not until the UmuD$'_2$C complex was purified under rigorous conditions, that it alone was found to possess translesion DNA synthesis capacity, and was named pol V [22]. At about the same time, the SOS inducible gene involved in untargeted mutagenesis, the DinB protein was also found to be a polymerase and assigned as pol IV [23]. Researchers had also been searching for eukaryotic polymerases involved in DNA damage bypass and had been focusing on the *RAD6* post-replication repair and mutagenesis pathway [24–26]. This search initially resulted in the discovery that *REV3* and *REV7* genes encoded a B family polymerase that was able to bypass cis-syn thymine dimers better than pol α, though not very efficiently [27].

A short time later it was found that *RAD30* gene, implicated in a non-mutagenic pathway, encoded a polymerase that was able to synthesize past thymine dimers much more efficiently than pol ζ and in a largely non-mutagenic manner [28]. This polymerase was named pol η and is a homolog of the *E. coli* UmuC protein. A homolog of pol η was also discovered in humans, defects in which were determined to be the cause of the genetic disease Xeroderma pigmentosum variant [29]. Since then a number of other homologs of the UmuC, DinB, and Rad30 proteins were discovered with differing abilities to synthesize opposite DNA photoproducts, but not past them [30,31]. These polymerases were collectively assigned to the Y polymerase family [32,33]. A second homolog of Rad30, RAD30B or pol ι was discovered in humans and though it is not to be able to insert efficiently opposite the cis-syn dimer, it can efficiently insert G opposite the 3′-T of the (6-4) product of TT [34,35]. It can not, however, extend opposite the 5′-T, but yeast pol ζ can [36]. As a result, it now appears that multiple polymerases might be involved in the bypass of DNA photolesions, and can be classified as inserters and/or extenders [37].

How and why some polymerases are able to bypass DNA photoproducts, while others are not, as well as how and why the nucleotides are selected for insertion opposite photoproducts, has become the subject of great interest. Answers to these questions are now forthcoming as a result of detailed kinetic and structural studies of polymerases and site-specific dipyrimidine photoproduct containing DNA. Knowledge of the specificity of these enzymes along with genetic data is also shedding light on which of these polymerases are involved in generating the observed UV-induced mutation spectra *in vivo*.

5.2 Structure and properties of dipyrimidine photoproducts

To understand the mechanisms by which photoproducts are bypassed by polymerases and their mutagenic consequences we must first consider what is known about the structure and properties of DNA photoproducts [13,38,39]. There are two major classes of primary photoproducts of double strand DNA induced by direct absorption of UVB light present in sunlight, the cis-syn cyclobutane dimers (CPDs) and the 5′-pyrimidine-(6-4)-pyrimidone, or simply (6-4) photoproducts (Figure 1A). The (6-4) photoproducts are not stable in sunlight and are photoisomerized by light centered at 320 nm (the UVA/UVB border) to their Dewar valence isomers [15,40–44], though at low doses of UVB light (6-4) products are found to predominate over their Dewar isomers [45]. It has been found by sequencing methods that CPDs form 20–40 times more abundantly than (6-4) products by simulated solar UV, and that the most abundant (6-4) product forms at TC sites [46]. Similar results were obtained by a mass spectrometry assay for UVB irradiation, though the amount of (6-4) product of TC appeared to be higher than produced by simulated solar UV light [45]. Studies of photoproduct distribution have also been carried out with monoclonal antibody assays and show that cis-syn

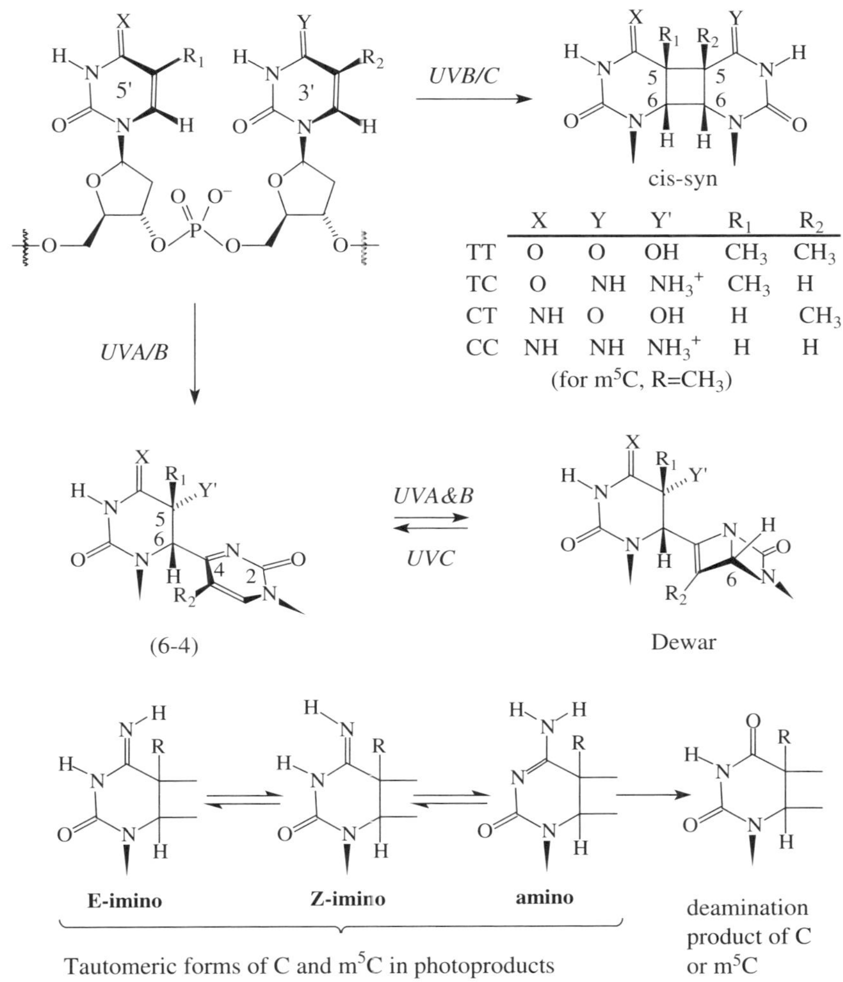

	X	Y	Y'	R_1	R_2
TT	O	O	OH	CH_3	CH_3
TC	O	NH	NH_3^+	CH_3	H
CT	NH	O	OH	H	CH_3
CC	NH	NH	NH_3^+	H	H

(for m^{5}C, R=CH_3)

Figure 1. Structure and reactions of primary DNA photoproducts of dipyrimidine sites in duplex B form DNA. (A) Initial UVB/C-induced formation of the cis-syn and (6-4) products and subsequent UVA/B induced isomerization of the (6-4) product to its Dewar valence isomer. (B) Spontaneous deamination reactions of C-containing photoproducts. C′s in either the 3′ or 5′ position of a cis-syn dimer can deaminate, but only the C in the 5′- position of a (6-4) or Dewar photoproduct can deaminate.

dimers predominate in natural or simulated sunlight [44,47,48]. Because the absorption spectrum of C is red shifted compared to T, the proportion of C-containing dimers is found to increase relative to TT dimers as the irradiation wavelength is increased [49]. Recently, it has been discovered that dipyrimidines containing a 5-methylated C are hotspots for cyclobutane dimer formation in sunlight because the absorption spectrum of methylated cytosine is red-shifted even more than cytosine [50,51].

C-containing CPDs have been known for a long time to spontaneously deaminate to U [52], while 5-methylC-containing dimers deaminate to T (Figure 1B)[43]. The half-life for deamination for C in a cis-syn dimer has been reinvestigated recently both in vitro and in vivo, and has been found to range from hours to days depending on the particular system, sequence context, and assay used [43,53–61] compared to thousands of years estimated for native C [62]. The more rapid deamination of C in a dimer can be attributed to loss of aromaticity of the cytosine ring due to saturation of the 5,6 double bond, thereby lowering the energetic barrier for the addition of water. The rapid rate of deamination of cytosine in cis-syn dimers makes it difficult to study their structure and properties, and in particular, which tautomeric form they adopt in DNA duplexes and in DNA polymerase active sites. Methylation of the 5 position of C slows down the rate of deamination of C in cis-syn dimers [43], presumably because the methyl sterically interferes with the addition of water.

The 5′-C of (6-4) products of CPy sequences also deaminate faster than C because their 5,6 double bond, like that of a CPD, has become saturated, though the half life of weeks is considerably longer than for a CPD [43]. The slower rate of deamination is presumably because of the inductive effect of the OH or NH_3^+ substituent at C5 position. Unlike the 3′-C of a cyclobutane dimer, which is otherwise identical to C except that the 5,6 double bond has become saturated, the 3′-C of a (6-4) product is incapable of a deamination event, because the amino group originally at the C4 position of C has been transferred to the C5 position of the 5′-pyrimidine ring. As a result the 3′-ring of the (6-4) product of PyC and PyT are otherwise identical in structure and properties, except for the presence of an H at C5 instead of a methyl group. As a consequence of tautomerization and deamination, the chemistry of C-containing dipyrimidine photoproducts becomes quite complex, and it most complex for CC sites as shown in Figure 2.

In the past few years, detailed physical and structural information has been obtained on photoproduct-containing duplexes in large part due to the availability of building blocks for the site-specific incorporation of photoproducts into oligonucleotides by standard phosphoramidite-based solid phase automated synthesis. Synthetic routes to phosphoramidite building blocks for the cis-syn cyclobutane dimer of TT [63–67] and TU [64] have been developed and optimized as well as for the (6-4) photoproduct of TT [68] and TC [69–71], though the yields are substantially less for making the (6-4) building blocks. Other methods for preparation of site-specific photoproduct-containing oligonucleotides such as the cis-syn dimer of TT [72,73], TU [74], UU [75], and TC [76], and the (6-4) products of TT [77,78] and TC [79] involve direct or sensitized irradiation followed by HPLC purification.

Analysis of 2D NMR data have led to solution state structures of cis-syn thymine dimer-containing duplexes opposite AA [80,81], and TA [82], as well as for the TT (6-4) photoproduct opposite AA [80] and GA [83], and for the TT Dewar product opposite AA [84,85]. The proposed solution state structure of the Dewar photoproduct [85] will have to be reinvestigated because the stereochemistry at C6 of the pyrimidone ring is opposite of that originally deduced for the dinucleotide

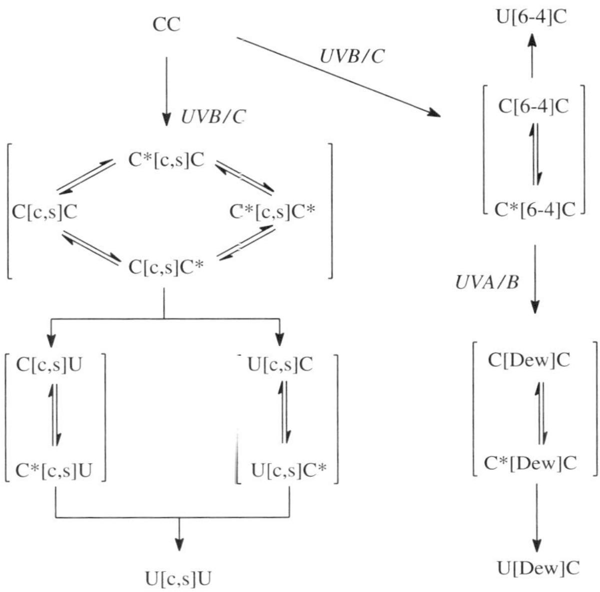

Figure 2. Photochemistry of a CC site illustrating the complexity of products that can ensue from the initial formation of the cis-syn dimer and (6-4) product. Structures without an asterisk refer to the amino tautomers and those with an asterisk refer to the *E* and *Z*-imino tautomers shown in Figure 1B. For dimer formation at a Cm^{5}C site, deamination would lead to U[c,s]m^{5}C, C[c,s]T and U[c,s]T.

photoproduct [15,40]. Most recently, the first crystal structure of a cis-syn thymine dimer-containing duplex has been solved [86] which differs from the solution structures, in that it has a 30° bend and a widened minor groove. How well this crystal structure corresponds to the solution structure of the duplex remains to be determined, however, as the bend angle of 30°, though in agreement with one solution study [87] is in disagreement with that of 7° deduced from another study [88,89]. None-the-less, both the solution NMR studies [80,81,90] and the X-ray study conclude that the 3′-T of the cis-syn dimer is oriented in a way to make a reasonably coplanar Watson-Crick base pair with an A, whereas the 5′-T is not (Figure 3). In the case of the (6-4) product of TT, the solution structure suggests that the 3′-pyrimidone ring pairs with a G by H-bonding between the N1H and the N2H of the G and the C2 carbonyl group of the pyrimidone ring, whereas the 5′-T can form a reasonably planar Watson-Crick base pair with A (Figure 3) [83]. The NMR study of the Dewar product indicates that the 3′-T forms H-bonds with G, but that the helix formed is more distorted than that formed with an A opposite the 3′-T [85].

Melting temperature studies show that cis-syn dimers of TT form the most stable DNA duplexes with AA, and are only 1.5 kcal/mole less stable that that those formed with a normal TT site [91]. Replacing a G opposite the 3′-T of a TT dimer

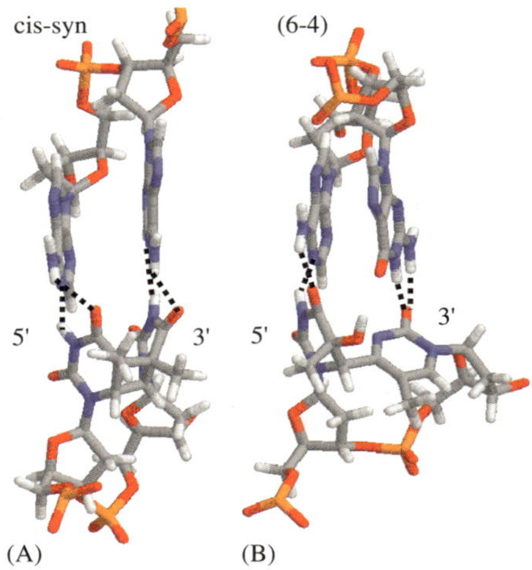

Figure 3. Solution state structures of the thermodynamically most stable base pairs for the cis-syn cyclobutane dimer and the (6-4) photoproduct of TT derived from 2D NMR data of DNA duplexes (PDB files 1TTD and 1CFL respectively). For the cis-syn thymine dimer, both Ts of the dimer pair with A, but the 5′-T forms a more distorted base pair than the 3′-T. For the (6-4) product, the 3′-T pairs best with G through H-bonding interactions between the O2 carbonyl of the pyrimidone ring and the N1 and N2 amino groups of G.

causes a further decrease in duplex stability of about 0.7 kcal/mole. Melting temperature studies on the (6-4) product of TT and its Dewar valence isomer reveal that duplexes with a G opposite the 3′-T are about 0.5 and 0.3 kcal more stable, respectively, than those with an A, though the G duplexes are about 4 kcal less stable than the dimer-containing duplexes [91,92]. For cis-syn, (6-4) and Dewar photoproducts of TT, the most stable duplex forms with an A opposite the 5′-T. A summary of the base pairing schemes for the cis-syn and (6-4) photoproducts of TT and by implication for TC, CT and CC sites and their deamination products is shown in Figure 4. Though there is no direct physical evidence of the base pairing properties of C in cis-syn dimers, or the 5′-C of (6-4) products because of their instability towards deamination, the selectivity of nucleotide insertion by DNA damage bypass polymerases suggests that they behave like a normal C to a large extent. A pairing is not shown for the 3′-base of the Dewar products because of questions concerning the stereochemistry of the C6 carbon used in modeling the NMR solution structure.

5.3 DNA polymerases

Polymerases have been classified into five major phylogenetic families, the A, B, C, D, Y and X. The A family is typified by *E. coli* polymerase I, the B family by pol α, the C family by *E. coli* pol III, the D family by Euryarchaeotic Pol II, the Y family by the umuC/DinB/RAD30 set of enzymes, and the X family by pol β (Table 1) [32,93]. For the purposes of this review, one can classify polymerases as being

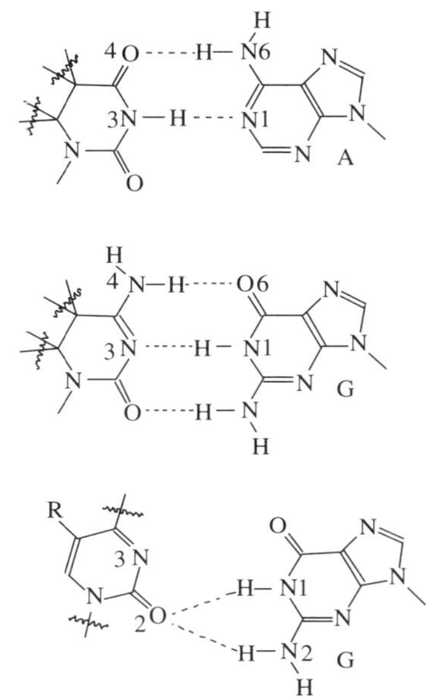

Figure 4. Schematic representations of the most stable base pairing interactions observed or expected for the 3′- and 5′- bases of the cis-syn cyclobutane and (6-4) products of dipyrimidine sites. Base pairing interactions involving C-containing dimers are surmised from the results of in vitro or in vivo photoproduct bypass experiments, and those involving the 5′-C of (6-4) products are expected to behave in the same way.

either replicative polymerases or DNA damage bypass polymerases based on their substrate specificity and fidelity. Polymerases whose primary function is to copy normal undamaged DNA efficiently and with high fidelity during the course of replication and repair, and which are rather intolerant of base modifications, will be classified as replicative polymerases [94]. Polymerases which have a much lower fidelity and efficiency that have been shown to efficiently bypass DNA damage will be classified as DNA damage bypass polymerases [30]. These bypass polymerases can be further categorized as inserters, because of their ability to insert nucleotides opposite DNA damage, or extenders, because of their ability to extend primers terminating opposite DNA damage [36,95].

For the most part it has been found that dipyrimidine photoproducts severely attenuate DNA synthesis by replicative polymerases, and when nucleotides are inserted opposite the 3′-pyrimidine, it is often an A, as it is for abasic sites [96]. The preference for inserting As opposite abasic sites has been termed the A-rule and has been attributed to the non-instructional behavior of an abasic site [97–99]. Because A is also often observed to be incorporated opposite dipyrimidine photoproducts by replicative polymerases, photoproducts were also originally categorized as non-instructional lesions. In contrast, *E. coli* pol V, the DNA damage bypass polymerase responsible for mutagenesis under SOS, is sensitive to the nature of the

Table 1. DNA polymerase families and representative replicative and DNA damage bypass polymerases[1]

Class	*E. coli*	Other Prokaryotic	Viral	Yeast	Human
A	pol I	Taq pol I	T7 DNA pol		
B	pol II	RB69 gp43	T4 gp	pol α	pol α
				pol δ	pol δ
				pol ζ	pol ζ
				(Rev3 + Rev7)	(REV3 + ?)
C	pol III				
X					pol β
Y	pol IV (DinB)	Dpo4			pol κ
	pol V ($UmuD'_2C$)			Rev1	REV1
				pol η (Rad30)	XP-V (RAD30)
					pol ι (RAD30B)

[1]Symbols α, alpha; β, beta; δ, delta; ζ, zeta; η, eta; ι, iota, κ, kappa.

3′-base of a photoproduct and inserts an A opposite the 3′-T of a cis-syn dimer, and a G opposite the 3′-pyrimidone of a TT suggesting that they are instructional [100,101]. Why dipyrimidine photoproducts behave in a non-instructional manner with some polymerases, and in an instructional manner with other polymerases can now be better understood on the basis of the structure and mechanism of the polymerases involved.

5.4 Structure and mechanism of replicative polymerases

To understand what happens when a replicative polymerase encounters a dipyrimidine photoproduct, one has to understand the structure and mechanism by which replicative polymerases incorporate nucleotides opposite undamaged templates [99,102,103]. Structures of replicative polymerases complexed to template primers and dideoxynucleotide triphosphates have revealed a most interesting and unexpected structural feature of the template primer that has important implications for the bypass of DNA photodamage. Contrary to what many have been led to expect from fanciful drawings of the mechanism of DNA replication published in papers and textbooks, the template for DNA synthesis is not held straight, but immediately bends by as much as 90° following the templating base, as shown for Taq polymerase [104] (Figure 5). In Taq polymerase, the bend is caused by the presence of the O-helix which flanks the nascent base pair and prevents the nucleotide to the 5′-side of the templating base to stack upon it and forces it out of the way. A bent template has been observed in a number of ternary complexes of DNA polymerases, such as the pol I or A family polymerases typified by T7 DNA polymerase [105], the pol α or B family polymerase RB69 [106], the pol X family polymerase pol β [107] and the pol Y family polymerase, Dpo4 [108] (Figure 6). A bend has also been observed in HIV reverse transcriptase [109].

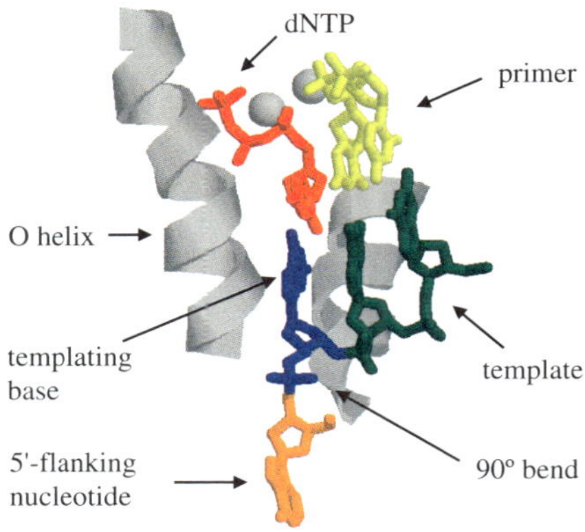

Figure 5. Structure of the ternary complex of Taq polymerase showing the location of the O-helix which causes the DNA template to bend out of the way (PDB file 3KT). The template strand is green, the primer strand is yellow, the ddNTP is red, the templating base is blue, and the base to the 5′-side of the templating base is orange.

The reason for the bend in the template becomes apparent when one considers that the function of replicative polymerases is to insure that the primer is elongated only when a Watson-Crick base pair forms between the templating nucleotide and the dNTP. To do this, replicative polymerases have evolved active sites that surround the nascent base pair and make use of steric fit rather than H-bonding to select for the Watson Crick base pairs and to exclude mismatched pairs [110–114]. As a result of contacts between the protein and the templating base, the template to the 5′-side is forced to bend out of the way. By examining of spacefilling structures of the Taq polymerase ternary complex [104] (Figure 7B) and ternary complexes of other replicative polymerases, one can readily appreciate that their active sites can not accommodate bases that are sterically modified to any great extent.

Early pre-steady state kinetic studies of the Klenow fragment and T7 DNA polymerases revealed that a step precedes phosphodiester bond formation that could be attributed to a conformational change [115–118]. Comparison of the structure of the T7 DNA polymerase ternary complex with the structure of the closely related Klenow fragment of *E. coli* polymerase I, revealed that the two polymerases had different conformations about the active site. The Klenow structure was suggested to represent an open conformation preceding the conformational step, and the T7 ternary complex to represent the closed conformation following the conformational step [102,105]. It wasn′t until a remarkable series of crystal structures of Taq polymerase, another pol I family polymerase, were obtained that the conformational changes could be more clearly visualized [104].

Crystal structures of Taq in various stages of primer elongation revealed that initial binding of the dNTP to the open form of the enzyme occurs without the assistance of the templating nucleotide, which lies outside of the active site (Figure 7A). In place of the templating base is a tyrosine, which stacks on the end of the duplex formed by the template and primer. The enzyme then undergoes

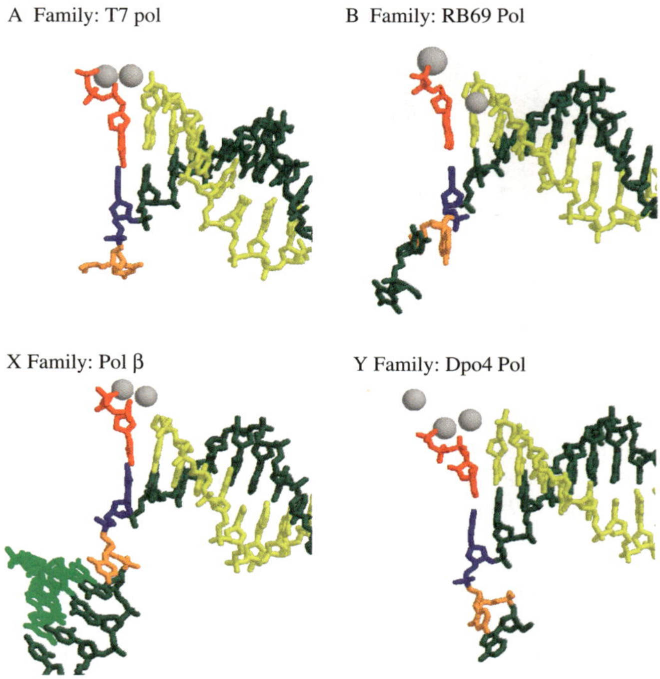

Figure 6. Structure of the DNA template primers in ternary complexes of DNA polymerases illustrating the bend in the template following the active site. Pol A family: T7 DNA polymerase (pol I family), Pol B family: RB69 (pol α family), Pol X family: pol β. and the pol Y family: Dpo4 (DinB family), PDB files 1T7P, 1IG9, 1PBZ and 1JX4, respectively. The template strand is green, the primer strand is yellow, the ddNTP is red, the templating base is blue, and the base to the 5′-side of the templating base is orange. In none of these structures is the base (orange) to the 5′-side of the templating base (blue) able to stack upon it, as would be necessary to accomodate a cis-syn dimer during insertion opposite the 3′-base of the dimer. All these polymerases would therefore be expected to insert nucleotides opposite the 3′-base of dipyrimidine photoproducts via a transient abasic site-like mechanism (Figure 9A).

a conformational change in which the tyrosine is displaced by the templating base, and the protein clamps down on the nascent base pair (Figure 7B). The steps in this process are shown schematically in Figure 8. Consideration of the Taq structures led to a base flipping proposal for how nucleotides were selected for insertion opposite undamaged DNA [103]. It was proposed that all four dNTPs would reversibly bind to the active site, but that the required conformational change leading to the closed conformation would only occur for the nucleotide that is complementary to the templating base. This conformational change was suggested to take place by a concerted mechanism in which the templating base flips into the active site at the same time that the O-helix clamps down on the nascent base pair. Phosphodiester bond formation is then activated by critical contacts that are made between the O-helix and the dNTP.

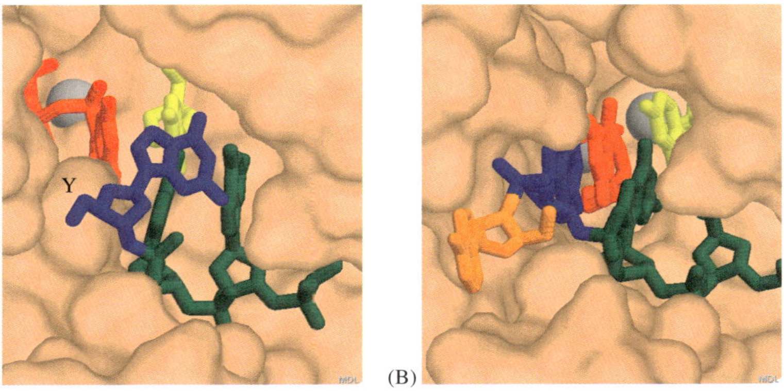

Figure 7. Structures of the binary and ternary complexes of Taq polymerase revealing the position of the templating base in the open and closed complexes, PDB files 2KT and 3KT respectfully. The template strand is green, the primer strand is yellow, the ddNTP is red, the templating base is blue, and the base to the 5′-side of the templating base is orange. The surface of the tyrosine that takes the place of the templating base in the open complex is indicated with a Y.

5.5 DNA synthesis opposite and past dipyrimidine photoproducts by replicative polymerases

The requirement that the templating base be able to flip into the active site of the polymerase suggests that if the structure of the templating base is altered by photoproduct formation, it may be unable to flip into the active site and trigger phosphodiester bond formation. In such a case, the rate of nucleotide insertion would be greatly attenuated and a competing 3′-5′ exonuclease activity would cause the polymerase to idle or stall at the site of the damaged base. In the absence of exonuclease activity, however, the enzyme might be able to slowly synthesize past the damaged base, with insertion taking place with the modified base outside of the active site, via a transient abasic-site like intermediate. The selectivity of nucleotide insertion in such a case would be expected to be the same as that for an abasic site, which is generally found to favor A for replicative polymerases (the A-rule) [97–99].

The transient abasic-site like mechanism was first proposed to account for the indiscriminant insertion of A opposite the 3′-base of different types of dipyrimidine photoproducts by the exonuclease deficient T7 DNA polymerase, Sequenase Version 2.0 [96,119]. Whereas the exonuclease proficient enzyme is unable to bypass cis-syn, (6-4) and Dewar photoproducts of TT, the exonuclease deficient enzyme can in the order cis-syn $<$ Dewar $<$ (6-4). Sequencing of the bypass products further revealed that A's were inserted opposite both the 3′ and 5′-Ts of the photoproducts despite substantial differences in the structure of the 3′-base of each photoproduct [96,119]. Such indiscriminate insertion opposite the 3′-T of the photoproducts is consistent with a mechanism in which the photoproduct is outside of the active side

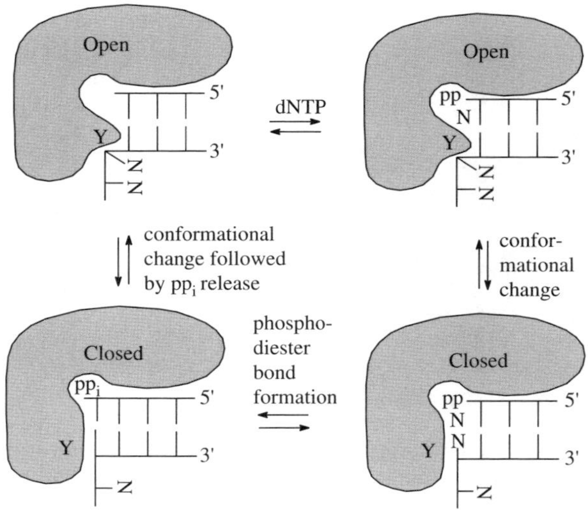

Figure 8. General scheme for nucleotide selection by pol I family polymerases as typified by Taq polymerase. The sequence of events involves initial binding of dNTP to the open complex in which tyrosine (Y) occupies the templating base site. This is followed by a conformational change involving concerted displacement of the tyrosine by the templating nucleotide and activation of the dNTP for phosphodiester bond formation which takes place next. The polymerase then undergoes another conformational change and in the process releases the inorganic pyrophosphate and translocates.

during insertion opposite the 3′-T as shown schematically in Figure 9A. A is also inserted opposite the 5′-T of these photoproducts, but it was suggested that this might be taking place with the photoproduct inside the active site and was being directed by the thymine-like structure and base pairing properties of the 5′-T.

Evidence in support of the proposed two step mechanism in which the photoproduct is outside the active site during insertion opposite the 3′-T and then inside during insertion opposite the 5′-T was obtained through the use of pyrene nucleotide [119]. Pyrene nucleotide has a large planar aromatic ring system that is almost the size of a base pair and which is selectively inserted by Klenow fragment and T7 DNA polymerase opposite abasic sites in preference to A [120]. In accord with the idea that the photoproducts can not flip into the active site during insertion opposite the 3′-T and therefore mimic the properties of an abasic site, pyrene nucleotide was found to be inserted in preference to A opposite the 3′-T of the cis-syn and (6-4) products as well as other dipyrimidine photoproducts [119]. On the other hand, pyrene nucleotide could not be inserted opposite the 5′-T of these photoproducts, indicating that these photoproducts were indeed inside the active site during the insertion step as originally proposed (Figure 9A) [96].

DNA synthesis arrest, and a transient abasic-site like mechanism for insertion opposite the 3′-T of a photoproduct is expected for any polymerase having an active site and mechanism that is similar to the pol I family polymerases. On a structural basis, pol β (X family) [121], and the pol α RB69 polymerase (B family) [106],

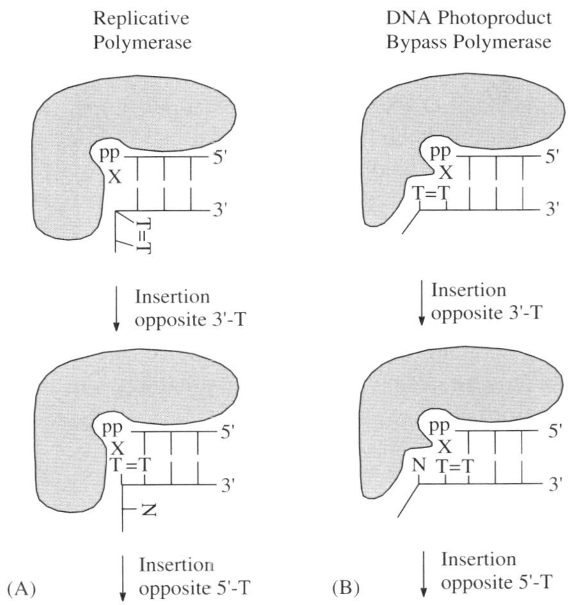

Figure 9. Diagrammatic representation of (A) a transient abasic site-like mechanism for bypass of dipyrimidine photoproducts for polymerases typified by T7 polymerase, and (B) a template-directed mechanism for bypass of dipyrimidine photoproducts by $UmuD'_2C$. In mechanism A, nucleotide insertion opposite the 3′-base of the photoproduct is independent of the base pairing properties because the photoproduct cannot enter the active site due to the presence of protein flanking the 5′-side of the templating base site. In mechanism B, nucleotide insertion is dependent on the base pairing properties of the 3′-base of the photoproduct because it can enter the active site due to the absence of an empty pocket flanking the 5′-side of the templating base site.

as well as HIV reverse transcriptase [122] all have the same type of constrained active site which results in bending of the template immediately following the templating nucleotide. One might also expect that other phylogenetically related polymerases, such as pol ζ [27], a B family polymerase involved in the bypass of DNA damage, might share similar structural features, which could explain why it cannot efficiently bypass thymine dimers on its own [36]. Recently, it has been found that although *E. coli* polymerase III is unable to bypass DNA photoproducts, it can bypass cis-syn dimers, but not (6-4) products, if the exonuclease activity of the epsilon subunit is suppressed [123,124].

5.6 Structure and mechanism of DNA damage bypass polymerases

A large number of DNA damage bypass polymerases have now been characterized and most belong to the Y family, which is also known as the umuC/DinB/Rad30 superfamily [30–33,125,126]. The best characterized of these enzymes from both a structural and mechanistic point of view is yeast polymerase η (pol η) for which a crystal structure and pre-steady state kinetic data have been obtained. Detailed analysis of the kinetics of DNA synthesis by yeast pol η opposite undamaged DNA has been interpreted to indicate the presence of a rate limiting conformational step preceding phosphodiester bond formation [127]. A crystal structure has been obtained for the pol η, which has been fit to the template primer from the T7 DNA polymerase ternary complex structure to give insight into the possible structure of this complex [128]. The model of the ternary complex reveals that the polymerase has a very open, unencumbered structure, which, unlike T7 polymerase, does not make contact with the 5′-face of the templating base (as can be seen in Figure 10A for a cis-syn dimer-containing template). It is not clear at the moment, whether or not the crystal structure of pol η obtained is in an open or closed conformation, or some other conformation.

Two crystal structures of ternary complexes of another Y family polymerase, *Sulfolobus solfataricus* P2 DNA polymerase IV (Dpo4), a DinB homolog, have been solved, and appear to be more similar to the T7 ternary complex than to the model for the pol η complex [108]. Like T7 polymerase and unlike pol η, Dpo4 polymerase contacts the 5′-face of the templating nucleotide and would not appear to be able to accommodate modifications to a base on this side (Figure 10B). Whereas the type I structure contains a perfectly matched template primer, the type II structure corresponds to the type I structure except that the primer is missing the 3′-terminal nucleotide. Another difference between the two structures is that the base to the 5′-side of the templating base in the type I structure is a purine that is oriented in an anti-parallel fashion, whereas the base in the type II structure is a

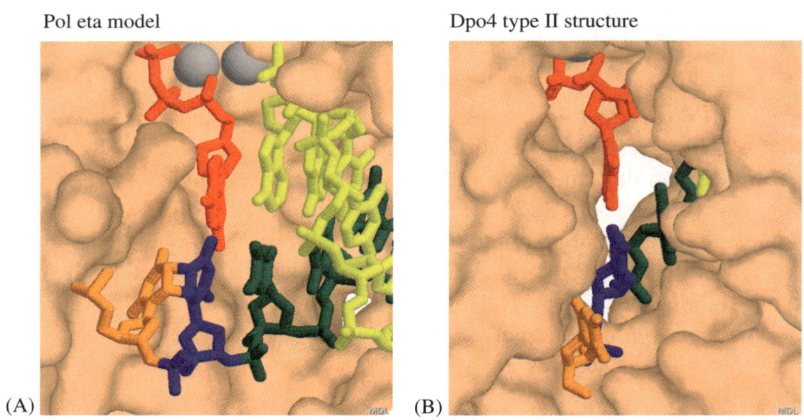

Figure 10. Structure of two Y family polymerases, (A) pol η fit to the template primer found in the T7 DNA polymerase structure following formation of a cyclobutane dimer between the templating base and the one flanking its 5′-side, and (B) Dpo4 complexed to a normal template primer showing the inability of the base flanking the 5′-side of the templating to stack on the templating base as would be required for cyclobutane dimer formation.

pyrimidine that is oriented in a parallel fashion, though neither of these bases is able to stack upon the templating base.

5.7 Bypass of dipyrimidine photoproducts by DNA damage bypass polymerases

E. coli pol V is the only enzyme known to date that is able to efficiently bypass both cis-syn and (6-4) photoproducts [101]. Because it inserts an A opposite the 3′-T of a cis-syn TT dimer and a G opposite the 3′-T of a (6-4) product, these photoproduct must be in the active site of the enzyme during the nucleotide selection step as depicted in schematically in Figure 9B. The observed insertion selectivities for A relative to G opposite the 3′-T of the cis-syn and (6-4) products is slightly more selective than what would be expected based on the experimentally determined thermodynamic stability of G relative to A in a DNA duplex or template primer (Table 2), indicating that fit might also be an important determinant of nucleotide insertion selectivity. Unfortunately there is no crystal structure of pol V at the moment from which to gain additional insight.

Yeast pol η, on the other hand, behaves in a very similar fashion to pol V, in that it can bypass a cis-syn thymine dimer and inserts A opposite the 3′-T of the cis-syn dimer [28] and a G opposite the 3′-T of the (6-4) product [129]. A model of the pol η crystal structure complexed to a dimer-containing template primer can explain how pol η is able to bypass a cis-syn thymine dimer with almost the same efficiency as normal DNA (Figure 10A) [128]. Unlike T7 DNA polymerase which makes contact with the 5′-side of the templating base, modeling has shown that pol η has a pocket at this side which is sufficiently large to accommodate the 5′-T of a thymine dimer during insertion opposite the 3′-T of the dimer (Figure 10A). As such, the 3′-T of the dimer could be in the active site and participate in the selection of the nucleotide for insertion (Figure 9B).

The nucleotide insertion selectivities for pol η [130] parallel the thermodynamic stability of nucleotides opposite these photoproducts [91,92] (Table 2) suggesting that nucleotide selection depends on the ability to form the most stable base pair.

Table 2. Selectivities for insertion of A:G opposite the 3′-T and 5′-T of the cis-syn and (6-4) photoproducts, compared to the thermodynamic stability of an A vs a G in a duplex

DNA Damage	pol V[1]	pol η[2] or (pol ζ)[3]	ΔG, 37°[4]	Thermodynamic selectivity
T(cis-syn)*T*	98:2	>99:1	−0.7	68:22
T(6-4)*T*	14:86	12:88	+0.5(0.6)[5]	44:56(38:62)[5]
T(cis-syn)T	95:5	>99:1	–	–
T(6-4)T	92:8	(>99:1)	–	–

[1]From V_{max}/K_m data [101].
[2]From V_{max}/K_m data [130].
[3]From V_{max}/K_m data [129].
[4]For a DNA dodecamer duplex with a central cis-syn or (6-4) photoproduct [91].
[5]Value in parentheses is for a primer template terminating in A or G opposite 3′-T at 25°C [92].

As was the case for pol V, the insertion selectivities are somewhat higher than predicted base on thermodynamics alone and may reflect an additional contribution from steric fit, though where this would come from is not clear given the open nature of the active site (Figure 10A). The homologous human pol η behaves in a similar fashion to the yeast enzyme [131], and must therefore have a similar active site structure that can accommodate the cis-syn dimer during insertion opposite the 3′-T of the dimer. In contrast, the homologous pol ι (RAD30B) is unable to insert opposite the 3′-T of the cis-syn dimer, but is able to insert an A opposite the 3′-T of a (6-4) product [36].

Unlike pol η, Dpo4 polymerase is not able to efficiently synthesize past a cis-syn dimer compared to undamaged DNA [132]. The inefficient rate of dimer bypass is consistent with the structure of the enzyme [108] which resembles that of a replicative DNA polymerase in that the enzyme contacts the 5′-side of the templating base and thus would prevent a dimer from being the active site during insertion opposite the 3′-base (Figure 10B). Though it was argued that a dimer should be able to fit because the T to the 5′-side of the templating base in the type II crystal structure of Dpo4 is aligned with the templating base as it would be in a dimer [108], a dimer cannot fit because the two bases are not stacked on top of each other as they would need to be in a dimer. This replicative polymerase-like feature of the active site could explain why some other Y family polymerases such as human pol ι [35,36,133] and human pol κ [34,134,135] are unable to insert opposite the 3′-T of thymine dimers efficiently.

It was also suggested that Dpo4 polymerase might be able to insert opposite the 3′-T of a dimer by a mechanism in which a nucleotide is inserted first opposite the 5′-T, followed by realignment and further extension of the primer [108]. For insertion to take place opposite the 5′-T of the dimer, the skipped nucleotide (the 3′-T) must be able to bulge out of the helix to allow for proper alignment of the primer opposite the 5′-T. This type of misalignment-mediated mechanism would not appear to work because the 3′-T of a dimer is covalently linked to the 5′-T, and therefore it would be unable to bulge out. The preferential insertion of A opposite the 3′-T is more likely to be the result of a transient abasic site-like mechanism (Figure 9A), which would be consistent with the indiscriminant insertion of A by Dpo4 observed opposite the 3′-T of a cis-syn dimer, (6-4) product and an abasic site [132].

5.8 UV-induced mutagenesis in *E. coli*

The major mutations produced by UV light in *E. coli* under SOS conditions are C→T and T→C mutations at dipyrimidine sites, as well as the tandem CC→TT mutations that are highly characteristic of UV light [136–139]. It had been known for a long time that the SOS inducible UmuC protein was involved in UV mutagenesis, and it was initially thought that its role might be in assisting pol III to synthesize past DNA photoproducts [140]. It wasn't until the rigorous purification of the UmuD$'_2$C complex and demonstration that it could bypass an abasic site in the presence of RecA was it realized that UmuD$'_2$C was a DNA damage bypass

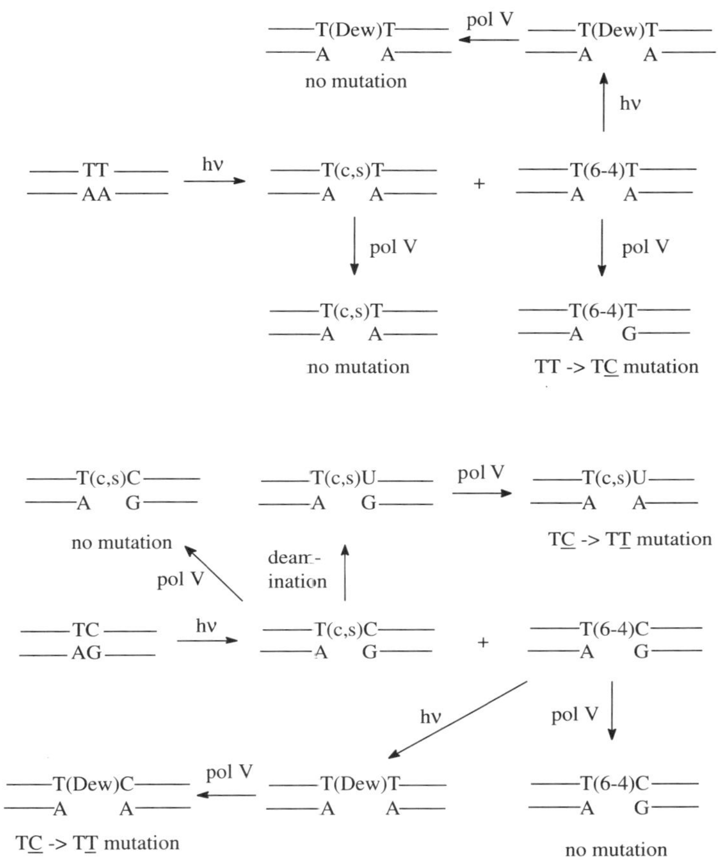

Figure 11. Observed and expected results of photoproduct bypass of the cis-syn, (6-4) and Dewar photoproducts of TT and TC by pol V that could explain the origin of the major UV-induced mutations observed in *E. coli* under SOS.

polymerase, now named pol V [22,141]. It was subsequently shown that the selectivity of nucleotide insertion opposite the cis-syn and (6-4) products of TT by reconstituted pol V closely matched the observed mutation spectra for these products in *E. coli* under SOS conditions [101]. The observed and predicted outcomes of the bypass of TT and TC cis-syn, (6-4) and Dewar photoproducts by pol V, as discussed below, are diagrammed in Figure 11, whereas the observed and expected outcomes for cis-syn dimers and (6-4) products for all sites are given in Tables 3 and 4.

Bypass of a TC cyclobutane dimer occurs in a relatively non-mutagenic manner in *E. coli* under SOS conditions [76], indicating that the 3′-C in a cyclobutane dimer adopts a tautomeric form identical to C, and presumably that the Cs in all C containing CPDs (TC, CT and CC) will behave in a similar manner (Figure 4). By analogy, it is likely that the 5′-C of CT and CC (6-4) products will also adopt

Table 3. Observed and expected principal outcomes from the bypass of cis-syn dimers by different polymerase systems. In making predictions, it is assumed that UmuD′$_2$C, pol η and pol ζ will treat a C or T in a dimer like the native base, and that pol δ will insert A opposite a 3′-C or T of a dimer via a transient abasic site-like mechanism. See text for details

	UmuD′$_2$C or pol η	**pol δ/pol ζ**
T[c,s]T	TT	TT
T[c,s]C	TC	T*T*
T[c,s]C via T[c,s]U	T*T*	T*T*
C[c,s]T	CT	CT
C[c,s]T via U[c,s]T	*T*T	*T*T
C[c,s]C	CC	C*T*
C[c,s]C via C[c,s]U	C*T*	C*T*
C[c,s]C via U[c,s]C	*T*C	*TC*
C[c,s]C via U[c,s]U	*TT*	*TT*

the tautomeric form corresponding to C, and if so, the 5′-C will be copied in a non-mutagenic manner. In contrast, the (6-4) TT product primarily causes a 3′-T→C mutation in *E. coli* under SOS [77] resulting from the preferential insertion of G opposite the 3′-T by pol V [101]. As would be expected from the structural similarity between the 3′-T and the 3′-C of TT and TC (6-4) products (Figure 1 & 4), G is inserted preferentially opposite the 3′-C of a TC (6-4) product, making bypass of the (6-4) TC photoproduct primarily non-mutagenic. The same is expected to hold true for the CT and CC (6-4) photoproducts. In contrast to the (6-4) product of TT, the Dewar product of TT is largely non-mutagenic in *E. coli* under SOS, and preferentially directs A opposite the 3′-T though it does produce a larger variety of minor mutations [77]. Because the Dewar pyrimidone ring system for T and C are identical expect for the presence of a methyl group, one would also expect the preferential insertion of A opposite the 3′-C of a TC Dewar product, leading to a 3′-C→T mutation.

Table 4. Observed and expected principal outcomes of the bypass of (6-4) photoproducts. In deriving predictions, it is assumed that UmuD′$_2$C and pol ζ will treat the 5′-base of the (6-4) products like the native base, and that UmuD′$_2$C and pol η will treat the 3′-T and 3′-C like a C. It is also assumed that pol ι and δ will treat the 3′-base of the (6-4) products like T. Though deamination of a 5′-C in a (6-4) product can take place, it is much slower than for cis-syn dimers and will be negligible during the same time period that cis-syn dimers deaminate. The 3′-C of (6-4) products is not subject to deamination. See text for details

	UmuD′$_2$C or pol η/pol ζ	**pol ι/pol ζ or pol δ/pol ζ**
T[64]T	T*C*	TT
T[64]C	TC	T*T*
C[64]T	C*C*	CT
C[64]T via U[64]T	*TC*	*T*T
C[c,s]C	CC	C*T*
C[c,s]C via U[c,s]C	*T*C	*TT*

Because C in a cis-syn dimer is not stable and deaminates to U, and because U has the same base pairing properties as T, it was proposed that UV-induced C→T mutations could be explained by the bypass of U-containing dimers resulting from deamination of C-containing dimers (Figure 1B & 2) [142]. Evidence that U-containing CPDs do indeed code as T in *E. coli* was shown later by mutagenesis studies in *E. coli* under SOS with site-specific photoproducts of TU and UU [74,75]. The origin of tandem CC→TT mutations that are a hallmark of UV mutagenesis [136] could thus be explained by a double deamination of a cis-syn dimer of CC to a UU dimer (Figure 2) which is then bypassed by pol V.

5.9 UV-induced mutagenesis in yeast

The UV mutation spectra of yeast is similar to that of *E. coli* [143] and humans [144–146], and consists mainly of C→T transitions at dipyrimidine sites. Differences have been found in the frequencies of mutations induced by individual photoproducts [145,147–149] and likely reflect differences in the enzymes and pathways involved in the various organisms. In yeast, bypass of DNA photoproducts involves the *RAD6* postreplication repair pathway, which appears to consist of mutagenic and non-mutagenic translesion synthesis pathways and a non-mutagenic DNA damage avoidance pathway [25,26,150]. The observed and predicted outcomes of the bypass of TT and TC cis-syn and (6-4) photoproducts and their deamination products by the various polymerases to be discussed are shown in Figure 12, whereas the observed and predicted outcomes for cis-syn and (6-4) products at all sites are given in Tables 3 and 4. Because little is known about how Dewar products interact with the yeast polymerases, no predictions are made for these products, but it is likely that they behave like they do in *E. coli* and preferentially direct A opposite the 3′-base.

One of the translesion synthesis pathways involves pol η which is encoded by the *RAD30* gene. As was already described above, this enzyme efficiently inserts As opposite both T's of a cis-syn thymine dimer in vitro [28] and there is both in vivo [151] and in vitro (Sun et al., unpublished results) evidence that this enzyme inserts preferentially G opposite the C of C-containing cis-syn dimers. Thus, pol η appears to behave just like *E. coli* pol V, and bypasses the initially formed cyclobutane dimers in a primarily non-mutagenic manner. U-containing dimers resulting from deamination of C-containing dimers (Figure 1 & 2) are also expected to behave like T-containing dimers, and direct the incorporation of A, resulting in C→T and CC→TT mutations. Thus, pol η, which has been assigned to a error-free or non-mutagenic pathway is also expected to contribute to an error prone or mutagenic deamination-bypass pathway given sufficient time for C-containing dimers to deaminate.

Attractive as the pol η-mediated cis-syn dimer deamination bypass model is for the formation of C→T mutations, it cannot explain why almost all UV-induced substitution mutations in yeast appear to require pol ζ [152], which is encoded by the *REV3* (catalytic subunit) and *REV7* genes [27]. Because pol ζ cannot efficiently insert nucleotides opposite the 3′-T of cis-syn or (6-4) products in vitro, but can

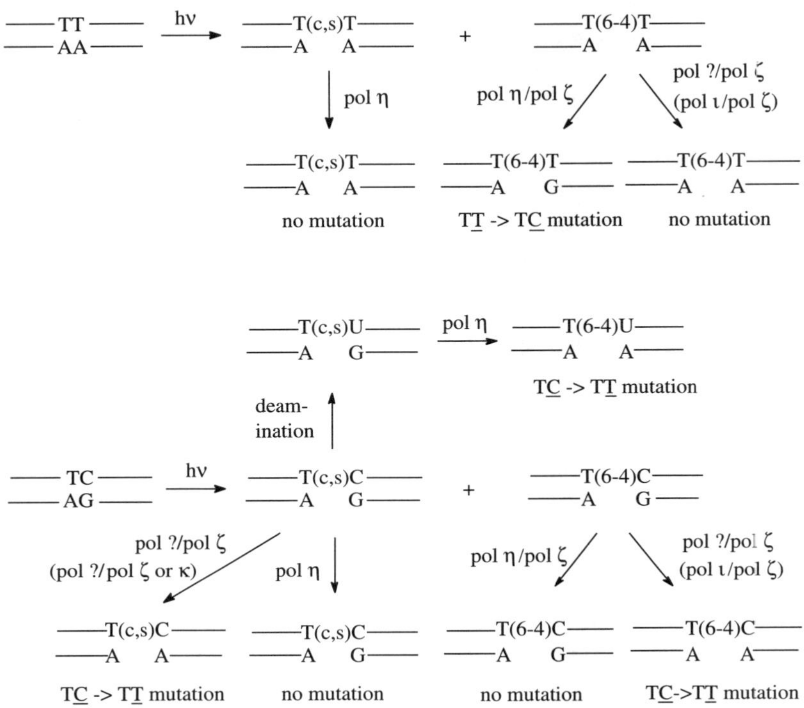

Figure 12. Observed and expected results of polymerase-mediated photoproduct bypass of the cis-syn and (6-4) photoproducts of TT and TC by polymerases of yeast and human cells that could explain the origin of the major UV-induced mutations in these cells. In yeast cells bypass is known to be mediated by pol η and pol ζ, and another as yet unidentified polymerase, that could be polymerase δ. In humans, bypass is also mediated by pol η, and presumably pol ζ. Pol ι may function to insert opposite the (6-4) product, and pol κ may play a role as an extender in the absence of, or in direct competition with, pol ζ or η. Another, as yet unidentified polymerase is needed to account for C→A mutations in the leading strand of XP-V cells that are presumably deficient in pol η, as well as another polymerase to account for transition mutations that form in the lagging strand. Pathways unique to human cells are given in parentheses.

extend primers terminating opposite the 3′-T, one or more other polymerases must be involved in inserting nucleotides opposite the 3′-pyrimidine of dipyrimidine photoproducts. A prime candidate for mutagenic insertion opposite the 3′-pyrimidine of (6-4) products is pol η which inserts G opposite the 3′-T of a TT (6-4) product with similar selectivity as observed for pol V (Table 2) [129]. This insertion product has been shown to be efficiently and accurately extended with A by pol ζ, resulting an overall TT→TC mutation [129]. Indeed, bypass of the (6-4) product of TT in yeast has been shown to lead to a TC mutation in a process that involves pol η [153], though not with the frequency predicted by in vitro studies [149]. The lower than expected frequency appears to be due to an alternate bypass pathway that does not involve pol η but some other polymerase which inserts A opposite the 3′-T of the (6-4) product [154]. Because the 3′-base of all (6-4) products have the same H-bonding properties and structure, except that for the presence of a methyl at C5 if derived from a T, or an H if derived from a C, they would be expected to direct nucleotide insertion in the same way (Figure 4). Therefore bypass of a TC (6-4) product mediated by pol η would not be expected to be mutagenic, but a TC→TT mutation would be expected to result from the alternate pathway. CT→TT and CC→TT mutations would not, however, be expected to be efficiently produced by this pathway, both because of the low frequency of (6-4) product induction at these sites, and because the 5′-C of (6-4) products deaminates very slowly, and is expected to code primarily as C.

One possibility for the polymerase involved in the alternate pathway that results in error-free or non-mutagenic bypass of a (6-4) TT product is polymerase δ, which is the first polymerase that would be expected to encounter a photoproducts during replication. Because it is a replicative polymerase of the pol B family, it is likely to be blocked by dipyrimidine photoproducts, but if insertion takes place, it would be expected to occur by a transient abasic site-like mechanism (Figure 9A). Such a mechanism would likely lead to the preferential insertion of A opposite the 3′-base of the photoproduct, given the known preference of calf thymus pol δ to insert A opposite abasic sites [155]. Because pol δ would be unlikely to extend any further, it would then dissociate and pol ζ would complete synthesis bypass, thereby resulting in non-mutagenic bypass of a (6-4) TT product and mutagenic bypass (3′-C→T) at a (6-4) TC site. A similar nucleotide insertion selectivity would be expected to occur opposite the 3′-C of a cis-syn dimer resulting in a mismatched primer that may not be as readily extended by pol η as a matched one [156] and thus require pol ζ for completion. This could explain the requirement for pol ζ in mutagenesis and why many of the C→T mutations observed in a UV-irradiated p53 gene transfected into yeast occur at a 3′-C [144,146]. The lower frequency of C → T mutations at the 5′-C may be because these mutations require deamination to occur, or because of pol ζ only infrequently inserts an A opposite a C in a dimer. The actual polymerase(s) that are involved in the alternate non-mutagenic pathway for the (6-4) TT product, and whether such a pathway exists for other (6-4) products and cis-syn dimers has yet to be determined.

5.10 UV-induced mutagenesis in humans

As noted before, the UV mutation spectrum in human cells is similar to both *E. coli* [143] and yeast [144,146], in that both C→T and CC→TT mutations are quite prevalent, as they are in the p53 gene of basal cell and squamous cell skin cancers [157–159]. Furthermore, it has been shown that PyCG sites are hotspots for both cis-syn dimer formation and C→T mutations induced by simulated sunlight [160]. The error free bypass of cis-syn thymine dimers in humans, like in yeast, has been established to involve the human analog of yeast Rad30, or human pol η, and defects in this enzyme have been established to be the cause of variant form of Xeroderma pigmentosum [29,161–164]. Xeroderma pigmentosum variant was linked to a defect in a DNA polymerase many years ago on the basis of the hypermutability of XP-V cells and arrest of DNA synthesis at pyrimidine dimers [165–167]. Later studies with SV40 vectors containing site-specific photoproducts in cell free extracts established that bypass of DNA photoproducts was inhibited in XP-V [168–172]. Ultimately, the polymerase that could complement the defect was isolated [29,163], and was found to be mutated or truncated in XP-V cells [29,162,164,173].

Human pol η has been shown to bypass the cis-syn thymine dimer almost as efficiently as an undamaged TT site [156,174]. Like the yeast enzyme, human pol η is also expected to bypass C-containing cis-syn dimers in a primarily non-mutagenic manner, and their deamination products in a mutagenic manner. Therefore, C→T and CC→TT mutations would be expected to result via pol η mediated bypass of U-containing cis-syn dimers resulting from the deamination of C-containing dimers (Figure 1 & 2). The observed and predicted pathways and outcomes of the bypass of TT and TC cis-syn and (6-4) photoproducts and their deamination products by pol η and other enzymes to be discussed are shown in Figure 12 along with the homologous yeast enzymes. The observed and predicted outcomes for cis-syn dimers and (6-4) products at all sites are given in Tables 3 and 4. We will not discuss the Dewar photoproduct because as yet there have been no experiments with either yeast or human polymerases

The human analog of Rev3 has been identified in human cells, but the homolog of Rev7 has not, and it may be that some other accessory protein to the catalytic Rev3 subunit is used to create a pol ζ-like enzyme. Thus it is expected that a pol ζ-like enzyme is also present in human cells and functions as an extender opposite DNA damage just as it does in yeast. Not surprisingly, human REV3 appears to be required for UV mutagenesis in human cells [175] though the extent of its role is not been established. As discussed for yeast, PyC→PyT mutations might arise from a pathway involving insertion of A opposite the 3′-U of a deaminated dimer by pol η, or opposite the 3′-C of a dimer or (6-4) product by another polymerase such as pol δ, followed by extension with pol ζ. CPy→TPy mutations would either require misinsertion of A opposite the 5′-C or deamination of the 5′-C, which would only be efficient for cis-syn dimers. PyT→PyC mutations would arise from insertion of G opposite the 3′-T of a (6-4) product by pol η, followed by extension by pol ζ. Use of photoreactivating enzymes specific for both cis-syn dimers and (6-4) products have shown that indeed both cis-syn dimers and (6-4) products are responsible for C→T mutations, but that only (6-4) products are responsible for T→C mutations [176].

In addition to the yeast Rad30 polymerase, humans have a homologous RAD30B polymerase, pol ι. This polymerase was found by two studies to be unable to bypass or even insert opposite a cis-syn dimer [35,36], but was found in another study to insert, albeit with low efficiency, with an unusual selectivity of T < G < A (42: 34: 24) [133]. All three studies found that pol ι is able to insert opposite the 3′-T of (6-4) TT photoproducts with a selectivity of A < G < T < C (78: 16: 5.5: 0.5) reported by one group [36] and A < T < G < C (50: 32: 18: 0.5) [133] by another group. Because the 3′-base of a TC (6-4) photoproduct is identical to that of the 3′-base of a TT (6-4) product, except for the presence of a methyl group, it should behave in an identical fashion to a TT (6-4) product. As such, bypass mediated by pol ι would be non-mutagenic at a TT (6-4) product, but mutagenic (C→T) at the 3′-C of a TC (6-4) product.

Another DNA damage bypass polymerase that has been found in humans is a member of the DinB or pol IV family [34,135,177,178] and has been assigned the name pol κ, though one of these groups had originally assigned it as pol θ [34]. Two independent studies have found that though this enzyme is unable to insert opposite the 3′-T of a cis-syn dimer [34,134,135] it can extend matched and mismatched bases terminating opposite the 3′-T of the dimer and may play a role as an extender in competition with pol ζ [95].

One of the interesting and peculiar features of XP-V that cannot yet be explained is that the UV-induced mutation spectrum is characterized by a high frequency of C→A and T→A transversions in a fibroblast cell line [179,180] and C→G transversions in a lymphoblast cell line [181]. In a more recent investigation, it was found that most of the C→A transversions produced in XP-V fibroblast cells were located in the leading strand of replication, whereas the lagging strand was characterized by the usual C→T transitions [182]. This strand bias implies that different polymerases are somehow involved in synthesizing opposite DNA photoproducts in the leading and lagging strands, though the polymerases that are involved, and how they become involved is still unknown. The only DNA damage bypass polymerase isolated so far that could explain the Py→A mutations observed for the leading strand is pol ι which has been found to preferentially insert T opposite the 3′-T of a cis-syn dimer of TT [133]. Though pol ι could account for the mutation spectrum, it would not appear to be a good candidate because it is extremely inefficient at inserting opposite the cis-syn dimer. It is also much more efficient in inserting opposite (6-4) products, and if it were involved, one would expect to see C→T mutations arising from insertion opposite the C of the (6-4) products of PyC. Pol ι could explain, however, the formation of C→T mutations at (6-4) PyC products in the lagging strand. More studies will be required to understand the biochemical basis for the curious UV-induced mutations observed in XP-V cells.

5.11 Conclusion

Many of the key players involved in mutagenesis in prokaryotic and eukaryotic systems have been identified, and that the structural and mechanistic basis for instructional and non-instructional behavior of photoproducts is becoming better understood. Which polymerases are actually recruited to bypass a photoproduct in a stalled replication fork and the mechanism by which they are recruited remains to be elucidated. Knowledge of structure and mechanism of how photoproducts are induced and then bypassed or avoided in vivo, together with how they are recognized and repaired, may make it possible to eventually explain and predict the types and frequencies of UV-induced mutations produced in vivo. Such knowledge could lead to a better understanding of factors that may predispose individuals to skin cancer and suggest ways to minimize the risk.

Acknowledgments

Graphic images of polymerase ternary complexes were prepared with the aid of Rasmol [183] and Protein Explorer [184,185]. This work was partly supported by NIH grant CA40463.

References

1. G.P. Pfeifer (1997). Formation and processing of UV photoproducts: Effects of DNA sequence and chromatin environment. *Photochem. Photobiol.*, **65**, 270–283.
2. H.S. Black, F.R. deGruijl, P.D. Forbes, J.E. Cleaver, H.N. Ananthaswamy, E.C. deFabo, S.E. Ullrich, R.M. Tyrrell (1997). Photocarcinogenesis: an overview. *J. Photochem. Photobiol. B*, **40**, 29–47.
3. D.E. Brash (1997). Sunlight and the onset of skin cancer. *Trends Genet.*, **13**, 410–414.
4. A. Sarasin (1999). The molecular pathways of ultraviolet-induced carcinogenesis. *Mutat. Res.*, **428**, 5–10.
5. N.M. Wikonkal, D.E. Brash (1999). Ultraviolet radiation induced signature mutations in photocarcinogenesis. *J. Investig. Dermatol. Symp. Proc.*, **4**, 6–10.
6. A.R. Lehmann (2000). Replication of UV-damaged DNA: new insights into links between DNA polymerases, mutagenesis and human disease. *Gene*, **253**, 1–12.
7. J.E. Cleaver (2000). Common pathways for ultraviolet skin carcinogenesis in the repair and replication defective groups of xeroderma pigmentosum. *J. Dermatol. Sci.*, **23**, 1–11.
8. J.E. Cleaver, E. Crowley (2002). UV damage, DNA repair and skin carcinogenesis. *Front. Biosci.*, **7**, d1024–1043.
9. R. Beukers, W. Berends (1961). Effects of ultraviolet irradiation on nucleic acids and their components. *Biochim. Biophys. Acta*, **49**, 181–189.
10. A. Wacker (1961). Alteration of pyrimidines by radiation *in vivo* and *in vitro*, *J. Chim. Phys.*, **58**, 1041–1045.
11. S.Y. Wang (1976). *Photochemistry and Photobiology of Nucleic Acids Volume I Chemistry*, Academic Press, Inc., USA.
12. S.Y. Wang (1976). *Photochemistry and Photobiology of Nucleic Acids Volume II Biology*, Academic Press, USA.
13. J. Cadet, P. Vigny (1990). The photochemistry of nucleic acids. In: H. Morrison (Ed), *Bioorganic Photochemistry* (pp. 1–272). John Wiley & Sons, USA.
14. A.J. Varghese, S.Y. Wang (1968). Thymine-thymine adduct as a photoproduct of thymine. *Science*, **160**, 186–187.
15. J.-S. Taylor, M.P. Cohrs (1987). DNA, light and Dewar pyrimidinones: The structure and biological significance of TpT3, *J. Am. Chem. Soc.*, **109**, 2834–2835.
16. J.E. Donnellan Jr., R.B. Setlow (1965). Thymine photoproducts but not thymine dimers found in uv-irradiated bacterial spores. *Science*, **149**, 308–310.
17. R.B. Setlow, P.A. Swenson, W.L. Carrier (1963). Thymine dimers and inhibition of DNA [deoxyribonucleic acid] synthesis by ultraviolet irradiation of cells. *Science*, **142**, 1464–1466.
18. M.D. Sutton, B.T. Smith, V.G. Godoy, G.C. Walker (2000). The SOS response: recent insights into umuDC-dependent mutagenesis and DNA damage tolerance. *Annu. Rev. Genet.*, **34**, 479–497.

19. B. Bridges (2000). DNA polymerases and SOS mutagenesis: can one reconcile the biochemical and genetic data?. *Bioessays*, **22**, 933–937.
20. R. Woodgate (2001). Evolution of the two-step model for UV-mutagenesis. *Mutat. Res.*, **485**, 83–92.
21. M. Gonzalez, R. Woodgate (2002). The "tale" of UmuD and its role in SOS mutagenesis. *Bioessays*, **24**, 141–148.
22. M. Tang, X. Shen, E.G. Frank, M. O'Donnell, R. Woodgate, M.F. Goodman (1999). UmuD′(2)C is an error-prone DNA polymerase, Escherichia coli pol V. *Proc. Natl. Acad. Sci. USA*, **96**, 8919–8924.
23. J. Wagner, P. Gruz, S.R. Kim, M. Yamada, K. Matsui, R.P. Fuchs, T. Nohmi (1999). The dinB gene encodes a novel E. coli DNA polymerase, DNA pol IV, involved in mutagenesis. *Mol. Cell*, **4**, 281–285.
24. J.P. McDonald, A.S. Levine, R. Woodgate (1997). The *Saccharomyces cerevisiae* RAD30 gene, a homologue of *Escherichia coli* dinB and umuC, is DNA damage inducible and functions in a novel error-free postreplication repair mechanism. *Genetics*, **147**, 1557–1568.
25. W. Xiao, B.L. Chow, S. Broomfield, M. Hanna (2000). The *Saccharomyces cerevisiae* RAD6 group is composed of an error-prone and two error-free postreplication repair pathways. *Genetics*, **155**, 1633–1641.
26. P. Cejka, V. Vondrejs, Z. Storchova (2001). Dissection of the functions of the *Saccharomyces cerevisiae* RAD6 postreplicative repair group in mutagenesis and UV sensitivity. *Genetics*, **159**, 953–963.
27. J.R. Nelson, C.W. Lawrence, D.C. Hinkel (1996). Thymine-thymine dimer bypass by yeast DNA polymerase zeta. *Science*, **272**, 1646–1649.
28. R.E. Johnson, S. Prakash, L. Prakash (1999). Efficient bypass of a thymine-thymine dimer by yeast DNA polymerase, pol eta. *Science*, **283**, 1001–1004.
29. C. Masutani, R. Kusumoto, F. Hanaoka (1999). The XPV (xeroderma pigmentosum variant) gene encodes human DNA polymerase eta. *Nature*, **399**, 700.
30. M.F. Goodman (2002). Error-prone repair DNA polymerases in prokaryotes and eukaryotes. *Annu. Rev. Biochem.*, **71**, 17–50.
31. Z. Wang (2001). Translesion synthesis by the UmuC family of DNA polymerases. *Mutat. Res.*, **486**, 59–70.
32. P.M. Burgers, E.V. Koonin, E. Bruford, L. Blanco, K.C. Burtis, M.F. Christman, W.C. Copeland, E.C. Friedberg, F. Hanaoka, D.C. Hinkle, C.W. Lawrence, M. Nakanishi, H. Ohmori, L. Prakash, S. Prakash, C.A. Reynaud, A. Sugino, T. Todo, Z. Wang, J.C. Weill, R. Woodgate (2001). Eukaryotic DNA polymerases: proposal for a revised nomenclature. *J. Biol. Chem.*, **276**, 43487–43490.
33. H. Ohmori, E.C. Friedberg, R.P. Fuchs, M.F. Goodman, F. Hanaoka, D. Hinkle, T.A. Kunkel, C.W. Lawrence, Z. Livneh, T. Nohmi, L. Prakash, S. Prakash, T. Todo, G.C. Walker, Z. Wang, R. Woodgate (2001). The Y-family of DNA polymerases. *Mol. Cell*, **8**, 7–8.
34. R.E. Johnson, S. Prakash, L. Prakash (2000). The human DINB1 gene encodes the DNA polymerase Polq. *Proc. Natl. Acad. Sci. USA*, **97**, 3838–3843.
35. Y. Zhang, F. Yuan, X. Wu, J.S. Taylor, Z. Wang (2001). Response of human DNA polymerase iota to DNA lesions. *Nucleic Acids Res.*, **29**, 928–935.
36. R.E. Johnson, M.T. Washington, L. Haracska, S. Prakash, L. Prakash (2000). Eukaryotic polymerases iota and zeta act sequentially to bypass DNA lesions. *Nature*, **406**, 1015–1019.
37. S. Prakash, L. Prakash (2002). Translesion DNA synthesis in eukaryotes: a one- or two-polymerase affair. *Genes Dev.*, **16**, 1872–1883.

38. J. Cadet, M. Berger, T. Douki, B. Morin, S. Raoul, J.L. Ravanat, S. Spinelli (1997). Effects of UV and visible radiation on DNA-final base damage. *Biol. Chem.*, **378**, 1275–1286.
39. J.L. Ravanat, T. Douki, J. Cadet (2001). Direct and indirect effects of UV radiation on DNA and its components. *J. Photochem. Photobiol. B*, **63**, 88–102.
40. J.-S. Taylor, D.S. Garrett, M.P. Cohrs (1988). Solution-state structure of the Dewar pyrimidinone photoproduct of thymidylyl-(3′[to]5′)-thymidine. *Biochemistry*, **27**, 7206–7215.
41. J.-S. Taylor, H.-F. Lu, J.J. Kotyk (1990). Quantitative conversion of the (6-4) photoproduct of TpdC to its dewar valence isomer upon exposure to simulated sunlight. *Photochem. Photobiol.*, **51**, 161–167.
42. T. Douki, L. Voituriez, J. Cadet (1991). Characterization of the (6-4) photoproduct of 2′-deoxycytidylyl-(3′→5′)-thymidine and of its Dewar valence isomer, *Photochem. Photobiol.*, **53**, 293–297.
43. T. Douki, J. Cadet (1994). Formation of cyclobutane dimers and (6-4) photoproducts upon far-UV photolysis of 5-methylcytosine-containing dinucleoside monophosphates. *Biochemistry*, **33**, 11942–11950.
44. P.H. Clingen, C.F. Arlett, L. Roza, T. Mori, O. Nikaido, M.H. Green (1995). Induction of cyclobutane pyrimidine dimers, pyrimidine(6-4)pyrimidone photoproducts, and Dewar valence isomers by natural sunlight in normal human mononuclear cells. *Cancer Res.*, **55**. 2245–2248.
45. T. Douki, J. Cadet (2001). Individual Determination of the Yield of the Main UV-Induced Dimeric Pyrimidine Photoproducts in DNA Suggests a High Mutagenicity of CC Photolesions. *Biochemistry*, **40**, 2495–2501.
46. J.H. Yoon, C.S. Lee, T.R. O'Connor, A. Yasui, G.P. Pfeifer (2000). The DNA damage spectrum produced by simulated sunlight. *J. Mol. Biol.*, **299**. 681–693.
47. N. Kobayashi, S. Katsumi, K. Imoto, A. Nakagawa, S. Miyagawa, M. Furumura, T. Mori (2001). Quantitation and visualization of ultraviolet-induced DNA damage using specific antibodies: application to pigment cell biology. *Pigment Cell Res.*, **14**, 94–102.
48. P.H. Clingen, C.F. Arlett, J. Cole, A.P. Waugh, J.E. Lowe, S.A. Harcourt, N. Hermanova, L. Roza, T. Mori, O. Nikaido et al. (1995). Correlation of UVC and UVB cytotoxicity with the induction of specific photoproducts in T-lymphocytes and fibroblasts from normal human donors. *Photochem. Photobiol.*, **61**, 163–170.
49. D.L. Mitchell, J. Jen, J.E. Cleaver (1992). Sequence specificity of cyclobutane pyrimidine dimers in DNA treated with solar (ultraviolet B) radiation. *Nucleic Acids Res.*, **20**, 225–229.
50. S. Tommasi, M.F. Denissenko, G.P. Pfeifer (1997). Sunlight induces pyrimidine dimers preferentially at 5-methylcytosine bases. *Cancer Res.*, **57**, 4727–4730.
51. R. Drouin, J.P. Therrien (1997). UVB-induced cyclobutane pyrimidine dimer frequency correlates with skin cancer mutational hotspots in p53. *Photochem. Photobiol.*, **66**, 719–726.
52. R.B. Setlow, W.L. Carrier, F.J. Bollum (1965). Pyrimidine dimers in uv-irradiated poly dI:dC. *Proc. Natl. Acad. Sci. USA*, **53**, 1111–1118.
53. D. Fix, R. Bockrath (1981). Thermal resistance to photoreactivation of specific mutations potentiated in E. coli B/r ung by ultraviolet light. *Mol. Gen. Genet.*, **182**, 7–11.
54. D. Fix (1986). Thermal resistance of UV-mutagenesis to photoreactivation in E. coli B/r uvrA ung: estimate of activation energy and further analysis. *Mol. Gen. Genet.*, **204**, 452–456.

55. D.G.E. Lemaire, B.P. Ruzsciska (1993). Kinetic analysis of the deamination reactions of cyclobutane dimers of thymidylyl-3′,5′-2′-deoxycytidine and 2′-deoxycytidylyl-3′,5′-thymidine. *Biochemistry*, **32**, 2525–2533.
56. T. Douki, J. Cadet (1992). Far-UV photochemistry and photosensitization of 2′-deoxycytidylyl-(3′-5′)-thymidine: isolation and characterization of the main photoproducts. *J. Photochem. Photobiol. B: Biol.*, **15**, 199–213.
57. I. Tessman, M.A. Kennedy, S.-K. Liu (1994). Unusual kinetics of uracil formation in single and double-stranded DNA by deamination of cytosine in cyclobutane pyrimdine dimmers. *J. Mol. Biol.*, **235**, 807–812.
58. Y. Barak, O. Cohen-Fix, Z. Livneh (1995). Deamination of cytosine-containing pyrimidine photodimers in UV-irradiated DNA. *J. Biol. Chem.*, **270**, 24174–24179.
59. W. Peng, B.R. Shaw (1996). Accelerated deamination of cytosine residues in UV-induced cyclobutane pyrimidine dimers leads to CC→TT transitions. *Biochemistry*, **35**, 10172–10181.
60. Y. Tu, R. Dammann, G.P. Pfeifer (1998). Sequence and time-dependent deamination of cytosine bases in UVB- induced cyclobutane pyrimidine dimers in vivo. *J. Mol. Biol.*, **284**, 297–311.
61. A. Burger, D. Fix, H. Liu, J. Hays, R. Bockrath (2003). In vivo deamination of cytosine-containing cyclobutane pyrimidine dimers in E. coli: a feasible part of UV-mutagenesis. *Mutat. Res.*, **522**, 145–156.
62. L.A. Frederico, T.A. Kunkel, B.R Shaw (1990). A sensitive genetic assay for the detection of cytosine deamination: determination of rate constants and the activation energy. *Biochemistry*, **29**, 2532–2537.
63. J.-S. Taylor, I.R. Brockie, C.L. O'Day (1987). A building block for the sequence-specific introduction of cis-syn thymine dimers into oligonucleotides. Solid-phase synthesis of TpT[c,s]pTpT. *J. Am. Chem. Soc.*, **109**, 6735–6742.
64. J.-S. Taylor, S. Nadji (1991). Unraveling the origin of the major mutation induced by ultraviolet light, the C-T transition at dTpdC sites. A DNA synthesis building block for the cis-syn cyclobutane dimer of dTpdU. *Tetrahedron*, **47**, 2579–2590.
65. T. Murata, S. Iwai, E. Ohtsuka (1990). Synthesis and characterization of a substrate for T4 endonuclease V containing a phophorodithioate linkage at the thymine dimer site. *Nucleic Acids Res.*, **18**, 7279–7286.
66. P. Ordoukhanian, J.S. Taylor (1997). Solid phase-supported thymine dimers for the construction of dimer- containing DNA by combined chemical and enzymatic synthesis: a potentially general method for the efficient incorporation of modified nucleotides into DNA. *Nucleic Acids Res.*, **25**, 3783–3786.
67. J.V. Kosmoski, M.J. Smerdon (1999). Synthesis and nucleosome structure of DNA containing a UV photoproduct at a specific site. *Biochemistry*, **38**, 9485–9494.
68. S. Iwai, M. Shimizu, H. Kamiya, E. Ohtsuka (1996). Synthesis of a phosphoramidite coupling unit of the pyrimidine (6-4) pyrimidone photoproduct and its incorporation into oligodeoxynucleotides. *J. Am. Chem. Soc.*, **118**, 7642–7643.
69. T. Mizukoshi, K. Hitomi, T. Todo, S. Iwai (1998). Studies on the chemical synthesis of oligonucleotides containing the (6-4) photoproduct of Thymine-Cytosine and its repair by (6-4) photolyase. *J. Am. Chem. Soc.*, **120**, 10634–10642.
70. S. Iwai, T. Mizukoshi, Y. Fujiwara, C. Masutani, F. Hanaoka, Y. Hayakawa (1999). Benzimidazolium triflate-activated synthesis of (6-4) photoproduct-containing oligonucleotides and its application. *Nucleic Acids Res.*, **27**, 2299–2303.
71. S. Iwai, T. Mizukoshi, K. Hitomi, T. Todo (1999). Chemical synthesis of oligonucleotides containing the (6-4) photoproduct at the thymine-cytosine site and its repair by (6-4) photolyase. *Nucleosides Nucleotides*, **18**, 1325–1327.

72. J. Kemmink, R. Boelens, T.M.G. Koning, R. Kaptein, G.A. van der Marel, J.H. van Boom (1987). Conformational changes in the oligonucleotide duplex d(GCGTTGCG)-d(CGCAACGC) induced by formation of a cis-syn thymine dimmer. *Eur. J. Biochem.*, **162**, 37–43.
73. S.K. Banerjee, R.B. Christensen, C.W. Lawrence, J.E. LeClerc (1988). Frequency and spectrum of mutations produced by a single cis-syn thymine-thymine cyclobutane dimer in a single-stranded vector. *Proc. Natl. Acad. Sci. USA*, **85**, 8141–8145.
74. N. Jiang, J.-S. Taylor (1993). In vivo evidence that uv-induced C→T mutations at dipyrimdine sites could result from the replicative bypass of cis-syn cyclobutane dimers or their deamination products. *Biochemistry*, **32**, 472–481.
75. P.E.M. Gibbs, C.W. Lawrence (1993). U-U and T-T cyclobutane dimers have different mutational properties. *Nucleic Acids Res.*, **21**, 4059–4065.
76. M.J. Horsfall, A. Borden, C.W. Lawrence (1997). Mutagenic properties of the T-C cyclobutane dimmer. *J. Bacteriol.*, **179**, 2835–2839.
77. J.E. LeClerc, A. Borden, C.W. Lawrence (1991). The thymine-thymine pyrimidine-pyrimidone(6-4) ultraviolet light photoproduct is highly mutagenic and specifically induces 3′ thymine-to-cytosine transitions in *Escherichia coli*. *Proc. Natl. Acad. Sci. USA*, **88**, 9685–9689.
78. C.A. Smith, J.-S. Taylor (1993). Preparation and characterization of a set of oligonucleotide 49-mers containing site-specific *cis-syn*, *trans-syn-I*, (6-4) and Dewar photoproducts of Thymidylyl(3′→5′)-thymidine. *J. Biol. Chem.*, **268**, 11143–11151.
79. M.J. Horsfall, C.W. Lawrence (1994). Accuracy of replication past the T-C (6-4) adduct. *J. Mol. Biol.*, **235**, 465–471.
80. J.-K. Kim, D. Patel, B.-S. Choi (1995). Contrasting structural impacts induced by cis-syn cyclobutane dimer and (6-4) adduct in DNA duplex decamers: Implication in mutagenesis and repair activity. *Photochem. Photobiol.*, **62**, 44–50.
81. K. McAteer, J. Jing, J. Kao, J.S. Taylor, M.A. Kennedy (1998). Solution-state structure of a DNA dodecamer duplex containing a cis-syn thymine cyclobutane dimer, the major UV photoproduct of DNA. *J. Mol. Biol.*, **282**, 1013–1032.
82. J.H. Lee, Y.J. Choi, B.S. Choi (2000). Solution structure of the DNA decamer duplex containing a 3′-T.T base pair of the cis-syn cyclobutane pyrimidine dimer: implication for the mutagenic property of the cis-syn dimmer. *Nucleic Acids Res.*, **28**, 1794–1801.
83. J.H. Lee, G.S. Hwang, B.S. Choi (1999). Solution structure of a DNA decamer duplex containing the stable 3′ T.G base pair of the pyrimidine(6-4)pyrimidone photoproduct [(6-4) adduct]: implications for the highly specific 3′ T→C transition of the (6-4) adduct. *Proc. Natl. Acad. Sci. USA*, **96**, 6632–6636.
84. J.H. Lee, G.S. Hwang, J.K. Kim, B.S. Choi (1998). The solution structure of DNA decamer duplex containing the Dewar product of thymidylyl(3→5′)thymidine by NMR and full relaxation matrix refinement. *FEBS Lett.*, **428**, 269–274.
85. J.H. Lee, S.H. Bae, B.S. Choi (2000). The Dewar photoproduct of thymidylyl(3′→5′)-thymidine (Dewar product) exhibits mutagenic behavior in accordance with the "A rule". *Proc. Natl. Acad. Sci. USA*, **97**, 4591–4596.
86. H. Park, K. Zhang, Y. Ren, S. Nadji, N. Sinha, J.S. Taylor, C. Kang (2002). Crystal structure of a DNA decamer containing a cis-syn thymine dimmer. *Proc. Natl. Acad. Sci. USA*, **99**, 15965–15970.
87. I. Husain, J. Griffith, A. Sancar (1988). Thymine dimers bend DNA. *Proc. Natl. Acad. Sci. USA*, **85**, 2558–2562.
88. C.-I. Wang, J.-S. Taylor (1991). Site-specific effect of thymine dimer formation on A-tract bending and its biological implications. *Proc. Natl. Acad. Sci. USA*, **88**, 9072–9076.

89. C.-I. Wang, J.-S. Taylor (1993). The trans-syn-I thymine dimer bends DNA by ≈22° and unwinds DNA by ≈15°. *Chem. Res. Toxicol.*, **6**, 519–523.
90. J. Kemmink, R. Boelens, T. Koning, G.A. van der Marel, J.H. van Boom, R. Kaptein (1987). 1H NMR study of the exchangeable protons of the duplex d(GCGT[]TGCG)-d(CGCAACGC) containing a thymine photodimer. *Nucleic Acids Res.*, **15**, 4645–4653.
91. Y. Jing, J.F.-L. Kao, J.-S. Taylor (1998). Thermodynamic and base-pairing studies of matched and mismatched DNA dodecamer duplexes containing cis-syn, (6-4) and Dewar photoproducts of TT. *Nucleic Acids Res.*, **26**, 3845–3853.
92. Y. Fujiwara, S. Iwai (1997). Thermodynamic studies of the hybridization properties of photolesions in DNA. *Biochemistry*, **36**, 1544–1550.
93. U. Hubscher, G. Maga, S. Spadari (2002). Eukaryotic DNA polymerases, *Annu. Rev. Biochem.*, **71**, 133–163.
94. T.A. Kunkel, K. Bebenek (2000). DNA replication fidelity. *Annu. Rev. Biochem.*, **69**, 497–529.
95. M.T. Washington, R.E. Johnson, L. Prakash, S. Prakash (2002). Human DINB1-encoded DNA polymerase kappa is a promiscuous extender of mispaired primer termini. *Proc. Natl. Acad. Sci. USA*, **99**, 1910–1914.
96. C.A. Smith, J. Baeten, J.S. Taylor (1998). The ability of a variety of polymerases to synthesize past site-specific cis-syn, trans-syn-II, (6-4), and Dewar photoproducts of thymidylyl-(3′→5′)-thymidine. *J. Biol. Chem.*, **273**, 21933–21940.
97. B.S. Strauss (1991). The 'A rule' of mutagen specificity: a consequence of DNA polymerase bypass of non-instructional lesions? *Bioessays*, **13**, 79–84.
98. B.S. Strauss (2002). The "A" rule revisited: polymerases as determinants of mutational specificity. *DNA Repair*, **1**, 125–135.
99. J.S. Taylor (2002). New structural and mechanistic insight into the A-rule and the instructional and non-instructional behavior of DNA photoproducts and other lesions. *Mutat. Res.*, **510**, 55–70.
100. C.W. Lawrence, S.K. Banerjee, A. Borden, J.E. LeClerc (1990). T-T cyclobutane dimers are misinstructive, rather than noninstructive, mutagenic lesions. *Mol. Gen. Genet.*, **222**, 166–168.
101. M. Tang, P. Pham, X. Shen, J.-S. Taylor, M. O'Donnell, R. Woodgate, M.F. Goodman (2000). Roles of E. coli DNA polymerases IV and V in lesion-targeted and untargeted SOS mutagenesis. *Nature*, **404**, 1014–1018.
102. S. Doublie, T. Ellenberger (1998). The mechanism of action of T7 DNA polymerase. *Curr. Opin. Struct. Biol.*, **8**, 704–712.
103. P.H. Patel, M. Suzuki, E. Adman, A. Shinkai, L.A. Loeb (2001). Prokaryotic DNA polymerase I: evolution, structure, and "base flipping" mechanism for nucleotide selection. *J. Mol. Biol.*, **308**, 823–837.
104. Y. Li, S. Korolev, G. Waksman (1998). Crystal structures of open and closed forms of binary and ternary complexes of the large fragment of Thermus aquaticus DNA polymerase I: structural basis for nucleotide incorporation. *EMBO J.*, **17**, 7514–7525.
105. S. Doublie, S. Tabor, A.M. Long, C.C. Richardson, T. Ellenberger (1998). Crystal structure of a bacteriophage T7 DNA replication complex at 2.2 A resolution. *Nature*, **391**, 251–258.
106. M.C. Franklin, J. Wang, T.A. Steitz (2001). Structure of the replicating complex of a pol alpha family DNA polymerase. *Cell*, **105**, 657–667.
107. M.R. Sawaya, R. Prasad, S.H. Wilson, J. Krout, H. Pelletier (1997). Crystal structures of human DNA polymerase á complexed with gapped and nicked DNA: Evidence for an induced fit mechanism. *Biochemistry*, **36**, 11205–11215.

108. H. Ling, F. Boudsocq, R. Woodgate, W. Yang (2001). Crystal structure of a Y-family DNA polymerase in action: a mechanism for error-prone and lesion-bypass replication. *Cell*, **107**, 91–102.
109. A. Jacobo-Molina, J. Ding, R.G. Nanni, A.D. Clark, X. Lu, C. Tantillo, R.L. Williams, G. Kamer, A.L. Ferris, P. Clark, A. Hizi, S.H. Hughes, E. Arnold (1993). Crystal structure of human immunodeficiency virus type 1 reverse transcriptase complexed with double-stranded DNA at 3.0 A resolution shows bent DNA. *Proc. Natl. Acad. Sci. USA*, **90**, 6320–6324.
110. D.L. Sloane, M.F. Goodman, H. Echols (1988). The fidelity of base selection by the polymerase subunit of DNA polymerase III holoenzyme. *Nucleic Acids Res.*, **16**, 6465–6475.
111. S. Moran, R.X.-F. Ren, E.T. Kool (1997). A thymidine triphosphate shape analog lacking Watson-Crick pairing ability is replicated with high sequence selectivity. *Proc. Natl. Acad. Sci. USA*, **94**, 10506–10511.
112. E.T. Kool (2001). Hydrogen bonding, base stacking, and steric effects in dna replication. *Annu. Rev. Biophys. Biomol. Struct.*, **30**, 1–22.
113. Y. Li, G. Waksman (2001). Crystal structures of a ddATP-, ddTTP-, ddCTP, and ddGTP- trapped ternary complex of Klentaq1: insights into nucleotide incorporation and selectivity. *Protein Sci.*, **10**, 1225–1233.
114. E.T. Kool (2002). Active site tightness and substrate fit in DNA replication. *Annu. Rev. Biochem.*, **71**, 191–219.
115. R.D. Kuchta, P. Benkovic, S.J. Benkovic (1988). Kinetic mechanism whereby DNA polymerase I (Klenow) replicates DNA with high fidelity. *Biochemistry*, **27**, 6716–6725.
116. M.E. Dahlberg, S.J. Benkovic (1991). Kinetic mechanism of DNA polymerase I (Klenow fragment): Identification of a second conformational change and evaluation of the internal equilibrium constant. *Biochemistry*, **30**, 4835–4843.
117. I. Wong, S.S. Patel, K.A. Johnson (1991). An induced-fit kinetic mechanism for DNA replication fidelity: direct measurement by single-turnover kinetics. *Biochemistry*, **30**, 526–537.
118. S.S. Patel, I. Wong, K.A. Johnson (1991). Pre-steady-state kinetic analysis of processive DNA replication including complete characterization of an exonuclease-deficient mutant. *Biochemistry*, **30**, 511–525.
119. L. Sun, M. Wang, E.T. Kool, J.S. Taylor (2000). Pyrene nucleotide as a mechanistic probe: evidence for a transient abasic site-like intermediate in the bypass of dipyrimidine photoproducts by T7 DNA polymerase. *Biochemistry*, **39**, 14603–14610.
120. T.J. Matray, E.T. Kool (1999). A specific partner for abasic damage in DNA. *Nature*, **399**, 704–708.
121. H. Pelletier, M.R. Sawaya, A. Kumar, S.H. Wilson, J. Kraut (1994). Structures of ternary complexes of rat DNA polymerase beta, a DNA template-primer, and ddCTP. *Science*, **264**, 1891–1903.
122. J. Ding, K. Das, Y. Hsiou, S.G. Sarafianos, A.D. Clark Jr., A. Jacobo-Molina, C. Tantillo, S.H. Hughes, E. Arnold (1998). Structure and functional implications of the polymerase active site region in a complex of HIV-1 RT with a double-stranded DNA template- primer and an antibody Fab fragment at 2.8 A resolution. *J. Mol. Biol.*, **284**, 1095–1111.
123. A. Borden, P.I. O'Grady, D. Vandewiele, A.R. Fernandez de Henestrosa, C.W. Lawrence, R. Woodgate (2002). Escherichia coli DNA polymerase III can replicate efficiently past a T- T cis-syn cyclobutane dimer if DNA polymerase V and

the 3′ to 5′ exonuclease proofreading function encoded by dnaQ are inactivated. *J. Bacteriol.*, **184**, 2674–2681.
124. D. Vandewiele, A. Borden, P.I. O'Grady, R. Woodgate, C.W. Lawrence (1998). Efficient translesion replication in the absence of *Escherichia coli* Umu proteins and 3′-5′ exonuclease proofreading function. *Proc. Natl. Acad. Sci. USA*, **95**, 15519–15524.
125. W.A. Beard, S.H. Wilson (2001). DNA lesion bypass polymerases open up. *Structure (Camb)*, **9**, 759–764.
126. E.C. Friedberg, P.L. Fischhaber, C. Kisker (2001). Error-prone DNA polymerases: novel structures and the benefits of infidelity. *Cell*, **107**, 9–12.
127. M.T. Washington, L. Prakash, S. Prakash (2001). Yeast DNA polymerase eta utilizes an induced-fit mechanism of nucleotide incorporation. *Cell*, **107**, 917–927.
128. J. Trincao, R.E. Johnson, C.R. Escalante, S. Prakash, L. Prakash, A.K. Aggarwal (2001). Structure of the catalytic core of S. cerevisiae DNA polymerase eta: implications for translesion DNA synthesis. *Mol. Cell*, **8**, 417–426.
129. R.E. Johnson, L. Haracska, S. Prakash, L. Prakash (2001). Role of DNA polymerase zeta in the bypass of a (6-4) TT photoproduct. *Mol. Cell Biol.*, **21**, 3558–3563.
130. M.T. Washington, R.E. Johnson, S. Prakash, L. Prakash (2000). Accuracy of thymine-thymine dimer bypass by saccharomyces cerevisiae DNA polymerase eta. *Proc. Natl. Acad. Sci. USA*, **97**, 3094–3099.
131. R.E. Johnson, M.T. Washington, S. Prakash, L. Prakash (2000). Fidelity of human DNA polymerase eta. *J. Biol. Chem.*, **275**, 7447–7450.
132. F. Boudsocq, S. Iwai, F. Hanaoka, R. Woodgate (2001). Sulfolobus solfataricus P2 DNA polymerase IV (Dpo4): an archaeal DinB- like DNA polymerase with lesion-bypass properties akin to eukaryotic poleta. *Nucleic Acids Res.*, **29**, 4607–4616.
133. A. Tissier, E.G. Frank, J.P. McDonald, S. Iwai, F. Hanaoka, R. Woodgate (2000). Misinsertion and bypass of thymine-thymine dimers by human DNA polymerase iota. *EMBO J.*, **19**, 5259–5266.
134. E. Ohashi, T. Ogi, R. Kusumoto, S. Iwai, C. Masutani, F. Hanaoka, H. Ohmori (2000). Error-prone bypass of certain DNA lesions by the human DNA polymerase kappa. *Genes Dev.*, **14**, 1589–1594.
135. Y. Zhang, F. Yuan, X. Wu, M. Wang, O. Rechkoblit, J.S. Taylor, N.E. Geacintov, Z. Wang (2000). Error-free and error-prone lesion bypass by human DNA polymerase kappa *in vitro*. *Nucleic Acids Res.*, **28**, 4138–4146.
136. C. Coulondre, J.H. Miller (1977). Genetic studies of the *lac* repressor. IV. Mutagenic specificity in the *lacI* gene of *Escherichia coli*. *J. Mol. Biol.*, **117**, 577–606.
137. J.E. LeClerc, N.L. Istock (1982). Specificity of UV mutagenesis in the *lac* promoter of *M13lac* hybrid phage DNA. *Nature*, **297**, 596–598.
138. J.E. LeClerc, N.L. Istock, B.R. Saran, R.J. Allen (1984). Sequence analysis of ultraviolet-induced mutations in M13lacZ hybrid phage DNA. *J. Mol. Biol.*, **180**, 217–237.
139. J.H. Miller (1985). Mutagenic specificity of ultraviolet light. *J. Mol. Biol.*, **182**, 45–68.
140. M. Rajagopalan, C. Lu, R. Woodgate, M. O'Donnell, M.F. Goodman, H. Echols (1992). Activity of the purified mutagenesis proteins UmuC, UmuD', and RecA in replicative bypass of an abasic DNA lesion by DNA polymerase III. *Proc. Natl. Acad. Sci. USA*, **89**, 10777–10781.
141. P. Pham, E.M. Seitz, S. Saveliev, X. Shen, R. Woodgate, M.M. Cox, M.F. Goodman (2002). Two distinct modes of RecA action are required for DNA polymerase V-catalyzed translesion synthesis. *Proc. Natl. Acad. Sci. USA*, **99**, 11061–11066.

142. J.-S. Taylor, C.L. O'Day (1990). Cis-syn thymine dimers are not absolute blocks to replication by DNA polymerase I of *Escherichia coli in vitro*. *Biochemistry*, **29**, 1624–1632.
143. H.C. Hsia, J.S. Lebkowski, P.-M. Leong, M.P. Calos, J.H. Miller (1989). Comparison of ultraviolet irradiation-induced mutagenesis of the lacI gene in *Escherichia coli* and in human 293 cells, *J. Mol. Biol.*, **205**, 103–113.
144. D.J. Moshinsky, G.N. Wogan (1997). UV-induced mutagenesis of human p53 in a vector replicated in Saccharomyces cerevisiae. *Proc. Natl. Acad. Sci. USA*, **94**, 2266–2271.
145. A. Inga, G. Scott, P. Monti, A. Aprile, A. Abbondandolo, P.A. Burns, G. Fronza (1998). Ultraviolet-light induced p53 mutational spectrum in yeast is indistinguishable from p53 mutations in human skin cancer. *Carcinogenesis*, **19**, 741–746.
146. D.J. Moshinsky, G.N. Wogan (2000). UV-induced mutagenesis of human p53: analysis using a double-selection method in yeast. *Environ. Mol. Mutagen.*, **35**, 31–38.
147. P.E.M. Gibbs, B.J. Kilbey, S.K. Banerjee, C.W. Lawrence (1993). The frequency and accuracy of replication past a thymine-thymine cyclobutane dimer are very different in *Saccharomyces-cerevisiae* and *Escherichia-coli*. *J. Bacteriol.*, **175**, 2607–2612.
148. C.W. Lawrence, P.E.M. Gibbs, A. Borden, M.J. Horsfall, B.J. Kilbey (1993). Mutagenesis induced by single UV photoproducts in *Escherichia-coli* and yeast. *Mutat. Res.*, **299**, 157–163.
149. P.E.M. Gibbs, A. Borden, C.W. Lawrence (1995). The T-T pyrimidine (6-4) pyrimidinone UV photoproduct is much less mutagenic in yeast than in *Escherichia coli*. *Nucleic Acids Res.*, **23**, 1919–1922.
150. S. Broomfield, T. Hryciw, W. Xiao (2001). DNA postreplication repair and mutagenesis in *Saccharomyces cerevisiae*. *Mutat. Res.*, **486**, 167–184.
151. S.L. Yu, R.E. Johnson, S. Prakash, L. Prakash (2001). Requirement of DNA polymerase eta for error-free bypass of UV-induced CC and TC photoproducts. *Mol. Cell Biol.*, **21**, 185–188.
152. C.W. Lawrence, V.M. Maher (2001). Mutagenesis in eukaryotes dependent on DNA polymerase zeta and Rev1p. *Philos. Trans. R Soc. Lond. B Biol. Sci.*, **356**, 41–46.
153. H. Zhang, W. Siede (2002). UV-induced T→C transition at a TT photoproduct site is dependent on *Saccharomyces cerevisiae* polymerase eta *in vivo*. *Nucleic Acids Res.*, **30**, 1262–1267.
154. A. Bresson, R.P. Fuchs (2002). Lesion bypass in yeast cells: Pol eta participates in a multi-DNA polymerase process. *EMBO J.*, **21**, 3881–3887.
155. D.J. Mozzherin, S. Shibutani, C.K. Tan, K.M. Downey, P.A. Fisher (1997). Proliferating cell nuclear antigen promotes DNA synthesis past template lesions by mammalian DNA polymerase delta. *Proc. Natl. Acad. Sci. USA*, **94**, 6126–6131.
156. M.T. Washington, R.E. Johnson, L. Prakash, S. Prakash (2001). Accuracy of lesion bypass by yeast and human DNA polymerase eta. *Proc. Natl. Acad. Sci. USA*, **98**, 8355–8360.
157. D.E. Brash, J.A. Rudolph, J.A. Simon, A. Lin, G.J. McKenna, H.P. Baden, A.J. Halperin, J. Ponten (1991). A role for sunlight in skin cancer: UV-induced p53 mutations in squamous cell carcinoma. *Proc. Natl. Acad. Sci. USA*, **88**, 10124–10128.
158. A. Ziegler, D.J. Leffell, S. Kunala, H.W. Sharma, M. Gailani, J.A. Simon, A.J. Halperin, H.P. Baden, P.E. Shapiro, A.E. Bale, D.E. Brash (1993). Mutation hotspots due to sunlight in the p53 gene on nonmelanoma skin cancers. *Proc. Natl. Acad. Sci. USA*, **90**, 4216-4220.
159. N. Dumaz, H.J. van Kranen, A. de Vries, R.J. Berg, P.W. Wester, C.F. van Kreijl, A. Sarasin, L. Daya-Grosjean, F.R. de Gruijl (1997). The role of UV-B light in skin

carcinogenesis through the analysis of p53 mutations in squamous cell carcinomas of hairless mice. *Carcinogenesis*, **18**, 897–904.

160. Y.H. You, G.P. Pfeifer (2001). Similarities in sunlight-induced mutational spectra of CpG-methylated transgenes and the p53 gene in skin cancer point to an important role of 5-methylcytosine residues in solar UV mutagenesis. *J. Mol. Biol.*, **305**, 389–399.
161. C. Masutani, M. Araki, A. Yamada, R. Kusumoto, T. Nogimori, T. Maekawa, S. Iwai, F. Hanaoka (1999). Xeroderma pigmentosum variant (XP-V) correcting protein from HeLa cells has a thymine dimer bypass DNA polymerase activity. *EMBO J.*, **18**, 3491–3501.
162. M. Yuasa, C. Masutani, T. Eki, F. Hanaoka (2000). Genomic structure, chromosomal localization and identification of mutations in the xeroderma pigmentosum variant (XPV) gene. *Oncogene*, **19**, 4721–4728.
163. A. Yamada, C. Masutani, S. Iwai, F. Hanaoka (2000). Complementation of defective translesion synthesis and UV light sensitivity in xeroderma pigmentosum variant cells by human and mouse DNA polymerase eta. *Nucleic Acids Res.*, **28**, 2473–2480.
164. B.C. Broughton, A. Cordonnier, W.J. Kleijer, N.G. Jaspers, H. Fawcett, A. Raams, V.H. Garritsen, A. Stary, M.F. Avril, F. Boudsocq, C. Masutani, F. Hanaoka, R.P. Fuchs, A. Sarasin, A.R. Lehmann (2002). Molecular analysis of mutations in DNA polymerase eta in xeroderma pigmentosum-variant patients. *Proc. Natl. Acad. Sci. USA*, **99**, 815–820.
165. V.M. Maher, L.M. Ouellette, R.D. Curren, J.J. McCormick (1976). Frequency of ultraviolet light-induced mutations is higher in xeroderma pigmentosum variant cells than in normal human cells. *Nature*, **261**, 593–595.
166. A.R. Lehmann (1979). The relationship between pyrimidine dimers and replicating DNA in UV-irradiated human fibroblasts. *Nucleic Acids Res.*, **7**, 1901–1912.
167. A.M. Cordonnier, R.P. Fuchs (1999). Replication of damaged DNA: molecular defect in xeroderma pigmentosum variant cells. *Mutat. Res.*, **435**, 111–119.
168. D.L. Svoboda, J.-M.H. Vos (1995). Differential replication of a single, UV-induced lesion in the leading or lagging strand by a human cell extract: Fork uncoupling or gap formation. *Proc. Natl. Acad. Sci. USA*, **92**, 11975–11979.
169. M. Cordeiro-Stone, L.S. Zaritskaya, L.K. Price, W.K. Kaufmann (1997). Replication fork bypass of a pyrimidine dimer blocking leading strand DNA synthesis. *J. Biol. Chem.*, **272**, 13945–13954.
170. D.L. Svoboda, L.P. Briley, J.M. Vos (1998). Defective bypass replication of a leading strand cyclobutane thymine dimer in xeroderma pigmentosum variant cell extracts. *Cancer Res.*, **58**, 2445–2448.
171. I. Ensch-Simon, P.M.J. Burgers, J.-S. Taylor (1998). Bypass of a site-specific cis-syn thymine dimer in an SV40 vector during *in vitro* replication by HeLa and XPV cell-free extracts. *Biochemistry*, **37**, 8218–8226.
172. J. Yao, K. Dixon, M.P. Carty (2001). A single (6-4) photoproduct inhibits plasmid DNA replication in xeroderma pigmentosum variant cell extracts. *Environ. Mol. Mutagen.*, **38**, 19–29.
173. P. Kannouche, B.C. Broughton, M. Volker, F. Hanaoka, L.H. Mullenders, A.R. Lehmann (2001). Domain structure, localization, and function of DNA polymerase eta, defective in xeroderma pigmentosum variant cells. *Genes Dev.*, **15**, 158–172.
174. R. Kusumoto, C. Masutani, S. Iwai, F. Hanaoka (2002). Translesion synthesis by human DNA polymerase eta across thymine glycol lesions. *Biochemistry*, **41**, 6090–6099.

175. P.E. Gibbs, W.G. McGregor, V.M. Maher, P. Nisson, C.W. Lawrence (1998). A human homolog of the *Saccharomyces cerevisiae* REV3 gene, which encodes the catalytic subunit of DNA polymerase zeta. *Proc. Natl. Acad. Sci. USA*, **95**, 6876–6880.
176. E. Otoshi, T. Yagi, T. Mori, T. Matsunaga, O. Nikaido, S.T. Kim, K. Hitomi, M. Ikenaga, T. Todo (2000). Respective roles of cyclobutane pyrimidine dimers, (6-4)photoproducts, and minor photoproducts in ultraviolet mutagenesis of repair-deficient xeroderma pigmentosum A cells. *Cancer Res.*, **60**, 1729–1735.
177. V.L. Gerlach, L. Aravind, E.C. Friedberg (1999). Human and mouse homologs of Escherichia coli DinB (DNA polymerase IV), members of the UmuC/DinB superfamily. *Proc. Natl. Acad. Sci. USA*, **96**, 11922.
178. V.L. Gerlach, W.J. Feaver, P.L. Fischhaber, E.C. Friedberg (2001). Purification and characterization of pol kappa, a DNA polymerase encoded by the human DINB1 gene. *J. Biol. Chem.*, **276**, 92–98.
179. Y.-C. Wang, V.M. Maher, J.J. McCormick (1991). Xeroderma pigmentosum variant cells are less likely than normal cells to incorporate dAMP opposite photoproducts during replication of UV-irradiated plasmids. *Proc. Natl. Acad. Sci. USA*, **88**, 7810–7814.
180. Y.C. Wang, V.M. Maher, D.L. Mitchell, J.J. McCormick (1993). Evidence from mutation spectra that the uv hypermutability of xeroderma-pigmentosum variant cells reflects abnormal, error-prone replication on a template containing photoproducts. *Mol. Cell Biol.*, **13**, 4276-4283.
181. H.L. Waters, S. Seetharam, M.M. Seidman, K.H. Kraemer (1993). Ultraviolet hypermutability of a shuttle vector propagated in xeroderma pigmentosum variant cells. *J. Invest. Dermatol.*, **101**, 744–748.
182. W.G. McGregor, D. Wei, V.M. Maher, J.J. McCormick (1999). Abnormal, error-prone bypass of photoproducts by xeroderma pigmentosum variant cell extracts results in extreme strand bias for the kinds of mutations induced by UV light. *Mol. Cell Biol.*, **19**, 147–154.
183. R.A. Sayle, E.J. Milner-White (1995). RASMOL: biomolecular graphics for all. *Trends Biochem. Sci.*, **20**, 374.
184. E. Martz (1999–2002). Protein Explorer Software http://proteinexplorer.org
185. E. Martz (2002). Protein Explorer: easy yet powerful macromolecular visualization. *Trends Biochem. Sci.*, **27**, 107–109.

Chapter 6

The Ogg1 protein of *Saccharomyces cerevisiae*: properties and biological functions

Serge Boiteux

Table of contents

Abstract

Oxidative DNA damage has been implicated in the etiologies of human pathologies such as cancer, neurodegenerative diseases and ageing. An oxidized DNA base, 8-oxo-7,8-dihydroguanine (8-oxoG) is abundantly produced in DNA exposed to free radicals and reactive oxygen species. The biological relevance of 8-oxoG has been unveiled by the study of two mutator genes in *E. coli*, *fpg* and *mutY*. Both genes code for DNA *N*-glycosylases that cooperate to prevent the mutagenic effects of 8-oxoG in DNA. In yeast and human cells, base excision repair of 8-oxoG is initiated by the Ogg1 protein, a DNA *N*-glycosylase/AP lyase which is the functional homologue of the bacterial *fpg* gene product. The inactivation of *OGG1* in yeast creates a mutator phenotype that is specific for the generation of GC to TA transversions. In this paper, we will review recent knowledge about the Ogg1 protein and repair of 8-oxoG in the yeast *Saccharomyces cerevisiae*, which can be used as a paradigm for DNA repair in eukaryotes.

6.1 Formation and biological consequences of 8-oxoguanine in DNA

Oxidative DNA damage results from the attack of sugar and base moieties by free radicals and reactive oxygen species (ROS) which are formed as byproducts of normal cell metabolism and during exposure to physical and chemical agents such as γ-irradiation, UVA-irradiation or H_2O_2 [1–3]. ROS such as $O_2^{\bar{}}$, OH•, $^1O_2$, NO•, RO• or ROO• damage DNA directly or after metal-catalyzed transformation in the cell [1–3]. Oxidative DNA damage includes DNA single-strand breaks (SSB) and double-strand breaks (DSB), base modifications, apurinic/apyrimidinic (AP) sites and [DNA-Protein]-cross-links [2,3]. A guanine lesion, 8-oxo-7,8-dihydroguanine (8-oxoG), is certainly the oxidized DNA base that has retained most of the attention in the last few years [4–6]. 8-OxoG is a lesion that does not efficiently block DNA synthesis catalyzed by DNA polymerases *in vitro* [4]. Furthermore, DNA polymerases can incorporate dAMP or dCMP opposite 8-oxoG in the template strand [4]. The coding properties of 8-oxoG suggest that this lesion could be mutagenic, yielding GC to TA transversions after two rounds of replication [4–6]. To counteract the deleterious action of 8-oxoG, living organisms have evolved DNA repair mechanisms. In *Escherichia coli*, 8-oxoG is primarily repaired by the base excision repair (BER) pathway using two DNA *N*-glycosylases: the Fpg protein excising 8-oxoG in damaged DNA and the MutY protein excising adenine residues incorporated by DNA polymerases opposite 8-oxoG [4–6]. Inactivation of *fpg* or *mutY* genes leads to a spontaneous mutator phenotype characterized by the exclusive increase in GC to TA transversions [4–6]. Furthermore, the double mutant *fpg mutY* exhibits a very strong GC to TA mutator phenotype indicating that Fpg and MutY may act synergistically to prevent mutagenesis [For reviews: 4–6]. These results in *E.coli* strongly suggest that the endogenous oxidative stress, due to cell metabolism, induces the formation of 8-oxoG in DNA, which, if not repaired, can be a major cause of genetic instability. Eukaryotic cells also possess DNA *N*-glycosylases that can prevent the deleterious action of 8-oxoG in DNA. The Ogg1 protein is the major 8-oxoG DNA *N*-glycosylase in *S.cerevisiae* and mammalian cells [6]. Finally, recent studies strongly suggest that the inactivation of DNA repair genes preventing the mutagenic effect of 8-oxoG, *MTH1*, *MYH1* and *OGG1*, results in cancer predisposition in man and mice [7–9].

6.2 Base excision repair of 8-oxoguanine in *S. cerevisiae*

In *Saccharomyces cerevisiae*, 8-oxoG is mostly repaired by the base excision repair (BER) pathway, which is initiated by the Ogg1 DNA *N*-glycosylase/AP lyase [10,11]. Ogg1 catalyzes the cleavage of the glycosylic bond between 8-oxoG and the sugar-phosphate backbone yielding an AP site in DNA. Ogg1 also possesses an AP lyase activity that can incise the phosphodiester bond immediately 3′ of the AP site yielding a 3′-dRP end [12]. Afterwards, the blocked 3′-dRP end is released by the phosphodiesterase activity of the yeast AP endonucleases (Apn1 or Apn2) or the 3′-flap endonuclease activity of Rad1-Rad10 [13]. Finally, the gap is filled in by a DNA polymerase and the nick is sealed by the action of a DNA ligase [14,15]. Here, we will focus on the first step of the BER process, which is catalyzed by the Ogg1 protein.

6.2.1 The Ogg1 protein: structure and expression

The *OGG1* gene of *Saccharomyces cerevisiae* is localized on chromosome XIII and codes for a 43-kDa protein, Ogg1 [10,11]. Comparison of the amino acid sequence of the Ogg1 protein with that of Fpg, its functional homolog in *E. coli*, does not reveal any obviously conserved region. In contrast, sequence alignment shows that Ogg1 is a member of a super-family of DNA repair proteins sharing a common ancestor gene with the Nth protein (endonuclease III) of *E.coli* [11]. The Ogg1 protein is conserved amongst eukaryotes such as yeast, plant, insect and mammals (Figure 1). However, organisms such as *Schizosaccharomyces pombe* and *Caenorhabditis elegans* do not possess an obvious Ogg1 protein homolog. The yeast Ogg1 protein possesses a bipartite nuclear localization signal (NLS) at the carboxy terminal-end and a mitochondrial localization signal (MTS) at the amino terminal-end. The cellular localization of Ogg1 was investigated using GFP-Ogg1 fusion proteins. The results show that the deletion of the last 63-amino acids at the C-terminus results in a mitochondrial localization whereas deletion of the first 11-amino acids from the N-terminus results in a nuclear localization [16]. Therefore, in yeast, the Ogg1 protein is present in both mitochondria and nucleus in physiological conditions.

The regulation of the *OGG1* gene has not been investigated in detail in *S.cerevisiae*. However, the expression of *OGG1* messenger RNA has been reported to be regulated in the course of the cell cycle with a peak of expression in late G1 phase [17]. In contrast, another study does not report induction of *OGG1* at a

```
Dm    1 -----------------------MKAVLQDR-G------------VIGLSLEECDLERTL
Sc    1 --------------------------MSYKFG-------------KLAINKSELCLANVL
At    1 MKRPRPTSQPSISSTVKPPLSPPVTPILKQKLH-R--TGTP-KWFPLKLTHTELTLPLTF
Hs    1 ---------------------MPARALLPRRMGHRTLASTPALWASIPCPRSELRLDLVL
                                  ..  . .             . . ..*. *   .

Dm   25 LGGQSFRWRSICDGNRTEYGGVVFNT----YWVLQQE--ESFITYEAY-G--TSSPLATK
Sc   22 QAGQSFRWIWDEKLNQYSTTMKIGQQEKYSVVILRQDE-ENEILEFVAVG---DCGNQDA
At   57 PTGQTFRWK---KTGAICYSGTIGPH----LVSLRQRPGDDAVSYCVH-----CSTSPKS
Hs   40 PSGQSFRWR---EQSPAHWSGVLADQ----VWTLTQT--EEQLHCTVYRGDKSQASRPTP
          **.***.        ... ..       .  * *   .  .   .  .

Dm   76 DYSSLISDYLRVDFDLKVNQKD-WLSKDDNFVKFLSKP--VRLLSQEPFENIFSFLCSQN
Sc   78 LKTHLM-KYFRLDVSLKHLFDNVWIPSDKAFAKLSPQG--IRILAQEPWETLISFICSSN
At  105 AELALL-DFLNAEISLAELWSD-FSKKDPRFGELARHLRGARVLRQDPLEQLIQFLCSSN
Hs   91 DELEAVRKYFQLDVTLACLYHH-WGSVDSHFQEVAQKFQGVRLLRQDPIEQLFSFICSSN
            .. .  ....* ..   .   *  *  .  .   .*.* *.* *  . .*.**.*

Dm  133 NNIKRISSMIEWFCATFGTKIGHFNGADAYTFPTINRFHDIPCEDLNAQLRAAKFGYRAK
Sc  135 NNISRITRMCNSLCSNFGNLITTIDGVAYHSFPTS---EELTSRATEAKLRELGFGYRAK
At  163 NNIARITKMVD-FVSSLGLHLGDIDGFEFHQFPSL----DRLSRVSEEEFRKAGFGYRAK
Hs  150 NNIARITGMVERLCQAFGPRLIQLDDVTYHGFPSL---QALAGPEVEAHLRKLGLGYRAR
        ***  **. *..  .  .* ..  ...  .. **..    ..      .. .*  ..****.

Dm  193 FIAQTLQEIQKK-------GGQNWFISL-KSMPFEKAREELTLLPGIGYKVADCICLMSM
Sc  192 YIIETARKLVNDKAEANITSDTTYLQSICKDAQYEDVREHLMSYNGVGPKVADCVCLMGL
At  218 YITGTVNALQAKP-----GGGNEWLLSL-RKVELQEAVAALCTLPGVGPKVAACIALFSL
Hs  207 YVSASARAILEEQ------GGLAWLQQL-RESSYEEAHKALCILPGVGTKVADCICLMAL
        ..  .  .         .. .. .. .  .....  *  ..*.* ***.*..*. .

Dm  245 GHLESVPVDIHIYRIAQNIYLPH-LTGQ-KNV----------TKKIYEEVSKH---FQKL
Sc  252 HMDGIVPVDVHVSRIAKRDYQISANKNHLKELRTKYNALPISRKKINLELDHIRLMLFKK
At  272 DQHSAIPVDTHVWQIATNILLPD-LAGA--KL----------TPKLHGRVAEA---FVSK
Hs  260 DKPQAVPVDVHMWHIAQRDYSWHPTTSQAKGP----------SPQTNKELGNF---FRSL
              .***.*...**  .          . .         . ..  ..      .

Dm  290 HGKYAGWAQAILFSADLS-Q---FQNTSTVACKK-KSNKKPKK-----------------
Sc  312 WGSYAGWAQGVLFSKEIGGTS--GSTTTGTIKKRKWDMIKETEAIVTKQMKLKVELSDLH
At  316 YGEYAGWAQTLLFIAELPAQKTLLQSFSQPINKLDESAEKLDLGRAKKILKLMQYLAPWQ
Hs  307 WGPYAGWAQAVLFSADLR-Q---SRHAQEPPAKRRKGSKGPEG-----------------
        .* ******..**....  .       .     *.     .

Dm      ----------------
Sc  370 IKEAKID---------
At  376 MSVVICYDHCLLRNQS
Hs      ----------------
```

Figure 1. Amino acid sequences of *Saccharomyces cerevisiae* (Sc), *Drosophila melanogaster* (Dm), *Arabidopsis thaliana* (At) and human (Hs) Ogg1 proteins. The amino acids identical in the four sequences are shown (*).

specific phase of the cell cycle [18]. Furthermore, the expression of *OGG1* is not stimulated nor repressed after exposure to a variety of DNA damaging agents DNA [18]. Table 1 summarizes some of the properties of the yeast and human Ogg1 proteins.

6.2.2 The Ogg1 protein: catalytic properties

In its active form, the Ogg1 protein is a monomer composed of 376 amino acids with a calculated pI of 8.83 (Table 1). The Ogg1 protein possesses a DNA *N*-glycosylase activity which excises 8-oxoG, Fapy-G and Methyl-Fapy-G residues

Table 1. Comparison of the properties of the yeast and human Ogg1 proteins

	yOgg1	**α-hOgg1[a]**	**β-hOgg1[b]**
Gene localization	Chromosome XIII	3p25	3p25
Protein (amino acids)	376	345	424
Molecular Mass (kDa)	42.8	38.8	47.2
Isoelectric point	8.83	8.89	6.52
Cellular localization			
Nucleus	Yes	Yes	No
Mitochondrion	Yes	No	Yes
RX-structure	No	Yes[c]	No
Enzymatic activities			
DNA glycosylase	Yes	Yes	Yes
Me-Fapy-G:C	Yes	Yes	Yes
Fapy-G:C	Yes	Yes	nd
8-oxo G:C	Yes	Yes	Yes
8-oxo G:A	No	No	No
8-oxo A:C	Yes	Yes	nd
AP lyase	Yes: β-elimination	Yes: β-elimination	Yes: β-elimination
AP:C	Yes	Yes	Yes
AP:A	No	No	nd
Mutant phenotype	mutator G:C to T:A	nd	nd
Complementation			
E.coli fpg mutY	Yes	Yes	Yes
S.cerevisiae ogg1	Yes	Yes	nd
DNA damage-inducible	No	nd	nd

nd: not determined.
[a]Nuclear version of the human Ogg1 protein.
[b]Mitochondrial version of the human Ogg1 protein.
[c]RX-structure, see [29].

in damaged DNA [10–12,19]. The Ogg1 protein efficiently releases 8-oxoG placed opposite a cytosine. In contrast, 8-oxoG placed opposite a guanine or an adenine is a very poor substrate for Ogg1 [10–12]. The Ogg1 protein also removes 8-oxo-7,8-dihydroadenine (8-oxoA) opposite a cytosine [20]. Therefore, Ogg1 exhibits a narrow substrate specificity compared with Fpg, which is also able to release several oxidized pyrimidines in oligodeoxyribonucleotides [21,22]. The Ogg1 protein is also endowed with an AP lyase activity that incises DNA at the 3′-side of an AP site in DNA *via* a β-elimination reaction [12]. Cleavage at AP sites by Ogg1 also requires the presence of a cytosine in the complementary strand opposite the lesion [12]. These results show that cytosine opposite the lesion is essential for its recognition and/or excision by Ogg1. The catalytic mechanism of the Ogg1 protein involves the formation of a transient covalent imino enzyme-DNA intermediate which is stabilized in the presence of the reducing agent NaBH4 [12]. Based on sequence analysis, lysine 241 of the yeast Ogg1 protein was expected to be directly involved in the catalytic mechanism. Indeed, the conversion of lysine 241 into glutamine (K241Q) abolishes both DNA *N*-glycosylase and AP lyase activities of Ogg1 [23]. However, the K241Q mutant of Ogg1 is still able to bind with a high affinity DNA containing 8-oxoG or AP site analogs [23]. It should be noted that the yeast Ogg1 and the nuclear human α-hOgg1 have identical substrate specificities and catalytic mechanisms (Table 1).

6.2.3 The Ogg1 protein: biological functions

The *OGG1* gene is not essential, its disruption does not affect the viability of a haploid strain [24]. Furthermore, *ogg1* mutants are not unusually sensitive to DNA damaging agents such as H_2O_2, γ-irradiation, methyl methanesulfonate and UVC-irradiation [24]. On the other hand, *ogg1* mutants exhibit a spontaneous mutator phenotype with enhanced frequencies for Can^R mutants and Lys^+ revertants compared to the wild type strain [24]. In fact, Ogg1-deficient strains specifically accumulate GC to TA transversions as demonstrated using the *cyc1* reversion assay [24] and sequencing of Can^R mutants [25]. Ogg1-deficient strains of *S.cerevisiae* also present an increased frequency of mitochondrial mutants [16]. The Ogg1-deficient strains accumulate deletions in mt-DNA yielding "petite" colonies [16]. These results in yeast strongly suggest that endogenous oxidative stress occurs in eukaryotes yielding 8-oxoG and presumably other mutagenic oxidative DNA damages. A recent study suggests that the *ogg1*-knockout mice present a marked predisposition to lung cancer [9].

6.3 Interplay of Ogg1 with other repair and replication pathways: an integrated model

Although critical, Ogg1-mediated BER is not the unique cellular defense against the deleterious action of 8-oxoG in DNA. In fact, *S.cerevisiae* uses at least three strategies to prevent 8-oxoG-induced mutagenesis: (i) Excision of 8-oxoG from damaged DNA by BER with Ogg1 as a major player (this review) and NER as a back up [26,27], (ii) Removal of the newly incorporated adenine opposite 8-oxoG by the MMR pathway [25], (iii) Incorporation of the "correct" cytosine opposite 8-oxoG by Pol η [28]. A comprehensive model is proposed, it shows the interplay of DNA repair and DNA replication mechanisms (Figure 2). The complexity of the model suggests that removal of 8-oxoG in DNA is critical for yeast. Our model also explains why Ogg1-deficient strains exhibit only a modest mutator phenotype. In Ogg1-deficient strains, mutagenesis is prevented by the action of NER on the 8-oxoG-containing strand and by the action of MMR and Pol η on the adenine-containing complementary strand. These last results raise the possibility for a role for 8-oxoG in the etiologies of pathologies associated with deficiencies in NER such as *Xeroderma pigmentosum* (XP), Pol η such as XP-Variant (XP-V) and MMR such as hereditary non-polyposis colon cancer (HNPCC).

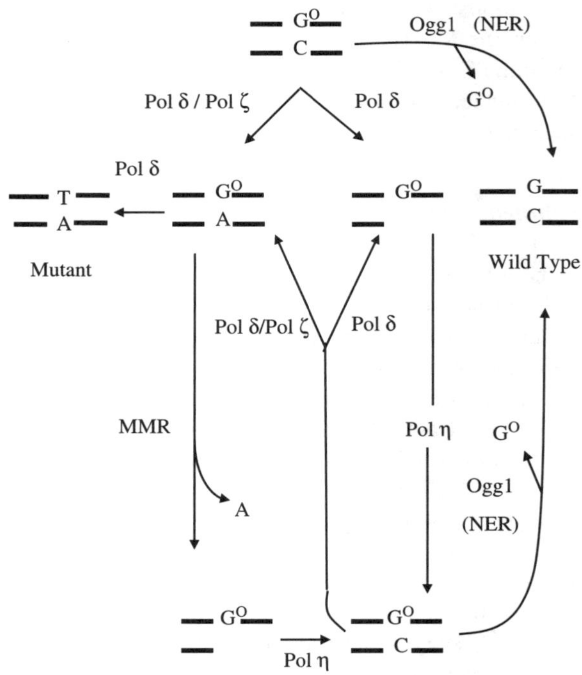

Figure 2. An integrated model for the repair of 8-oxoG in *S.cerevisiae*. In this model, mutagenesis by 8-oxoG (G = 0) is prevented at several levels: i) Excision of 8-oxoG from the DNA before replication by BER mediated by Ogg1 or NER as a back up system. ii) Pol δ can pause before 8-oxoG permitting Pol η to incorporate a "correct" cytosine opposite 8-oxoG generating a substrate for the BER and NER pathways. iii) Alternatively, Pol δ can incorporate adenine opposite 8-oxoG yielding the G = 0:A mispair, which is elongated by the action of Pol ζ yielding a GC to TA mutations after a second round of replication. iv) However, the mutagenic G = 0:A mispair is a substrate of MMR that can release the adenine opposite 8-oxoG, permitting incorporation of a "correct" cytosine by Pol η, followed by excision of 8-oxoG by BER and NER. This model shows that several repair pathways can act synergistically to prevent the mutagenic effect of 8-oxoG in eukaryotes.

Acknowledgements

This work was supported by the Centre National de la Recherche Scientifique (CNRS) and the Commissariat à l'Energie Atomique (CEA).

References

1. B. Halliwell, J.M.C. Gutteridge (1989). Free Radicals in Biology and Medicine. Second edition. Clarendon Press, Oxford.
2. J. Cadet, M. Berger, T. Douki, J.-L. Ravanat (1997). Oxidative Damage to DNA: Formation, measurement and biological significance. *Rev. Physiol. Biochem. Pharmacol.*, **131**, 1–87.

3. M. Dizdaroglu (1991). Chemical determination of free-radical induced damage to DNA. *Free Rad. Biol. Med.*, **10**, 225–242.
4. A.P. Grollman, M. Moriya (1993). Mutagenesis by 8-oxoguanine: an enemy within. *Trends in Genet.*, **9**, 246–249.
5. M.L. Michaels and J.H. Miller (1992). The GO system protects organisms from the mutagenic effect of the spontaneous lesion 8-hydroxyguanine (7,8-dihydro-8-oxoguanine). *J. Bacteriol.*, **174**, 6321–6325.
6. S. Boiteux and J.P. Radicella (1999). Base excision repair of 8-hydroxyguanine protects DNA from endogenous oxidative stress. *Biochimie* **81**, 59–67.
7. T. Tsusuki, A. Egashira, H. Igarashi, T. Iwakuma, Y. Nakatsuru, Y. Tominaga, H. Kawate, K. Nakao, K. Nakaitura, F. Ide, S. Kura, Y. Nakabeppu, M. Katsuki, T. Ishikawa, M. Sekigushi (2001). Spontaneous tumorigenesis in mice defective in the MTH1 gene encoding 8-oxo-dGTPase. *Proc. Natl. Acad. Sci. USA*, **98**, 11456–11461.
8. N. Al-Tassan, N.H. Chmiel, J. Maynard, N. Fleming, A.L. Livingston, G.T. Williams, A.K. Hodges, D.R. Davies, S.S. David, J.R. Sampson, J.P. Cheadle (2002). Inherited variants of MYH associated with somatic GC to TA mutations in colorectal tumors. *Nature Genetics*, **30**, 227–232.
9. K. Sakumi, Y. Tominaga, M. Furuichi, Ping Xu, T. Ysuzuki, M. Sekiguchi, and Y. Nakabeppu (2003). Ogg1 knockout-associated lung tumorigenesis and its suppression by Mth1 gene disruption. *Cancer Res.*, **63**, 902–905.
10. P. Auffret van der Kemp, D. Thomas, R. Barbey, R. De Oliveira, R. and S. Boiteux (1996). Cloning and expression in *E. coli* of the *OGG1* gene of *S. cerevisiae* which codes for a DNA glycosylase that excises 7,8-dihydro-8-oxoguanine and 2,6-diamino-4-hydroxy-5-N-methylformamidopyrimidine. *Proc. Natl. Acad. Sci. USA*, **93**, 5197–5202.
11. H.M. Nash, S.D. Bruner, O.D. Scharer, T. Kawate, T.A. Addona, E. Spooner, W.S. Lane, and G.L. Verdine (1996). Cloning of a yeast 8-oxoguanine DNA glycosylase reveals the existence of a base excision repair protein superfamily. *Current Biol.*, **6**, 968–980.
12. P.M. Girard, N. Guibourt, and S. Boiteux (1997). The Ogg1 protein of *S. cerevisiae*: a 7,8-dihydro-8-oxoguanine DNA glycosylase/AP lyase whose lysine 241 is a critical residue for catalytic activity. *Nucleic Acids Res.*, **25**, 3404–3411.
13. M. Guillet, and S. Boiteux (2002). Endogenous DNA abasic sites cause cell death in the absence of Apn1, Apn2 and Rad1-Rad10 in *Saccharomyces cerevisiae, EMBO J.*, **21**, 2833–2841.
14. S. Boiteux, and J.P. Radicella (2001). The human *OGG1* gene: Structure, functions, and its implication in the process of carcinogenesis. *Arch. Biochem. Biophys.*, **377**, 1–8.
15. J.H.J. Hoeijmakers (2001). Genome maintenance mechanisms for preventing cancer. *Nature*, **411**, 366–374.
16. K.K. Singh, B. Sigala, H.A. Sikder, and C. Schwimmer (2001). Inactivation of *Saccharomyces cerevisiae OGG1* DNA repair gene leads to an increased frequency of mitochondrial mutants. *Nucleic Acids Res.*, **29**, 1381–1388.
17. J.R. Cho, M.J. Campbell, E.A. Winzeler, L. Steinmetz, A. Conway, L. Wodocka, T.G. Wolfsberg, A.E.Gabrielian, D. Landsman, D.J. Lockhart, and R.W. Davis (1998). A genome-wide transcriptional analysis of the mitotic cell cycle. *Molecular Cell*, **2**, 65–73.
18. S.A. Jelinsky, P. Estep, G. Church, and L.D. Samson (2000). Regulatory networks revealed by transcriptional profiling of damaged *Saccharomyces cerevisiae* cells: Rpn4 links base excision repair with proteasomes. *Mol. Cell Biol.*, **20**, 8157–8167.
19. B. Karahalil, P.-M. Girard, S. Boiteux, and M. Dizdaroglu (1998). Substrate specificity of the Ogg1 protein of *Saccharomyces cerevisiae*: Excision of guanine lesions produced in DNA by ionizing radiation- or hydrogen peroxide/metal ion-generated free radicals. *Nucleic Acids Res.*, **26**, 1228–1232.

20. P.-M. Girard, C. D'Ham, J. Cadet, and S. Boiteux (1998). Opposite base-dependent excision of 7,8-dihydro-8-oxoadenine by the Ogg1 protein of *Saccharomyces cerevisiae*. *Carcinogenesis* **19**, 1299–1305.
21. C. D'Ham, A. Romieu, M. Jaquinod, D. Gasparutto, and J. Cadet (1998). Excision of 5,6-dihydrothymine and 5-hydroxycytosine from defined sequence oligonucleotides by *Escherichia coli* endonuclease III and Fpg protein: Kinetic and mechanistic aspects. *Biochemistry*, **38**, 3335–3344.
22. D. Gasparutto, M. Ait-Abbas, M. Jaquinod, S. Boiteux, and J. Cadet (2000). Repair and coding properties of 5-hydroxy-5-methylhydantoin nucleosides inserted into DNA oligomers. *Chem. Res. Toxicol.* **13**, 575–584.
23. N. Guibourt, B. Castaing, P. Auffret van der Kemp, and S. Boiteux (2000). Catalytic and DNA binding properties of the Ogg1 protein of *Saccharomyces cerevisiae*: Comparison between the wild-type and the K241R and K241Q active site mutant proteins. *Biochemistry*, **39**, 1716–1724.
24. D. Thomas, A.D. Scott, R. Barbey, M. Padula, and S. Boiteux (1997). Inactivation of *OGG1* increases the incidence of GC–>TA transversions in *Saccharomyces cerevisiae*: evidence for endogenous oxidative damage to DNA in eukaryotic cells. *Mol. Gen. Genet.*, **254**, 171–178.
25. T.T. Ni, G.T. Marsischky, and R.D. Kolodner (1999). MSH2 and MSH6 are required for the removal of Adenine misincorporated opposite 8-oxo-guanine in *S. cerevisiae*. *Molecular Cell*, **4**, 439–444.
26. R.L. Swanson, N.J. Morey, P.W. Doetsch, and S. Jinks-Robertson (1999). Overlapping specificities of base excision repair, nucleotide excision repair, recombination and translesion synthesis pathways for DNA base damage in *Saccharomyces cerevisiae*. *Mol. Cell Biol.* **19**, 2929–2935.
27. A.D. Scott, M. Neishabury, D.H. Jones, S.H. Reed, S. Boiteux, and R.D. Waters (1999). Spontaneous mutation, oxidative DNA damage and the roles of base and nucleotide excision repair in the yeast *Saccharomyces cerevisiae*. *Yeast*, **15**, 205–218.
28. L. Haracska, Sung-Lim Yu, R.E. Johnson, L. Prakash, and S. Prakash (2000). Efficient and accurate replication in the presence of 7,8-dihydro-8-oxoguanine by DNA polymerase η. *Nature Genetics*, **25**, 458–461.
29. S.D. Brunner, D.P.G. Norman and G.L. Verdine (2000). Structural basis for recognition and repair of the endogenous mutagen 8-oxoguanine in DNA. *Nature*, **403**, 859–866.

Chapter 7

The role of a yeast homologue of the human phosphatase activator hPTPA in the cellular response to oxidative DNA damage

Dindial Ramotar, Jocelyn David and Elliot Drobetsky

Table of contents

7.1 Introduction

Organisms respond to environmental stress in large part by triggering phosphorylation-dependent signal transduction pathways, which are tightly regulated by multiple kinases and phosphatases. These pathways are often initiated at the cell surface and culminate in the expression of protective genes involved in DNA repair, cell cycle arrest, and apoptosis. While the roles of many of the kinases involved in cellular stress-induced signal transduction have been elucidated, far less is known about the role, regulation, and even the identity of the phosphatases. Many phosphatases are known to have a single catalytic subunit, as well as one or more regulatory subunits that impose diverse functional roles. In this chapter, we focus on the yeast *Saccharomyces cerevisiae yPTPA1* gene, encoding a putative phosphatase activator, which may play a role in signaling the oxidative stress response. The yeast yPtpa1 protein shares significant identity with a human protein PTPA1, which was isolated based on its ability to stimulate the weak phosphotyrosyl phosphatase activity of the serine/threonine phosphatase PP2A. To date, neither the biological function, nor the precise pathway in which the human hPTPA executes its role, is known.

Recent studies from yeast may shed light into the possible function of this putative phosphatase activator. Yeast mutants lacking yPtpa1 protein are hypersensitive to the chemical carcinogen 4-nitroquinoline-1-oxide (4-NQO), as well as to ultraviolet A (UVA; 320–400-nm), both potent inducers of oxidative stress. The precise mechanism by which yPtpa1 confers resistance to 4-NQO and UVA is not known, but several experiments have revealed that this protein is required to modulate gene expression. It is possible that yPtpa1 belongs to a signaling pathway which senses oxidative stress and triggers gene expression. As such, hPTPA may play a similar protective mechanism against oxidative stress in human cells, i.e., in order to forestall multistage carcinogenesis. Below, we first provide a brief account of oxygen free radical production and its cellular consequences, followed by our current understanding of the biological role of yPtpa1.

7.2 Reactive oxygen species

7.2.1 Sources and consequences

Oxidative stress arises when there is an imbalance between production and scavenging of intracellular reactive oxygen species (ROS) such as superoxide anion ($O_2^{\bullet -}$), singlet oxygen, and hydroxyl radical (OH) [1–4]. ROS are highly reactive molecules, which have one or more unpaired electrons and, as such, are stabilized through the attack and removal of electrons from other sources (i.e., concomitant oxidation of the donor molecule). ROS are produced by *environmental sources*, e.g., γ-rays, natural sunlight (UVB + UVA), and antitumour drugs such as bleomycin [2,5–10]. They are also produced by *endogenous sources*, e.g., in humans through (i) mitochondrial respiration, (ii) the inflammatory response (activated neutrophils and macrophages), and (iii) the dissociation of H_2O_2 *via* the Fenton reaction to form $^{\bullet}OH$ [2,11].

ROS are associated with several human disorders, including atherosclerosis, neurodegenerative diseases such as amyotrophic lateral sclerosis, Down's syndrome, Alzheimer's disease, and the normal aging process [12–17]. It is therefore critical that all organisms retain the capacity to detoxify ROS. A primary line of defense consists of various antioxidants, e.g., gluthathione and vitamins, as well as enzymes, e.g., superoxide dismutase and catalase that minimize ROS levels in cells [12,18]. Depending on the genetic state and age of a cell these defense mechanisms can be overwhelmed by increased levels of ROS, which in turn can directly inactivate proteins or enzymes, damage membranes through lipid peroxidation, and induce highly-genotoxic DNA adducts [19–21]. The manner in which cells cope with the latter type of damage is of particular interest to the authors, and below we further describe some important ROS-induced oxidative DNA lesions, and the enzymes that initiate their repair.

7.2.2 ROS-induced DNA lesions

(i) Base damage: ROS induce a variety of modified bases, e.g., thymine glycol, the ring-closed 8-oxoguanine, and 5-formyluracil [22,23]. These modified bases must be removed from duplex DNA, otherwise they may cause DNA polymerase to insert an incorrect nucleotide thereby creating permanent mutations [24–27]. For example, 8-oxoguanine very frequently mispairs with adenine resulting in G to T transversion mutations, while 5-formyluracil can mispair with cytosine causing T to G transversions [23,28,29]. Thus, specific mutations can serve as molecular signatures to help establish the nature of a DNA lesion induced by a particular treatment [29].

(ii) Sugar damage: ROS also damage the sugar moiety of DNA to create both oxidized apurinic/apyrimidinic (AP) sites, as well as single strand breaks that are terminated with 3′-blocked ends [5,6]. Unrepaired AP sites are targets for mutagenesis because they lack template information, while the 3′-blocked ends prevent the action of DNA polymerase [30–32]. Thus, the maintenance of DNA integrity by dividing cells requires efficient mechanisms to ensure the proper removal of oxidized DNA lesions.

7.2.3 Repair of oxidized DNA bases via base-excision repair

Prokaryotes and eukaryotes possess multiple repair pathways to combat the deleterious effects of oxidative insult to cellular DNA [33,34]. Oxidized bases such as 8-oxoguanine and thymine glycol are removed through the base excision repair pathway (BER) [33]. In the initiation step of BER, damaged bases are removed by a highly conserved group of enzymes called DNA glycosylases, which are specific for single or a class of adducts [35,36]. For example, in *E. coli*, 8-oxoguanine and other oxidized purines are removed by formamidopyrimidine DNA glycosylase (Fpg), for which functional homologues also exist in yeast and mammalian cells [37]. Cells that lack Fpg are hypersensitive to H_2O_2, and, under normal growth conditions, show a high rate of signature G to T transversion mutations as a direct consequence [38,39]. All glycosylases such as Fpg hydrolyze the N-glycosylic bond between a damaged base and the deoxyribose sugar to produce an unmodified AP site as a secondary lesion [33,40].

Some glycosylases can also cleave the resulting AP sites by a β-elimination reaction (β-lyase) to produce a 3′-terminal blocked end, such as, the 3′-α, β-unsaturated aldehyde produced by endonuclease III that removes thymine glycol [41]. The blocked end must then be removed by 3′-diesterase in order to allow DNA replication to proceed (see below). Otherwise, AP sites are incised by hydrolytic AP endonucleases, which cleave the phosphodiester bond immediately 5′ to the AP site, producing a free 3′-OH for DNA repair synthesis and a 5′-deoxyribose-phosphate that must also be removed [33]. Many AP endonucleases also contain 3′-diesterase activity that removes 3′-blocked termini, such as those created by ROS-induced sugar fragmentation and by the β-lyase activity of some DNA glycosylases [6,31,33,40]. To date, every organism examined possesses bifunctional AP endonucleases/3′-diesterases. The bifunctional AP endonucleases/3′-diesterases comprise two distinct families, exemplified by *Escherichia coli* endonuclease IV (Endo IV) and exonuclease III (Exo III), and these enzymes show specificity on the removal of oxidative DNA lesions [33,34,42,43].

Following cleavage of the AP site, DNA polymerase β is recruited to replace the single nucleotide, and ligase seals the nick. In mammalian cells, the 5′-deoxyribose-phosphate is removed by the 5′-deoxyribose-phosphodiesterase activity of the N-terminal region of DNA polymerase β [44]. In some cases, DNA polymerase β or ε or δ can replace at least 5 to 8 nucleotides causing the 5′-deoxyribose-phosphate strand to be displaced creating a flap-structure which is cleaved by the Fen1 endonuclease [45]. The nick is then sealed by ligase.

7.3 The carcinogen 4-nitroquinoline-1-oxide induces oxidative DNA lesions

A few years ago, we set out to search for new factors that are involved in protecting cells against the model carcinogen 4-nitroquinoline-1-oxide (4-NQO). This agent has been widely employed in mammalian systems as a paradigm for DNA damage-induced carcinogenesis. To exert its neoplastic effect, 4-NQO must first undergo metabolic activation to the proximate carcinogen 4-hydroxyaminoquinoline-1-oxide (4-HAQO), which, following acylation, reacts with DNA to form stable quinoline-purine monoadducts, i.e., at the exocyclic N2 and N6 positions of guanine and adenine, respectively [46–48]. In bacteria, yeast, and mammalian cells, these genotoxic "bulky" DNA lesions are processed largely by the nucleotide excision repair (NER) pathway, which recognizes and creates two incisions, i.e., on either side of the DNA strand containing the helix-distorting lesion, to create a gap [49,50]. This single-stranded gap is restored to the original double stranded form by the DNA polymerases δ and ε and ligase. Thus, 4-NQO-induced "bulky lesions" are removed in a manner analogous to classical dipyrimidine photoproducts generated by the model DNA damaging agent 254 nm UV light [49,50]. Thus, mutants deficient in NER are hypersensitive to the genotoxic effects of 254 nm UV, as well as 4-NQO [49,50]. The apparent strong similarity in modes of cellular processing for 254 nm UV- and 4-NQO-induced DNA damage in diverse prokaryotic and eukaryotic systems has often resulted in categorization of the latter agent as "UV-mimetic" [51]. However, this characterization may be inappropriate, as several recent investigations have clearly demonstrated that 4-NQO, unlike 254 nm UV, can generate a substantial degree of intracellular oxidative stress [52,53]. Indeed, it was shown that 4-NQO is a potent inducer of the *E. coli* soxR/S regulon that responds to intracellular superoxide imbalances: SoxR is activated by superoxide anion and triggers the expression of the transcriptional activator SoxS, which in turn induces the expression of several proteins including the AP endonuclease, endonuclease IV [52,53]. It is believed that 4-NQO undergoes redox cycling to produce superoxide anion, which can be further converted into genotoxic reactive oxygen species (e.g., singlet oxygen and hydroxyl radicals) that engender modified bases and DNA strand breaks [52,54]. In fact, it has been shown *in vitro* that activated 4-HAQO can generate the highly premutagenic DNA lesion 8-oxoguanine, as well as DNA strand breaks, in the presence of Cu(II) [54,55]. It is therefore not surprising that bacterial mutants lacking either Fpg or AP endonucleases are hypersensitive to the mutagenic effects of 4-NQO, but not of 254 nm UV [52,56].

7.4 Yeast as a model to study 4-NQO induced oxidative DNA lesions

Based on the demonstrated ability of 4-NQO to induce oxidative stress that contributes to cell killing and mutagenesis in bacteria, it is very possible that this agent may act similarly in eukaryotic cells. As such, we decided to use the budding yeast *Saccharomyces cerevisiae* as a model organism to identify novel eukaryotic genes implicated in the cellular response to 4-NQO-induced oxidative DNA damage. Yeast has served as an outstanding experimental model for delineating the mechanisms underlying many biological processes including gene regulation, cell cycle control, and DNA repair [57]. The latter notion is substantiated by the elegant discovery that yeast DNA mismatch repair-deficient mutants essentially recapitulate the genetic defects recovered in non-polyposis heriditary colon cancer cells (e.g., carrying germ line alterations in the human DNA mismatch repair genes hMSH2 and hMLH1) [58,59].

In addition, yeast offers several advantages, such as a versatile genetic system that can be easily manipulated, e.g., genes can be readily deleted in a single step [60]. This organism grows on simple medium under aerobic and anaerobic conditions, and can maintain episomal genes in single or multicopies. The complete sequencing of the yeast genome has allowed rapid identification of genes and their putative structural and functional homologues in other organisms. Thus, the ability to combine the strengths of molecular biology, genetics, biochemistry, and cell biology, and to manipulate the yeast growth environment, all serve to make *S. cerevisiae* an ideal system to study oxidative stress.

7.5 Isolation of 4-NQO sensitive yeast mutants

We initially began to search for yeast mutants that were hypersensitive to oxidative DNA damage induced by 4-NQO. This was achieved by treating a wild-type haploid yeast strain DBY747 with a high concentration of the alkylating agent methyl methane sulfonate (0.3% for 1 hr), and the survivors were screened for drug sensitivity on rich medium solid agar plates containing 0.4 μg/ml of 4-NQO [61]. This concentration of 4-NQO permits 90% survival of the wild-type strain, and colonies that do not survive this drug concentration were scored as mutants [61]. Mutants showing cross sensitivity to 254 nm UV were deemed defective in the NER pathway known to repair UV-induced dipyrimidine photoproducts, as well as 4-NQO induced "bulky" DNA adducts, and therefore eliminated from the analysis. This screening procedure led to the isolation of five mutants, EBY1, EBY10, EBY20, EBY30, and EBY40 that were sensitive to 4-NQO, but not to 254 nm UV. These five mutants are likely defective in a process that responds to oxidative stress caused by 4-NQO, as they show parental resistance to 4-NQO in the absence of oxygen, i.e., under anaerobic conditions.

7.6 EBY1 is defective in the yPTPA1 gene required for repairing 4-NQO induced DNA lesions

We decided to systematically characterize each of the mutants beginning with EBY1. Cloning by complementation of the 4-NQO sensitivity of EBY1 revealed that this mutant was defective in a gene that we designated *yPTPA1,* as it encoded a protein sharing 40% identity with the human phosphotyrosyl phosphatase activator, hPTPA [61,62]. The human hPTPA was originally isolated as a protein that stimulates the phosphotyrosyl phosphatase activity of the mammalian serine/threonine phosphatase PP2A, suggesting that yPtpa1 may act in yeast to activate a phosphatase. Interestingly, the human PTPA gene is located on chromosome 9q34, and this location has been designated a hot spot for melanoma and leukemia [63].

To investigate the role of yPtpa1p in 4-NQO induced oxidative stress, we deleted the chromosomal copy of *yPTPA1* in the wild-type haploid strain DBY747, and showed that the resulting *yptpa1Δ* mutant (strain EBY2) was 6-fold more sensitive to 4-NQO [61]. We further demonstrated that the EBY1 mutant bore a single mutation in the *yPTPA1* gene, providing absolute confirmation that the 4-NQO hypersensitivity was due to a defect in this gene [61]. A more striking observation is that the *yptpa1Δ* mutant was 5-fold more sensitive to rigorously-purified UVA wavelengths, a major component of terrestrial sunlight known to produce ROS and oxidative DNA damage. To date, this is the first reported UVA hypersensitive yeast mutant, and thus provides an important tool to study the mechanism of UVA defense in eukaryotic cells. This mechanism is critical as UVA is known to induce mutations in mammalian cells, and furthermore exposure to UVA has been linked specifically to the development of malignant melanoma [64,65]. Further analysis revealed that the *yptpa1Δ* mutant was also hypersensitive to diamide, but only modestly sensitive to H_2O_2. The *yptpa1Δ* mutant is not sensitive to 254 nm UV or the alkylating agent methyl methane sulfonate (MMS) neither of which produce ROS [61]. These observations are consistent with a role for yPtpa1 specifically in oxidative stress response.

We further demonstrated that the *yptpa1Δ*mutant exhibited a marked deficiency in the recovery of high molecular weight DNA upon exposure to 4-NQO revealing a defect at the level of DNA repair [61]. The *yptpa1Δ* mutant also displayed a 3-fold increase in the spontaneous mutation rate, which was enhanced as much as 13- and 17-fold when treated with 4-NQO and UVA, respectively, as compared to the parent [61]. The mutation rate was not enhanced in the *yptpa1Δ* mutant when challenged with 254 nm UV or MMS. Collectively, these data (i) strongly support a role for active oxygen intermediates in determining the genotoxic outcome of 4-NQO exposure, and (ii) suggest a novel mechanism in yeast involving yPtpa1-mediated activation of a phosphatase that participates in the repair of oxidative DNA damage, implying that hPTPA may exert a similar function in humans [64,65]. In fact, studies have shown that phosphatases are involved in DNA repair and oxidative stress response [66,67]. For example, inhibition of PP2A phosphatase in human cell extracts prevents the repair of UV damaged DNA by the nucleotide excision DNA repair pathway [66], and expression of the CL100 serine/threonine phosphatase is rapidly induced by UVA [68,69]. However, the exact role of these phosphatases in DNA repair and UVA response remains unclear.

7.7 yPtpa1 interacts with a chaperone, Ydj2, and a phosphatase, Sit4, to mediate 4-NQO and UVA resistance

It was logical to assume that yPtpa1 could exert a role in the DNA damage response by stimulating one of the PP2A phosphatases in yeast. Since yeast has at least eleven PP2A members (Pph21, Pph22, Pph3, **Sit4**, Ppg1, Glc7, Cna1, Cna2, Ppz1, Ppz2, and Sal6) [70], it was difficult to predict which of these could be activated by yPtpa1. We therefore used yPtpa1 as bait to trap the relevant phosphatase using the yeast two hybrid system [71,72]. Surprisingly, we did not isolate a phosphatase, but identified a previously uncharacterized protein, herein designated Ydj2, which shared 39% identity with Ydj1 (*Y*east homolog of *D*na*J*) [73]. The Ydj-family consists of at least 15 members Ydj1, **Ydj2** (our work), Scj1, Xdj1, Cai1, Caj1, Mdj1, Mdj2, Sec63, Jem1, Sis1, and four additional members whose functions are unknown. We confirmed that Ydj2 interacts with yPtpa1 both *in vivo* and *in vitro*. Furthermore, we found that Ydj2 is essential for cell viability and is localized only to the nucleus, as determined using a Gfp-Ydj2 fusion (I. Brodeur and D. Ramotar, unpublished). We do not know the role of Ydj2 in yeast, but it could function to disaggregate or assemble protein complexes in the nucleus. In fact, many of the Ydj members function to assemble and disassemble protein complexes and to transport proteins to the endoplasmic reticulum and mitochondria [73–76]. So far, Ydj members have no documented role as phosphatases.

Is there any connection between yeast DnaJs and cellular phosphatases? A hint in this respect was provided by a study showing that overproduction of the Ydj member, Sis1 was capable of restoring activity to a mutant form of Sit4 phosphatase [77]. Thus, it was reasonable to assume that yPtpa1-Ydj2 complex could activate Sit4. If so, a *sit4Δ* mutant should be as sensitive as a *yptpa1Δ* mutant to 4-NQO/UVA. We therefore deleted the *SIT4* gene from the parent strain DBY747 and found that the resulting *sit4Δ* mutant was indeed sensitive to 4-NQO/UVA and to the same extent as the *yptpa1Δ* mutant (D. Ramotar, unpublished). Moreover, a *yptpa1Δ* sit4Δ double mutant was constructed and found to be equally sensitive to 4-NQO/UVA, as either the *yptpa1Δ* or *sit4Δ* single mutant. The lack of any additive 4-NQO/UVA sensitivity in the absence of both yPtpa1 and Sit4 strongly suggests that these two proteins participate in the same pathway. One might expect that overproduction of Sit4 may restore 4-NQO and UVA resistance to the *yptpa1Δ* mutant, but this was not the case. Thus, yPtpa1 is absolutely required for a specific function of Sit4 in 4-NQO/UVA resistance. While we were unable to show that yPtpa1 and Sit4 interact *in vivo*, a very recent study re-isolated yPtpa1 as a mutation that caused lethality in cells lacking the Cla4 kinase (a kinase without a specific function) [78]. This latter study also demonstrated that yPtpa1 immunoprecipitated the Sit4 phosphatase, thus confirming our genetic findings that yPtpa1 and Sit4 interact. We are in the process of testing (i) whether yPtpa1 stimulates Sit4 phosphatase activity in a Ydj2-dependent manner, and (ii) whether yPtpa1 and Sit4 interaction is mediated by Ydj2.

7.8 The Sit4 phosphatase forms multiple complexes and play a role in gene expression

Sit4 forms a complex with Tap42 in the TOR signaling pathway (see below) [79]. Moreover, Sit4 is also associated with at least three related high molecular weight proteins (*S*it4 *a*ssociated *p*roteins), Sap190, Sap185, and Sap155 [80]. Sit4 forms one complex with Sap190/185 and another with Sap155, and each complex is believed to perform a different function [80]. Sap190/185 can compete with Sap155 for binding to Sit4, raising the possibility that other factors could compete with Sap proteins to alter Sit4 specificity or cellular location [80]. Thus, in response to oxidative stress, yPtpa1, with the aid of Ydj2, could displace the Sap proteins from Sit4 to create an activated yPtpa1-Sit4 complex with specialized function [70]. In fact, in yeast, the serine/threonine PP1 phosphatase catalytic subunit Glc7 participates in diverse cellular functions, including glycogen biosynthesis, synaptonemal complex formation, and sporulation, which are dictated by specific regulatory subunits Gac1, Red1, and Gip1, respectively [81,82].

One of the functions of Sit4 is to execute a role in gene expression. This protein was originally characterized as required for expression of the *HIS4* gene, and later, for the normal expression of the cyclins *CLN1*, *CLN2*, and *HCS26* mRNA during late G1 phase of the cell cycle [83–85]. However, it remains unclear how Sit4 influences gene expression.

7.9 yPtpa1 plays a role in gene expression

7.9.1 yPtpa1 down-regulates a subset of ribosomal protein genes in response to rapamycin

Thus far, we have learned that yPtpa1 interacts with Sit4, perhaps in the presence of Ydj2, and that the complex is required to mediate resistance against 4-NQO/UVA (presumably by activating gene expression). We have direct evidence showing that yPtpa1 is required to modulate gene expression. The first evidence came from the finding that Sit4 interacts with the phosphorylated form of Tap42 [79]. The Sit4-Tap42 complex is a component of the TOR (*T*arget *O*f *R*apamycin)-kinase signaling pathway that can respond to the immunosuppressant rapamycin. Under normal growth conditions, it is believed that the TOR signaling pathway leads to the recruitment of the histone acetylase, Esa1 to the promoters of a subset of ribosomal protein genes, which then acetylate histone H4 and thus open the chromatin structure to allow gene expression [86]. In rapamycin-exposed cells, the Tor kinase is unable to phosphorylate Tap42, preventing its association with Sit4, leading to a dramatic reduction in the expression of several ribosomal protein genes including *RPS5, 6A, 26A, 28A, RPL3, 9, 10, 25, 29*, and *32* [79,87]. Under this rapamycin treatment condition, it is believed that chromatin structure at the ribosomal protein gene promoters is in a repressed state (likely through a mechanism involving deacetylation of histones).

A mutant allele of *TAP42* (*tap42–1*) was hyper-resistant to rapamycin, but deletion of the *SIT4* gene manifested no such phenotype [79,87]. It was postulated that regulation, rather than loss, of the phosphatase activity confers rapamycin resistance. Because yPtpa1 is associated with Sit4, we tested if *yptpa1Δ* mutants were also hyper-resistant to rapamycin. Interestingly, the *yptpa1Δ* mutants were indeed hyper-resistant to rapamycin, as compared to the isogenic parents. As expected, overproduction of Sit4 in the *yptpa1Δ* mutant did not revert the strain to rapamycin-sensitivity. Moreover, the expression of a *subset* of the ribosomal protein genes was not completely shut down in the *yptpa1Δ* mutants, as compared to the parent, in response to rapamycin (100 ng/ml for 2 hr). For example, *RPS26A* mRNA remains expressed (but not fully), while that of *RPL9A* was shut down in the *yptpa1Δ* mutant, as compared to the parent. This unpublished finding demonstrates for the first time that yPtpa1 has a role in the TOR signaling pathway, and its absence allows expression of key genes involved in rapamycin resistance. We have absolutely no knowledge yet if yPtpa1 and Tap42 confer different specificity on Sit4, and how the complex of yPtpa1-Sit4-Ydj2 may influence gene expression, i.e., in a negative way in response to rapamycin.

7.9.2 yPtpa1 is required to express the YCF1 gene in response to arsenite

The observation that *yptpa1Δ* mutants manifests different responses to two stresses (i.e., sensitivity to 4-NQO/UVA, but resistance to rapamycin), prompted a search for yet additional phenotypes. Among a large number of drugs and metal ions tested, we found that *yptpa1Δ* mutant was sensitive to arsenite. In yeast, two efflux pumps are known to detoxify arsenite, the vacuolar pump Ycf1 and the plasma membrane pump Acr3, and these in turn are positively regulated by the AP-1 like transcriptional activators Yap1 and Yap8, respectively [88,89]. We first examined the expression level of *YCF1* in four strains: the parent, *yptpa1Δ*, yap1*Δ*, and a *yptpa1Δ* yap1*Δ* double mutant, when cells were exposed to arsenite (0.1 mM/3 hr). While *YCF1* mRNA was clearly induced 20-fold in the parent, it was only induced ~10-fold in the *yptpa1Δ* and *yap1Δ* mutant. No induction was detected in the *yptpa1Δ* yap1*Δ* double mutant, indicating that yPtpa1 and Yap1 play independent role in *YCF1* expression. The absence of *YCF1* mRNA expression in the double mutant is consistent with the hypersensitivity to arsenite. A multicopy plasmid pADH-YCF1, expressing the *YCF1* gene under the control of the constitutive promoter *ADH*, fully expressed the *YCF1* mRNA, and restored to the *yptpa1Δ yap1Δ* double mutant (slightly more than) parental resistance to arsenite. We note that *YCF1* mRNA stability is not affected by loss of yPtpa1. Moreover, the stability of various fusion proteins including Ycf1-HA, Sgs1-HA (a helicase), Rad53-HA (a cell cycle checkpoint kinase), and Imp2-HA (a transcriptional co-activator) are also not affected by the absence of yPtpa1. Taken together, these data suggest that yPtpa1 is required to promote expression of the *YCF1* gene in response to arsenite exposure.

7.10 Model

In summary, yPtpa1 is required to (i) "shut-down" gene expression of a subset of ribosomal protein genes in response to rapamycin, and (ii) "promote" expression of the *YCF1* gene in response to arsenite stress. It is also possible that yPtpa1 promotes gene expression in response to 4-NQO/UVA stresses. A simple model (Figure 1) to explain all the above findings is that a yPtpa1-Sit4 complex has the ability to regulate expression of a subset of genes, presumably by modulating

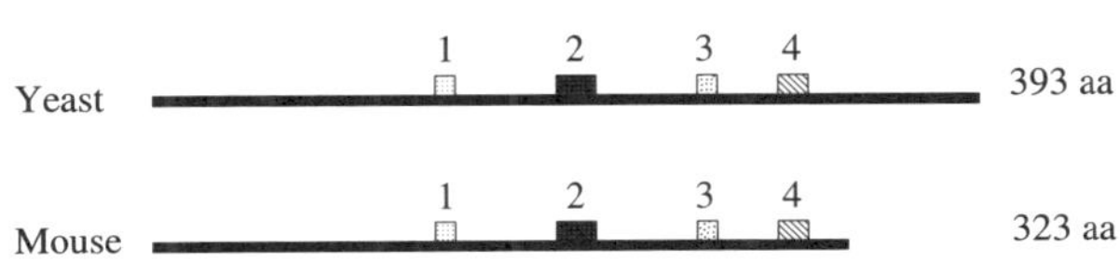

Figure 1. Comparison of the yeast *S. cerevisiae* and mouse PTPA proteins. Four regions shown in boxes correspond to highly conserved motifs. mPTPA lacks a C-terminal region.

chromatin structure through a process involving phosphorylation/dephosphorylation. In the case of 4-NQO or UVA exposure, the yPtpa1-Sit4 complex may activate a kinase that phosphorylates histones (H2A and or H3) to directly open the chromatin structure thus allowing gene expression (Figure 1). This model is supported by several studies documenting the need for chromatin remodeling in order to promote DNA repair. In one study, it was shown that the DNA damaging agents bleomycin and MMS activate the Mec1 kinase in yeast, leading to direct phosphorylation of serine 129 of histone H2A [90]. A mutation (S129A) that prevents phosphorylation of H2A causes cells to be hypersensitive to both bleomycin and MMS (*yptpa1Δ* is not sensitive to these agents) [90]. It is likely that that phosphorylation of H2A may be required to relax the chromatin to either allow gene expression to facilitate repair, or to permit access of repair proteins and other factors directly to the DNA lesions. Two additional studies showed that remodeling of the chromatin is required for DNA repair. Mutations in the INO80 ATPase chromatin remodeling complex of yeast and the TIP60 histone acetylase of mammalian cells cause sensitivity to MMS and γ-rays, respectively [91,92].

According to our model, the dual requirement of yPtpa1 and Yap1 to express the *YCF1* gene (arsenite efflux pump) may require both phosphorylation and acetylation of histones at the *YCF1* promoter. It was shown that the Yap1-related transcriptional activator, Gcn4 recruits the histone acetylase Gcn5 to acetylate histone H3 on lysine 14 at gene promoters [93]. More importantly, histone H3 is also phosphorylated at serine 10 and this is required for transcriptional activation, as well as to promote acetylation of lysine 14 by histone acetylase to optimize transcription [94].

7.11 Perspectives

We predict that the yPtpa1-Sit4 complex acts by modifying the chromatin structure at the promoter region of a subset of genes. In response to 4-NQO or UVA, this complex may lead to the expression of at least one gene involved in the repair of the oxidative DNA lesions. Alternatively, the yPtpa1-Sit4 complex could have a genome-wide effect, thereby remodeling the chromatin to allow the entry of DNA repair proteins to repair 4-NQO or UVA-induced DNA lesions. So far, we observed no differences in the expression of several known DNA repair or DNA damage-inducible genes including *OGG1* (involved in base-excision repair), *RAD1* (involved in nucleotide excision repair), *RAD52* (involved in recombinational repair), and *RNR2* (induced in response to DNA damaging agents). It is possible that an unknown gene specifically involved in the repair of 4-NQO/UVA-induced oxidative DNA lesions is not properly expressed in the *yptpa1Δ* mutant. Several approaches are underway to isolate this unknown gene, particularly using reverse genetics to identify the defective gene in the four remaining 4-NQO sensitive mutants discussed earlier (EBY10, EBY20, EBY30, and EBY40). These mutants are not complemented either by the *yPTPA1* or *SIT4 gene*, and thus bear mutations in other genes. Unlike *yptpa1Δ* mutants, the mutant EBY20 is not hyper-resistance to rapamycin. Thus, EBY20 is uniquely sensitive to 4-NQO/UVA. One possibility is that expression of the EBY20-gene is under the regulation of the yPtpa1-Sit4 complex. As such, the mRNA level of the EBY20-gene should be drastically reduced in the *yptpa1Δ* or *sit4Δ* mutant, and should not be induced in response to 4-NQO exposure. Another possibility is that the EBY20-gene may encode a protein whose activity could be regulated by phosphorylation/dephosphorylation by the yPtpa1-Sit4 complex. If the EBY20-gene product is a direct target for Sit4, it should be dephosphorylated in the parent strain, but not in the *sit4Δ* mutant in response to 4-NQO. Several additional questions can be asked about the EBY20-gene product: (i) Does it act alone or in combination with other proteins to repair 4-NQO-induced DNA lesions? (ii) Is it part of the same biochemical pathway as yPtpa1, Ydj2, and Sit4? and (iii) How important is this protein in preventing 4-NQO-induced mutations? An alternative approach to find the yPtpa1-target gene is to use DNA microarrays. A preliminary microarray performed under normal growth conditions revealed that several genes were upregulated or downregulated in the *yptpa1Δ* mutant as compared to the parent (H. Lee and D. Ramotar, unpublished). Experiments are currently in progress to determine if there are specific genes affected in the *yptpa1Δ* mutant in response to 4-NQO or UVA exposure. It is expected that deletion of any yPtpa1-target genes would confer sensitivity to 4-NQO and display an inability to repair damaged DNA.

It is noteworthy that we have also made substantial progress towards delineating the biological role of the mammalian PTPA. As a first step, we isolated the mouse mPTPA cDNA using the hPTPA cDNA as a probe (J. David et al., in preparation). The mPTPA predicted amino acid sequence is nearly identical (95.6%) to that of human hPTPA, suggesting a strong structural conservation. Preliminary data revealed that when mPTPA cDNA was engineered to express in yeast, it substituted for yPtpa1 in 4-NQO and UVA resistance. However, the mPTPA did not restore rapamycin sensitivity to the *yptpa1Δ* mutant. This finding might not be surprising as the C-terminal end of yPtpa1 has an extension of 70 amino acid residues, which is absent form mPTPA, and may be required for its role in rapamycin response (Figure 2). It would be interesting to attach the C-terminal region of yPtpa1 onto mPTPA and test if this heterologous fusion protein mPTPA-C-ter will restore parental rapamycin sensitivity to the *yptpa1Δ* mutant.

We have also used the mPTPA cDNA to screen a murine 129/Sv genomic library, which resulted in a single clone bearing an 11–Kb DNA fragment. Partial PCR mapping revealed that the clone contained exons, 5, 6, 7, 8, and 9 (encompassing 66% of the mPTPA coding region) within 6-kb of DNA. It is noteworthy that the human gene is composed of 10 exons and 9 introns that span 60-Kb of DNA. We are in the process of using the 11-Kb genomic fragment to construct a loxP targeting vector to make conditional $ptpa^{-/-}$ null mice [95]. We are especially planning to trigger loxP-induced deletion of exon 5 to 7, a region which is clearly required for function. For example, exon 7 contains a highly conserved segment (EPAGSQGVWGL*DD*^{206}) found in yPtpa1, mPTPA, and hPTPA, and a mutation D205G completely abrogated yPtpa1 function in yeast. The D205G substitution was identified as a mutation that prevented interaction with Ydj2. Thus, deletion of exon 5 to 7 should abolish mPTPA function in mice. We expect that murine embryonic fibroblasts (MEFs) derived from $ptpa^{-/-}$ mice

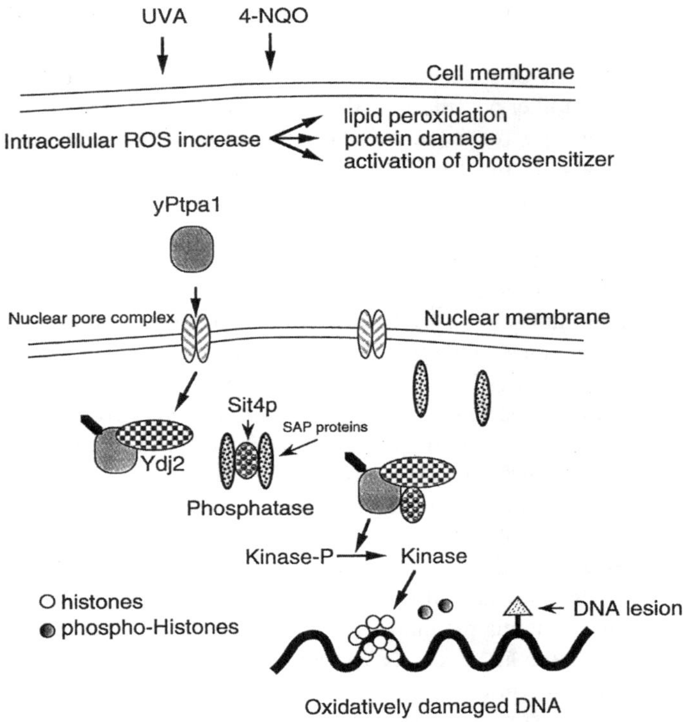

Figure 2. In response to 4-NQO/UVA, yPtpa1 becomes activated and interacts with Ydj2 and Sit4. The yPtpa1-Ydj2-Sit4 complex activates a kinase by dephosphorylation, which then phosphorylates histones (e.g., H3) to cause their release from the DNA to promote gene expression and or allow DNA repair. During this process, Ydj2 may prevent Sit4 interaction with other proteins.

will be (i) sensitive to UVA, (ii) unable to repair UVA-induced DNA lesions, and (iii) accumulate UVA-induced mutations at endogenous loci, e.g., *hprt* [19]. As such, we also predict that these mice will be susceptible to UVA-induced skin tumors. If indeed $ptpa^{-/-}$ mice exhibit enhanced sensitivity to UVA coupled with predisposition to cutaneous tumourigenesis, individuals with a defective hPTPA allele may similarly be at risk of developing sunlight-induced skin cancer.

Acknowledgements

This work was funded by a grant from the Canadian Institute of Health Research to D.R.

References

1. B. Halliwell, J.M. Gutteridge (1990). Role of free radicals and catalytic metal ions in human disease: an overview. *Methods Enzymol.,* **186**, 1–85.

2. J.A. Imlay, S. Linn (1988). DNA damage and oxygen radical toxicity. *Science,* **240**, 1302–1309.
3. R.S. Sohal (1997). Mitochondria generate superoxide anion radicals and hydrogen peroxide. *Faseb J.,* **11**, 1269–1270.
4. M. Saran W. Bors (1990). Radical reactions in vivo–an overview. *Radiat. Environ. Biophys.,* **29**, 249–262.
5. F. Hutchinson (1985). Chemical changes induced in DNA by ionizing radiation. *Prog. Nucleic Acid Res. Mol. Biol.,* **32**, 1[illegible]5–154.
6. C. von Sonntag (1991). The chemistry of free-radical-mediated DNA damage. *Basic Life Sci.,* **58**, 287–317; discussion 317–321.
7. S.M. Keyse, R.M. Tyrrell (1990). Induction of the heme oxygenase gene in human skin fibroblasts by hydrogen peroxide and UVA (365 nm) radiation: evidence for the involvement of the hydroxyl radical. *Carcinogenesis,* **11**, 787–791.
8. R.A. Bennett, P.S. Swerdlow, L.F. Povirk (1993). Spontaneous cleavage of bleomycin-induced abasic sites in chromatin and their mutagenicity in mammalian shuttle vectors. *Biochemistry,* **32**, 3188–3195.
9. J.W. Kozarich, L. Worth, Jr., B.L. Frank, D.F. Christner, D.E. Vanderwall, J. Stubbe (1989). Sequence-specific isotope effects on the cleavage of DNA by bleomycin. *Science,* **245**, 1396–1399.
10. E. Kvam, R.M. Tyrrell (1997). Induction of oxidative DNA base damage in human skin cells by UV and near visible radiation. *Carcinogenesis,* **18**, 2379–2384.
11. Y.C. Chong, G.H. Heppner, L.A. Paul, A.M. Fulton (1989). Macrophage-mediated induction of DNA strand breaks in target tumor cells. *Cancer Res.,* **49**, 6652–6657.
12. R.L. Jackson (1993). Anti-oxidants for the treatment and the prevention of atherosclerosis. *Biochem. Soc. Trans.,* **21**, 650–651.
13. H.X. Deng, A. Hentati, J.A. Tainer, Z. Iqbal, A. Cayabyab, W.Y. Hung, E.D. Getzoff, P. Hu, B. Herzfeldt, R.P. Roos, et al. (1993). Amyotrophic lateral sclerosis and structural defects in Cu,Zn superoxide dismutase [see comments]. *Science,* **261**, 1047–1051.
14. J. Busciglio, B.A. Yankner (1995). Apoptosis and increased generation of reactive oxygen species in Down's syndrome neurons in vitro. *Nature,* **378**, 776–779.
15. B. Halliwell (1992). Reactive oxygen species and the central nervous system. *J. Neurochem.,* **59**, 1609–1623.
16. B.N. Ames, M.K. Shigenaga, T.M Hagen (1993). Oxidants, antioxidants, and the degenerative diseases of aging. *Proc. Natl. Acad. Sci. USA,* **90**, 7915–7922.
17. R.S. Sohal, S. Agarwal, A. Dubey W.C. Orr (1993). Protein oxidative damage is associated with life expectancy of houseflies. *Proc. Natl. Acad. Sci. USA,* **90**, 7255–7259.
18. I. Fridovich (1997). Superoxide anion radical (O2-.), superoxide dismutases, and related matters. *J. Biol. Chem.,* **272**, 18515–18517.
19. B.S. Berlett, E.R. Stadtman (1997). Protein oxidation in aging, disease, and oxidative stress. *J. Biol. Chem.,* **272**, 20313–20316.
20. D. Steinberg (1997). Low density lipoprotein oxidation and its pathobiological significance. *J. Biol. Chem.,* **272**, 20963–20966.
21. E.S. Henle, S. Linn (1997) Formation, prevention, and repair of DNA damage by iron/hydrogen peroxide. *J. Biol. Chem.,* **272**, 19095–19098.
22. H. Kasai, P.F. Crain, Y. Kuchino, S. Nishimura, A. Ootsuyama, H. Tanooka (1986). Formation of 8-hydroxyguanine moiety in cellular DNA by agents producing oxygen radicals and evidence for its repair. *Carcinogenesis,* **7**, 1849–1851.

23. Q.M. Zhang, H. Sugiyama, I. Miyabe, S. Matsuda, I. Saito, S. Yonei (1997). Replication of DNA templates containing 5-formyluracil, a major oxidative lesion of thymine in DNA. *Nucleic Acids Res.,* **25**, 3969–3973.
24. H. Ide, L.A. Petrullo, Z. Hatahet, S.S. Wallace (1991). Processing of DNA base damage by DNA polymerases. Dihydrothymine and beta-ureidoisobutyric acid as models for instructive and noninstructive lesions. *J. Biol. Chem.,* **266**, 1469–1477.
25. T.R. O'Connor, S. Boiteux, J. Laval (1988). Ring-opened 7-methylguanine residues in DNA are a block to in vitro DNA synthesis. *Nucleic Acids Res.,* **16**, 5879–5894.
26. M.L. Wood, A. Esteve, M.L. Morningstar, G.M. Kuziemko, J.M. Essigmann, M. Dizdaroglu, E. Gajewski (1992). Genetic effects of oxidative DNA damage: comparative mutagenesis of 7,8-dihydro-8-oxoguanine and 7,8-dihydro-8-oxoadenine in Escherichia coli. Mechanistic studies of ionizing radiation and oxidative mutagenesis: genetic effects of a single 8-hydroxyguanine (7-hydro-8-oxoguanine) residue inserted at a unique site in a viral genome. *Nucleic Acids Res.,* **20**, 6023–6032.
27. T.J. McBride, B.D. Preston, L.A. Loeb (1991). Mutagenic spectrum resulting from DNA damage by oxygen radicals. *Biochemistry,* **30**, 207–213.
28. S. Shibutani, M. Takeshita, A.P. Grollman (1991). Insertion of specific bases during DNA synthesis past the oxidation-damaged base 8-oxodG. *Nature,* **349**, 431–434.
29. B.A. Kunz, E.S. Henson, H. Roche, D. Ramotar, T. Nunoshiba, B. Demple (1994). Specificity of the mutator caused by deletion of the yeast structural gene (APN1) for the major apurinic endonuclease. *Proc. Natl. Acad. Sci. USA,* **91**, 8165–8169.
30. L.A. Loeb, B.D. Preston (1986). Mutagenesis by apurinic/apyrimidinic sites. *Annu Rev Genet,* **20**, 201–230.
31. W.D. Henner, S.M. Grunberg, W.A. Haseltine (1983) Enzyme action at 3′ termini of ionizing radiation-induced DNA strand breaks. *J. Biol. Chem.,* **258**, 15198–15205.
32. D. Ramotar, S.C. Popoff, B. Demple (1991). Complementation of DNA repair-deficient Escherichia coli by the yeast Apn1 apurinic/apyrimidinic endonuclease gene. *Mol. Microbiol.,* **5**, 149–155.
33. B. Demple, L. Harrison (1994). Repair of oxidative damage to DNA: enzymology and biology. *Annu. Rev. Biochem.,* **63**, 915–948.
34. D. Ramotar (1997). The apurinic-apyrimidinic endonuclease IV family of DNA repair enzymes. *Biochem. Cell Biol.,* **75**, 327–336.
35. Q.M. Zhang, I. Miyabe, Y. Matsumoto, K. Kino, H. Sugiyama, S. Yonei (2000). Identification of repair enzymes for 5-formyluracil in DNA. Nth, Nei, and MutM proteins of Escherichia coli. *J. Biol. Chem.,* **275**, 35471–35477.
36. V.A. Bohr, G.L. Dianov (1999). Oxidative DNA damage processing in nuclear and mitochondrial DNA, *Biochimie,* **81**, 155–160.
37. J.P. Radicella, C. Dherin, C. Desmaze, M.S. Fox, S. Boiteux (1997). Cloning and characterization of hOGG1, a human homolog of the OGG1 gene of *Saccharomyces cerevisiae*. *Proc. Natl. Acad. Sci. USA,* **94**, 8010–8015.
38. M.L. Michaels, C. Cruz, A.P. Grollman, J.H. Miller (1992). Evidence that MutY and MutM combine to prevent mutations by an oxidatively damaged form of guanine in DNA. *Proc. Natl. Acad. Sci. USA,* **89**, 7022–7025.
39. D. Thomas, A.D. Scot, R. Barbey, M. Padula, S. Boiteux (1997). Inactivation of OGG1 increases the incidence of G. C−>T. A transversions in Saccharomyces cerevisiae: evidence for endogenous oxidative damage to DNA in eukaryotic cells. *Mol. Gen. Genet.*, **254**, 171–178.
40. K. Sakumi, M. Sekiguchi (1990). Structures and functions of DNA glycosylases. *Mutat. Res.*, **236**, 161–172.

41. A. Mazumder, J.A. Gerlt, M.J. Absalon, J. Stubbe, R.P. Cunningham, J. Withka, P.H. Bolton (1991). Stereochemical studies of the beta-elimination reactions at aldehydic abasic sites in DNA: endonuclease III from Escherichia coli, sodium hydroxide, and Lys-Trp-Lys. *Biochemistry*, **30**, 1119–1126.
42. G. Barzilay, I.D. Hickson (1995). Structure and function of apurinic/apyrimidinic endonucleases. *Bioessays*, **17**, 713–719.
43. A.A. Ischenko, M.K. Saparbaev (2002). Alternative nucleotide incision repair pathway for oxidative DNA damage. *Nature* **415**, 183–187.
44. J.A. Feng, C.J. Crasto, Y. Matsumoto (1998). Deoxyribose phosphate excision by the N-terminal domain of the polymerase beta: the mechanism revisited. *Biochemistry*, **37**, 9605–9611.
45. R. Prasad, G.L. Dianov, V.A. Bohr, S.H. Wilson (2000). FEN1 stimulation of DNA polymerase beta mediates an excision step in mammalian long patch base excision repair. *J. Biol. Chem.*, **275**, 4460–4466.
46. T. Sugimura, H. Otake, T. Matsushima (1968). Single strand scissions of DNA caused by a carcinogen, 4-hydroxylaminoquinoline 1-oxide. *Nature*, **218**, 392.
47. S. Galiegue-Zouitina, B. Bailleul, Y.M. Ginot, B. Perly, P. Vigny, M.H. Loucheux-Lefebvre (1986). N2-guanyl and N6-adenyl arylation of chicken erythrocyte DNA by the ultimate carcinogen of 4-nitroquinoline 1-oxide. *Cancer Res.*, **46**, 1858–1863.
48. M. Tada (1976). Main binding sites of the carcinogen, 4-nitroquinoline 1-oxide in nucleic acids. *Biochim. Biophys. Acta*, **454**, 558–566.
49. S. Prakash, P. Sung, L. Prakash (1993). DNA repair genes and proteins of *Saccharomyces cerevisiae*. *Annu. Rev. Genet.*, **27**, 33–70.
50. J.H. Hoeijmakers (1993). Nucleotide excision repair. II: From yeast to mammals. *Trends Genet.*, **9**, 211–217.
51. I.C. Felkner, F. Kadlubar (1968). Parallel between ultraviolet light and 4-nitroquinoline-1-oxide sensitivity in *Bacillus subtilis*. *J. Bacteriol.*, **96**, 1448–1449.
52. T. Nunoshiba, B. Demple (1993). Potent intracellular oxidative stress exerted by the carcinogen 4-nitroquinoline-N-oxide. *Cancer Res.*, **53**, 3250–3252.
53. B. Demple, C.F. Amabile-Cuevas (1991). Redox redux: the control of oxidative stress responses. *Cell*, **67**, 837–839.
54. K. Yamamoto, S. Inoue, S. Kawanishi (1993). Site-specific DNA damage and 8-hydroxydeoxyguanosine formation by hydroxylamine and 4-hydroxyaminoquinoline 1-oxide in the presence of Cu(II): role of active oxygen species. *Carcinogenesis*, **14**, 1397–1401.
55. K. Kohda, M. Tada, H. Kasai, S. Nishimura, Y. Kawazoe (1986). Formation of 8-hydroxyguanine residues in cellular DNA exposed to the carcinogen 4-nitroquinoline 1-oxide. *Biochem. Biophys. Res. Commun.*, **139**, 626–632.
56. J. Ruiz-Laguna, R.R. Ariza, M.J. Prieto-Alamo, S. Boiteux, C. Pueyo (1994). Fpg protein protects Escherichia coli K-12 from mutation induction by the carcinogen 4-nitroquinoline 1-oxide. *Carcinogenesis*, **15**, 425–429.
57. D. Botstein, S.A. Chervitz, J.M. Cherry (1997). Yeast as a model organism [comment]. *Science*, **277**, 1259–1260.
58. R. Fishel, M.K. Lescoe, M.R. Rao, N.G. Copeland, N.A. Jenkins, J. Garber, M. Kane, R. Kolodner (1993). The human mutator gene homolog MSH2 and its association with hereditary nonpolyposis colon cancer *Cell*, **75**, 1027–1038.
59. C.E. Bronner, S.M. Baker, P.T. Morrison, G. Warren, L.G. Smith, M.K. Lescoe, M. Kane, C. Earabino, J. Lipford, A. Lindblom (1994). et al., Mutation in the DNA mismatch repair gene homologue hMLH1 is associated with hereditary non-polyposis colon cancer. *Nature*, **368**, 258–261.

60. R. Rothstein (1991). Targeting, disruption, replacement, and allele rescue: integrative DNA transformation in yeast. *Methods Enzymol.*, **194**, 281–301.
61. D. Ramotar, E. Belanger, I. Brodeur, J.Y. Masson, E.A. Drobetsky (1998). A yeast homologue of the human phosphotyrosyl phosphatase activator PTPA is implicated in protection against oxidative DNA damage induced by the model carcinogen 4-nitroquinoline 1-oxide. *J. Biol. Chem.*, **273**, 21489–21496.
62. X. Cayla, C. Van Hoof, M. Bosch, E. Waelkens, J. Vandekerckhove, B. Peeters, W. Merlevede, J. Goris (1994). Molecular cloning, expression, and characterization of PTPA, a protein that activates the tyrosyl phosphatase activity of protein phosphatase 2A. *J. Biol. Chem.*, **269**, 15668–15675.
63. C. Van Hoof, M.S. Aly, A. Garcia, X. Cayla, J.J. Cassiman, W. Merlevede, J. Goris (1995). Structure and chromosomal localization of the human gene of the phosphotyrosyl phosphatase activator (PTPA) of protein phosphatase 2A. *Genomics*, **28**, 261–272.
64. E.A. Drobetsky, J. Turcotte, A. Chateauneuf (1995). A role for ultraviolet A in solar mutagenesis. *Proc. Natl. Acad. Sci. USA*, **92**, 2350–2354.
65. J. Moan, A. Dahlback, R.B. Setlow (1999). Epidemiological support for an hypothesis for melanoma induction indicating a role for UVA radiation. *Photochem. Photobiol.*, **70**, 243–247.
66. R.R. Ariza, S.M. Keyse, J.G. Moggs, R.D. Wood (1996). Reversible protein phosphorylation modulates nucleotide excision repair of damaged DNA by human cell extracts. *Nucleic Acids Res.*, **24**, 433–440.
67. A. Yacoub, M.R. Kelley, W.A. Deutsch (1997). The DNA repair activity of human redox/repair protein APE/Ref-1 is inactivated by phosphorylation. *Cancer Res.*, **57**, 5457–5459.
68. S.M. Keyse, E.A. Emslie (1992). Oxidative stress and heat shock induce a human gene encoding a protein-tyrosine phosphatase. *Nature*, **359**, 644–647.
69. M. Fiscella, H. Zhang, S. Fan, K. Sakaguchi, S. Shen, W.E. Mercer, G.F. Vande Woude, P.M. O'Connor, E. Appella (1997). Wip1, a novel human protein phosphatase that is induced in response to ionizing radiation in a p53-dependent manner. *Proc. Natl. Acad. Sci. USA*, **94**, 6048–6053.
70. M.J. Stark (1996). Yeast protein serine/threonine phosphatases: multiple roles and diverse regulation. *Yeast*, **12**, 1647–1675.
71. E.M. Phizicky, S. Fields (1995). Protein-protein interactions: methods for detection and analysis. *Microbiol. Rev.*, **59**, 94–123.
72. P. James, J. Halladay, E.A. Craig (1996). Genomic libraries and a host strain designed for highly efficient two-hybrid selection in yeast. *Genetics*, **144**, 1425–1436.
73. D.P. Atencio, M.P. Yaffe (1992). MAS5, a yeast homolog of DnaJ involved in mitochondrial protein import. *Mol. Cell Biol.*, **12**, 283–291.
74. H. Blumberg, P. Silver, P.A. Silver (1991). A split zinc-finger protein is required for normal yeast growth. A homologue of the bacterial heat-shock gene DnaJ that alters protein sorting in yeast. *Gene*, **107**, 101–110.
75. S. Nishikawa, T. Endo (1997). The yeast JEM1p is a DnaJ-like protein of the endoplasmic reticulum membrane required for nuclear fusion. *J Biol Chem*, **272**, 12889–12892.
76. T. Zhong, M.M. Luke, K.T. Arndt (1996). Transcriptional regulation of the yeast DnaJ homologue SIS1. *J. Biol. Chem.*, **271**, 1349–1356.
77. M.M. Luke, A. Sutton, K.T. Arndt (1991). Characterization of SIS1, a *Saccharomyces cerevisiae* homologue of bacterial dnaJ proteins. *J. Cell Biol.*, **114**, 623–638.
78. D.A. Mitchell, G.F. Sprague, Jr. (2001) The phosphotyrosyl phosphatase activator, Ncs1p (Rrd1p), functions with Cla4p to regulate the G(2)/M transition in *Saccharomyces cerevisiae*. *Mol. Cell Biol.*, **21**, 488–500.

79. Y. Jiang, J.R. Broach (1999). Tor proteins and protein phosphatase 2A reciprocally regulate Tap42 in controlling cell growth in yeast. *EMBO J.*, **18**, 2782–2792.
80. M.M. Luke, F. Della Seta, C.J. Di Como, H. Sugimoto, R. Kobayashi, K.T. Arndt (1996). The SAP, a new family of proteins, associate and function positively with the SIT4 phosphatase. *Mol. Cell Biol.*, **16**, 2744–2755.
81. J.S. Stuart, D.L. Frederick, C.M. Varner, K. Tatchell (1994). The mutant type 1 protein phosphatase encoded by glc7–1 from *Saccharomyces cerevisiae* fails to interact productively with the GAC1-encoded regulatory subunit. *Mol. Cell Biol.*, **14**, 896–905.
82. J. Tu, W. Song, M. Carlson (1996). Protein phosphatase type 1 interacts with proteins required for meiosis and other cellular processes in Saccharomyces cerevisiae. *Mol. Cell Biol.*, **16**, 4199–4206.
83. K.T. Arndt, C.A. Styles, G.R. Fink (1989). A suppressor of a HIS4 transcriptional defect encodes a protein with homology to the catalytic subunit of protein phosphatases. *Cell*, **56**, 527–537.
84. M.J. Fernandez-Sarabia, A. Sutton, T. Zhong, K.T. Arndt (1992). SIT4 protein phosphatase is required for the normal accumulation of SWI4, CLN1, CLN2, and HCS26 RNAs during late G1. *Genes Dev.*, **6**, 2417–2428.
85. J. Sidorova, L. Breeden (1993). Analysis of the SWI4/SWI6 protein complex, which directs G1/S-specific transcription in Saccharomyces cerevisiae. *Mol. Cell Biol.*, **13**, 1069–1077.
86. J.L. Reid, V.R. Iyer, P.O. Brown, K. Struhl (2000). Coordinate regulation of yeast ribosomal protein genes is associated with targeted recruitment of Esa1 histone acetylase. *Mol. Cell*, **6**, 1297–1307.
87. T. Powers, P. Walter (1999). Regulation of ribosome biogenesis by the rapamycin-sensitive TOR-signaling pathway in *Saccharomyces cerevisiae*. *Mol. Biol. Cell*, **10**, 987–1000.
88. M. Ghosh, J. Shen, B.P. Rosen (1999). Pathways of As(III) detoxification in Saccharomyces cerevisiae. *Proc. Natl. Acad. Sci. USA*, **96**, 5001–5006.
89. N. Bouganim, J. David, R. Wysocki, D. Ramotar (2001). Yap1 overproduction restores arsenite resistance to the ABC transporter deficient mutant ycf1 by activating ACR3 expression. *Biochem. Cell Biol.*, **79**, 441–448.
90. J.A. Downs, N.F. Lowndes, S.P. Jackson (2000). A role for *Saccharomyces cerevisiae* histone H2A in DNA repair. *Nature*, **408**, 1001–1004.
91. X. Shen, G. Mizuguchi, A. Hamiche, C. Wu (2000). A chromatin remodelling complex involved in transcription and DNA processing. *Nature*, **406**, 541–544.
92. T. Ikura, V.V. Ogryzko, M. Grigoriev, R. Groisman, J. Wang, M. Horikoshi, R. Scully, J. Qin, Y. Nakatani (2000). Involvement of the TIP60 histone acetylase complex in DNA repair and apoptosis. *Cell*, **102**, 463–473.
93. M.H. Kuo, E. vom Baur, K. Struhl, C.D. Allis (2000) Gcn4 activator targets Gcn5 histone acetyltransferase to specific promoters independently of transcription. *Mol. Cell*, **6**, 1309–1320.
94. W.S. Lo, R.C. Trievel, J.R. Rojas, L. Duggan, J.Y. Hsu, C.D. Allis, R. Marmorstein, S.L. Berger (2000). Phosphorylation of serine 10 in histone H3 is functionally linked *in vitro* and in *vivo* to Gcn5-mediated acetylation at lysine 14. *Mol. Cell*, **5**, 917–926.
95. J.L. Liu, S. Yakar, D. LeRoith (2000). Mice deficient in liver production of insulin-like growth factor I display sexual dimorphism in growth hormone-stimulated postnatal growth. *Endocrinology*, **141**, 4436–4441.

Chapter 8

DNA repair in RNA polymerase I transcribed genes

Antonio Conconi, Vyacheslav A. Bespalov, Deirdre Fahy and Michael J. Smerdon

Table of contents

Abstract

DNA damage is repaired by different mechanisms like nucleotide excision repair (NER) and base excision repair (BER). NER subpathways include transcription-coupled repair (TCR), which rapidly repairs actively transcribing genes, and global genomic repair (GGR), which repairs non-transcribed sequences. Until recently, it was believed that TCR operated exclusively on RNA polymerase II transcribed genes, with the RNA polymerase II transcription factor TFIIH providing a link between transcription and repair.

The ribosomal genes, transcribed by RNA polymerase I, exist as multiple copies in tandem array. A portion of these genes is free of nucleosomes and is transcriptionally active while the remaining genes are assembled into nucleosomes and are transcriptionally inactive. In general, repair of various DNA lesions in total rDNA was shown to be slower than in RNA polymerase II transcribed genes. Moreover, TCR was not observed in ribosomal genes. However, recent measurements of strand specific repair in separated active and inactive ribosomal genes demonstrate that TCR of RNA polymerase I transcribed genes occurs in yeast. This observation, and the recent findings that TFIIH is also a component of the RNA polymerase I transcription complex, sheds new light on DNA repair of ribosomal genes.

8.1 Introduction

DNA repair of a variety of lesions has been investigated in the genome as a whole and in specific genes. The majority of these studies were done for genes transcribed by RNA polymerase II, the protein coding genes.

Repair of ribosomal genes, which are present in multiple copies in the cell and are transcribed by RNA polymerase I, represents an interesting system to understand the interplay of chromatin, transcription and DNA repair. In fact, transcribing and silent ribosomal genes coexist in the same cell as non-nucleosomal and nucleosomal populations respectively, thus presenting two different substrates for DNA repair mechanisms. Also, RNA polymerase I transcription is distinct from RNA polymerase II transcription, albeit with some common features. Therefore, studies on DNA repair of ribosomal genes may reveal new insights on the mechanisms that couple transcription with excision repair.

Though repair of a wide range of DNA lesions in ribosomal genes is summarized below, emphasis is given to repair of UV damage. Moreover, this review focuses on transcription-coupled repair (TCR), one of the two subpathways of nucleotide excision repair (NER). Finally, to provide a context for the recent description of TCR in RNA polymerase I transcribed genes in yeast, we have outlined TCR in RNA polymerase II transcribed genes in the first part of this chapter. In the last section, we speculate on the mechanisms of RNA polymerase I TCR in yeast and revisit the possible existence of RNA polymerase I TCR in mammalian cells.

8.2 Transcription coupled DNA repair

8.2.1 DNA repair

Damage in DNA, ranging from small base (or sugar) modifications to bulky chemical adducts and double-strand breaks (reviewed in [1]), can block normal DNA processing mechanisms and cause cell death, mutation or cancer (reviewed in [2]). Cells have evolved different DNA repair pathways that specifically remove a variety of lesions from DNA (reviewed in [3,4]). BER removes nucleotides containing small modifications like oxidized, deaminated or alkylated bases. On the other hand, lesions that induce severe distortions of the DNA helix, like bulky DNA adducts and UV-photoproducts, [cyclobutane pyrimidine dimers (CPDs) and pyrimidine-(6-4)-pyrimidone photoproducts], are repaired by NER (reviewed in [2,5]).

In eukaryotes, DNA repair is modulated by DNA packaging in chromatin (reviewed in [6,7]). Most of the genome is transcriptionally inactive and these regions are characterized by stable arrays of nucleosomes (reviewed in [8]). On the other hand, transcriptionally active regions have modified (or absent) nucleosomes and are more permissive to DNA processing including repair (reviewed in [9,10]).

The repair of UV-photoproducts by NER involves distinct steps: recognition of the lesion, incisions upstream and downstream of the lesion in the damaged strand, removal of the oligonucleotide containing the lesion, filling in of the gap by DNA synthesis and ligation of the newly synthesized patch (reviewed in [11]). Although it is likely that removal of photodimers from the inactive and active regions of the genome are done by similar molecular processes, recognition of DNA damage in the inactive chromatin appears to differ from that in the active chromatin.

In the current model, NER is divided into two sub-pathways: 1) global genome repair (GGR), responsible for removing lesions from non-transcribed DNA, including sequences in the non-transcribed strand (NTS) of expressed genes; and 2) transcription-coupled repair (TCR), responsible for removing lesions from the transcribed strands (TS) of active genes (reviewed in [2,12]). In addition to the core NER enzymes that are employed by both GGR and TCR, there are genes specifically required for repair of either the transcribed or the non-transcribed DNA strands (reviewed in [2,13]). For example, the product of the XPC (*Xeroderma pigmentosum* group C) gene is involved in the early steps of GGR, like recognition of lesions and recruitment of the core NER factors to the damaged sites. Thus, a non-functional XPC gene causes a deficiency in GGR and only lesions in the TS of active genes are repaired in these cells. In contrast, and in addition to ongoing transcription, the products of the CSA and CSB genes (Cockayne syndrome group A and group B, respectively) are required for repair of lesions in the TS (reviewed in [13,14]). Therefore, in CS cells only lesions in non-transcribed DNA are repaired.

8.2.2 Transcription-coupled repair

Despite numerous studies done to elucidate the mechanisms of TCR, it is still unclear how this process occurs at the molecular level. *In vitro* systems were successfully employed to describe most of the events occurring in GGR. However, attempts made to reproduce TCR *in vitro* have failed. On the other hand, a number of interesting concepts have emerged from *in vivo* studies that have analyzed TCR in RNA polymerase II-transcribed genes. Moreover, these observations may shed light on TCR processes in RNA polymerase I- and III-transcribed genes.

8.2.2.1 Definition of transcription-coupled repair

The concept of transcription-coupled repair was first proposed after discovering that the rate of removal of CPDs varies throughout the genome (reviewed in [12]). Initially it was observed that CPDs present in the expressed *DHFR* (dihydrofolate reductase) gene in CHO (Chinese hamster ovary) cells were repaired more efficiently than those present in a silent sequence downstream of the gene [15]. Later, it was shown that the TS of active genes is rapidly repaired and this explained the original findings of the efficient removal of CPDs from active DNA regions [12,16]. Since the discovery of TCR in mammalian cells and *E. coli* [15,17], TCR has been found in most organisms analyzed (reviewed in [12]). Furthermore, TCR was observed throughout the cell cycle [18] and *in situ* (e.g., mouse skin [19]). Finally, the consequence of fast removal of CPDs from the TS and the biological importance of TCR is evidenced by the UV-induced mutation spectra of mammalian cells, showing a bias for mutations in the NTS [20–22].

Analysis of repair of the complementary strands of active genes has demonstrated the direct involvement of transcription in fast removal of DNA lesions. In these experiments, use of strand-specific probes in Southern blot analysis has shown that the TS is repaired faster than the NTS of a variety of class II genes (reviewed in [23]). Moreover, the use of α-amanitin, an inhibitor of RNA polymerase II elongation, abolished strand-specific repair in human and CHO cells [24–26]. In yeast, efficient repair of the TS of active genes was abolished in RNA polymerase II temperature sensitive mutants at the restrictive temperatures [27,28]. Thus, the designation 'transcription-coupled repair' infers that ongoing transcription is necessary for preferential removal of DNA damage from the TS.

Until recently it was thought that only genes transcribed by RNA polymerase II, and not those transcribed by RNA polymerases I or III, are targets of TCR (reviewed in [12,13]). As discussed below, at least in yeast we now know that TCR also occurs in RNA polymerase I-transcribed genes [29–31]. Nevertheless, it is important to remember that TCR is observed only where RNA polymerases are actively elongating.

8.2.2.2 Transcription factor TFIIH

The discovery that subunits of TFIIH are required in DNA repair established the crucial role of this transcription factor in both TCR and GGR. In addition to RNA polymerase II, at least five basal transcription factors (TFIIB, D, E, F and H) are required to initiate transcription at the minimal promoter (TATA box and initiation site) [32]. The multi-subunit protein composite TFIIH is recruited into the preinitiation complex at the promoter by the interaction with TFIIE and RNA polymerase II. The two largest subunits of TFIIH, XPB and XPD (*Xeroderma pigmentosum* groups B and D), are helicases having opposite polarity and

DNA-dependent ATPase activity. Following formation of the preinitiation complex, the DNA helicase activity associated with TFIIH catalyzes unwinding of DNA at the transcriptional start site. These DNA helicases play an important role in transcription during the step of 'promoter escape' by RNA polymerase II (reviewed in [32,33]). Finally, during the transition from the initiation to the elongation step, TFIIH is released, a process occurring after the transcription machinery reaches +30 [34].

NER also involves the formation of an open DNA bubble (of 24~32 bp) around the lesion. The creation of this unwound DNA region is ATP dependent and the early stages of the DNA opening require the two helicases of TFIIH (XPB and XPD), together with XPC. During NER, TFIIH may also play a role in distinguishing the damaged strand from the non-damaged strand if, for example, translocation along a DNA strand by TFIIH DNA helicases is interrupted by encountering a lesion (reviewed in [11,33]).

8.2.2.3 DNA lesions and arrest of transcription

Blockage of transcription by DNA damage has the potential for inducing changes in gene expression, with deleterious effects on the organism. However, the existence of DNA repair mechanisms linked to transcription allow for rapid recovery of gene expression (reviewed in [13,14]).

Although the connections between transcription and repair are not completely understood, stalled RNA polymerases at damage sites may be recognition signals for recruitment of repair proteins (reviewed by [12]). Importantly, it was recently shown that TCR is not limited to NER. In fact, lesions that are repaired by BER, and capable of blocking RNA polymerases, are also substrates for TCR [35,36]. Thus, it is possible that most DNA lesions which block the elongation of RNA polymerases are removed by repair processes that are coupled to transcription (reviewed in [14,37]).

Whether stalled RNA polymerases are the only signals required to recruit the repair machinery to the TS of active genes is still an open question [see 12]. As pointed out by Tornaletti and Hanawalt [13], if this is indeed the case, then how do cells distinguish between RNA polymerases that are stalled at DNA lesions and those arrested at natural DNA sequence pause sites?

8.2.2.4 Removal of stalled RNA polymerases

Upon induction of DNA damage, stalled RNA polymerases at lesion sites must be displaced to make DNA damage accessible to repair enzymes (reviewed in [12,14]). In *E. coli* it was shown that a 'transcription coupling repair factor' (TCRF), the product of the *mfd* gene, binds to and releases a blocked RNA polymerase at a lesion [38]. Furthermore, the Mfd protein interacts with the excision repair complex, promoting DNA repair [39]. In human cells, the product of the CSB gene has been implicated in the process of recognition and release of blocked RNA polymerases [40]. Indeed, it is possible that CSB together with XPB, XPD and XPG are responsible for the displacement of stalled RNA polymerases (reviewed by [14]). In support of this hypothesis are the *in vivo* studies by Le Page et al. [36], showing that transcription prevents DNA repair in CSB and XPG-CS mutants but not in normal cells. These results suggest that without TCR-specific factors, stalled transcription complexes cannot be displaced and DNA repair is inhibited.

Another poorly understood process is the fate of stalled RNA polymerases after they have been displaced by TCR factors. In *E. coli*, it was shown that during TCR the transcription complex is aborted [39]. In eukaryotic cells, genes are several kilobases long and avoidance of abortion of partially completed transcripts of large genes would be of great advantage to cells (reviewed in [13]). However, some data suggest that transcription complexes are degraded following RNA polymerase arrest at DNA lesions. For example, in UV-irradiated human cells there is evidence that CSA and CSB are required for polyubiquitination and targeting for proteosomal degradation of RNA polymerase II molecules [41,42]. In yeast, when the lesion cannot be rapidly removed, it appears that Def1 in complex with Rad26 (the counterpart of human CSB) enables ubiquitination and proteolysis of RNA polymerase II [43].

8.2.3 Transcription-coupled repair in the yeast Saccharomyces cerevisiae

The NER pathway has been remarkably conserved during the evolution from yeast to man. Therefore, genetic and biochemical studies in yeast have made important contributions to our current understanding of the mechanisms involved in NER in eukaryotes (for comprehensive reviews of NER in yeast, see [3,44]).

8.2.3.1 NER in yeast

In yeast the Rad1/10 and Rad2 endonucleases, the Rad4/23 complex, the Rpa protein, the Rad14 DNA-damage recognition protein and TFIIH (containing the Rad3 and Rad25 helicases) are necessary and sufficient for incision of damaged DNA *in vitro* [45]. These proteins are referred to as the core NER proteins. In addition, Rad26 is involved in TCR (see below), whereas *rad7* and *rad16* mutant cells are deficient in removal of CPDs from non-transcribed DNA, including the silent mating type loci, NTS of active genes and promoter sequences. In these mutants TCR is not affected and the TS of active genes is efficiently repaired. Thus, Rad7 and Rad16 appear to be necessary for GGR *in vivo*. In contrast to strains carrying mutations in one of the core NER proteins, *rad7* and *rad16* mutants are only moderately sensitive to UV light. Therefore, Rad7/16 and the human XPC cells have a similar phenotype. However, Rad4 is the yeast homologue of XPC, even though *rad4* and XPC do not share the same phenotype since *rad4* mutants are defective in repair of both DNA strands [23].

8.2.3.2 TCR in yeast

As is found in higher eukaryotes, the kinetics of removal of CPDs in yeast can vary throughout the genome, with the TS of active genes generally being repaired faster than non-transcribed DNA. In yeast, strand-specific repair was first observed in the *URA3* gene [46]. Subsequently, TCR was found to be dependent on ongoing transcription, since efficient repair of the transcribed strand of active genes was abolished at the restrictive temperature in RNA polymerase II mutants [27,28]. Moreover, strand-specific repair of the yeast *GAL7* gene was shown to be dependent on the induction of transcription [28].

8.2.3.3 The roles of RAD26 and RAD28 in TCR

Like CSB in human cells and the TCRF in *E. coli* [47], Rad26 may help recruit NER proteins to DNA lesions, where RNA polymerase II is stalled. Based on sequence homology, Rad26 and CSB belong to the Swi2/Snf2 family of proteins, which have been implicated in remodeling protein-DNA interactions during DNA repair (reviewed in [10,48]). This suggests that Rad26 could help remove stalled RNA polymerase II from the lesion during TCR, as described for the TCRF in *E. coli*. The *RAD28* gene is the yeast homologue of CSA but unlike the human CSA cells, the *rad28* mutants are not deficient in TCR.

Information on the role of Rad26 was obtained when TCR was followed in *rad26* mutants. In these cells, TCR occurs in a region comprising the first ~50 bp downstream of the transcription start site [49], and similar results were also found in CSB cells [50]. This DNA region (i.e., first 30 bp) corresponds to the beginning of RNA polymerase II elongation and TFIIH is not yet released from the transcription complex [34]. Thus, the inverse correlation between the requirement of Rad26 in TCR and the presence of TFIIH may suggest that the rate-limiting step of TCR is the recruitment of TFIIH to the lesion. If so, Rad26 could facilitate the recruitment of TFIIH to the damage site.

Despite the similarities described above, there are some differences between CSB and *rad26* cells (reviewed in [23]). Like human CSB cells, yeast mutants lacking the *RAD26* gene are TCR deficient. However, in contrast to the characteristic UV-sensitivity of CSB cells, *rad26* mutants and *wt* strains show similar resistance to UV irradiation. To explain this lack of UV-sensitivity, Verhage et al. [29] proposed the existence of a TCR process that is Rad26-independent. In fact, Rad26-independent TCR was observed when *rad7, rad16-rad26* double mutants were analyzed (reviewed in [23]). From these experiments, it appears that the requirement for Rad26 in TCR differs from gene to gene. For example, Rad26-independent TCR was not found in the *URA3* gene downstream of position +50, whereas Rad26-independent TCR occurred in the *RPB2* gene. Interestingly, in these *rad26* double mutants the TS of the induced *GAL7* gene (cells grown in galactose) was almost completely repaired by Rad26-independent TCR, while the repressed *GAL7* gene (cells grown in glucose) was essentially not repaired [29]. These findings suggest that TCR can operate independently of Rad26, and that ongoing transcription can give rise to TCR in yeast. The existence of Rad26-independent TCR was further supported by the deletion of the *SPT4* gene, which suppresses the *rad26* phenotypic defect by reactivating TCR. Furthermore, these data suggest that Rad26 acts as an elongation factor that enables TCR [51]. Most recently, mutants with deletions in RNA polymerase II subunits were analyzed and the results support the existence of Rad26-dependent and Rad26-independent TCR [52]. In this work it was shown that there is a role for one RNA polymerase II subunit (Rpb4) in Rad26 mediated TCR and for another polymerase subunit (Rpb9) in Rad26 independent TCR. Finally, a novel and rather different function for Rad26 was proposed. It was found that Rad26 promotes the passage of RNA polymerase II through DNA lesions induced by the alkylating agent methyl methanesulfonate [53]. Thus, the exact function of Rad26 is not yet understood.

8.3 Ribosomal genes

In comparison to the continuously growing number of studies on DNA repair in RNA polymerase II-transcribed genes, very little is known about DNA repair in RNA polymerase I-transcribed genes. Before discussing our current knowledge on DNA repair in rDNA, the molecular biology of ribosomal genes will be briefly summarized (for extensive reviews, see [54,55]).

8.3.1 Organization of ribosomal genes

8.3.1.1 rDNA tandem repeats

Tandem arrays of multiple copies of rDNA are found in the nucleoli of eukaryotic cells. The nucleolus is a dense chromatin region composed of rDNA, RNA polymerase I, rRNA and assembling ribosomes (reviewed in [56,57]). The yeast ribosomal genes are present in ~150 copies that are organized in a single tandem repeat on chromosome XII (Figure 1A), and one unit is ~9.1 kb long (Figure 1B). RNA polymerase I is responsible for transcription of the rRNA 35S precursor which is processed into the mature 25S, 18S and 5.8S rRNA present in ribosomes.

8.3.1.2 rDNA chromatin structure

The chromatin structure of RNA polymerase I-transcribed rDNA is well characterized (reviewed in [8,58,59]), and in most organisms (lower eukaryotes, plants and animals), only a fraction of ribosomal genes is transcriptionally active. Consequently, at the chromatin level there are two rDNA populations. One population is assembled into arrays of nucleosomes and is transcriptionally

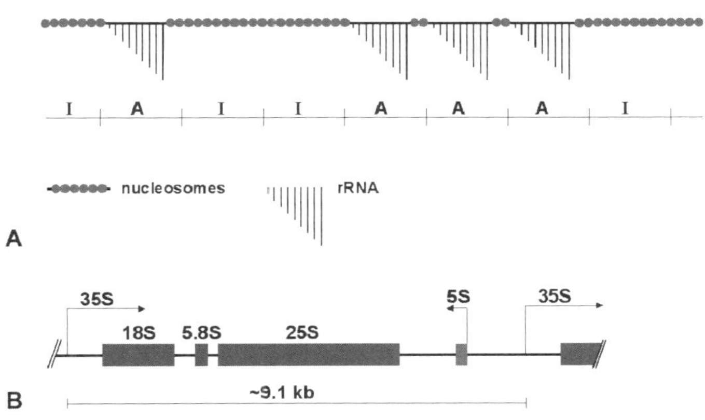

Figure 1. Ribosomal DNA transcription unit and ribosomal DNA chromatin. **A.** Schematic diagram representing active (A) and inactive (I) genes in rDNA chromatin. **B.** Active rDNA transcription unit. The 35S transcript corresponds to the rRNA precursor transcribed by RNA polymerase I and processed into mature 18S, 5.8S and 25S rRNAs. The 5S rDNA, present in the intergenic spacers, is transcribed by RNA polymerase III and in the opposite direction of RNA polymerase I transcription.

inactive, while the other population does not contain canonical nucleosomes and is either transcriptionally active or in a pre-active state (reviewed in [60]) (Figure 1A).

Psoralen photo-crosslinking has been used to separate nucleosomal from non-nucleosomal rDNA. The principle of this technique is that non-nucleosomal rDNA chromatin binds more psoralen than nucleosomal rDNA chromatin. As a result, the two rDNA populations can be separated by gel electrophoresis, since psoralen-crosslinked fragments from active (and pre-active) ribosomal genes have a slower migration than fragments from inactive genes (reviewed in [60,61]) (Figure 2). With this assay, it was found that the fraction of non-nucleosomal rDNA varies markedly among organisms and cell types (e.g., from ~20% to ~70%; [60,62,63,64]). In exponentially growing yeast cells, *in vivo* and *in vitro* psoralen analysis revealed that ribosomal genes are present in two chromatin structures (Figure 2B; [30,63]). Furthermore, it was shown that the non-nucleosomal rDNA units are randomly distributed in the rDNA locus and that the average number of non-nucleosomal ribosomal genes ranges from 1 to 3 copies per 'active cluster' [64]; schematically described in Figure 1A).

8.3.2 Transcriptional regulation of ribosomal genes

8.3.2.1 Transcription of rDNA during cell growth

Growing cells require continuous synthesis of ribosomes, a process that is mainly limited by the rate of synthesis of rRNA. Furthermore, in response to changes in cell growth stimuli, transcription of rDNA appears to influence cell proliferation by regulating the production of ribosomes. For example, signals like amino acid starvation, toxic lesions, aging, cancer and viral infections can modulate the transcription of ribosomal genes. There is also a correlation between cell differentiation and decrease of rDNA transcription. Moreover, synthesis of rRNA is controlled by the cell cycle and is arrested during mitosis.

It appears that the "key players" in growth-dependent expression of ribosomal genes are the transcription factor TIF-IA (inactivated in quiescent cells), phosphorylation of the regulatory protein UBF and structural modification of RNA polymerase I (dephosphorylation abolishes transcription initiation).

8.3.2.2 The RNA polymerase I machinery

Each class of RNA polymerase (I, II and III) employs a distinct set of transcription factors that are assembled into transcription initiation complexes at promoters. The mammalian RNA polymerase I is formed by at least 11 subunits (two large structural peptides and a number of smaller subunits) and a similar composition was described for the RNA polymerase I in yeast.

The RNA polymerase II transcription complex is composed of general initiation factors, RNA polymerase II and accessory factors. Many of these components are preassembled in the absence of DNA into a large structure called the holoenzyme, which is the form of RNA polymerase II that responds to transcriptional activators. Since the mechanisms of transcription initiation are

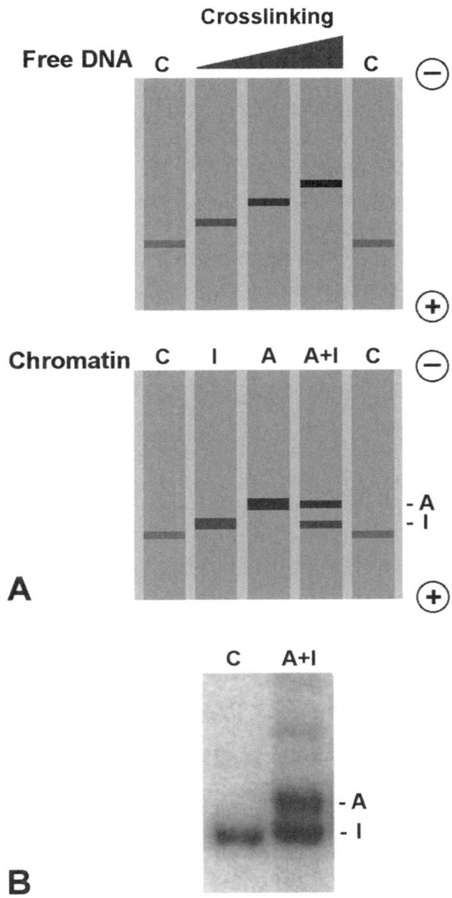

Figure 2. The psoralen assay. **A.** Upper panel: Schematic diagram of psoralen-crosslinked DNA fragments following gel electrophoresis. The crosslinked DNA fragments migrate slower on gels than non-crosslinked DNA fragments. The greater the number of psoralen crosslinks in a DNA fragment, the more its migration is retarded during native agarose gel electrophoresis, as compared to non-crosslinked DNA (C; control). Lower panel: Psoralen-crosslinked fragments from active ribosomal genes have a slower migration than fragments from inactive genes. DNA from inactive (I) nucleosomal rDNA contains psoralen crosslinks only between nucleosomes, and its migration on gels is only slightly retarded as compared to non-crosslinked control DNA (C; control). DNA from active rDNA chromatin (A) is heavily crosslinked and its migration is strongly retarded. If active and inactive ribosomal genes are present in the same cell (A+I), both A and I bands are observed. **B.** Analysis by gel electrophoresis of psoralen-crosslinked rDNA chromatin in exponentially growing yeast cells. Cells were crosslinked with psoralen and total DNA was extracted, digested with restriction enzyme (e.g., *Eco*RI), separated by gel electrophoresis, blotted and hybridized with the rDNA probe. The results show that active (A-band) and inactive (I-band) ribosomal genes are present in exponentially growing cells (A+I). C; control DNA isolated from non-crosslinked cells.

similar for all three classes of nuclear RNA polymerases, it is possible that RNA polymerase I also operates as a holoenzyme. Although it has been proposed that the RNA polymerase I transcription complex is formed by a 'stepwise assembly' mechanism, recent results suggest that cellular RNA polymerase I may indeed exist as a large complex with associated transcription factors [65]. In fact, RNA polymerase I holoenzyme containing the transcription factors TIF-IA, TIF-IB, UBF and limiting amounts of TIF-IC, was shown to be transcriptionally active after addition of template DNA.

Very recently it was reported that TFIIH plays an essential role in RNA polymerase I transcription [66]. During the characterization of the polypeptides present in the RNA polymerase I holoenzyme, it was found that TFIIH, a basal RNA polymerase II transcription factor (see Section 1.2.2), is also an integral component of the RNA polymerase I holoenzyme. Significantly, *in vitro* transcription systems lacking TFIIH were reactivated by addition of exogenous TFIIH and it was shown that part of the cellular TFIIH is localized within the nucleolus, at sites where rDNA is actively transcribed. TFIIH is associated with subpopulations of RNA polymerase I and TIF-IB, which is responsible for promoter selectivity and cooperates with UBF in formation of the pre-initiation complexes, by recruiting RNA polymerase I, TIF-IA and TIF-IC to rDNA promoters.

From these data it seems that during RNA polymerase I transcription, unlike the case of RNA polymerase II transcription, TFIIH is not required for the formation of the open complex and synthesis of the first phosphodiester bonds. Instead, it could play a role in transcription elongation. The requirement for ATP is a major difference between the function of TFIIH in RNA polymerase II- and I-transcription. Transcription by RNA polymerase II requires ATP for both the formation of an open complex and promoter clearance. In contrast, RNA polymerase I does not require ATP for transcription initiation. This indicates that neither the helicase nor the protein kinase of TFIIH is required for rDNA transcription.

These results could provide a molecular link between RNA polymerase I transcription and TCR of active ribosomal genes. The fact that TCR is not restricted to RNA polymerase II, but also occurs in RNA polymerase I-transcribed genes in yeast, gives further support to the biological significance of TFIIH in RNA polymerase I transcription.

8.4 Transcription-coupled repair in the RNA polymerase I-transcribed ribosomal genes

8.4.1 DNA damage and repair in ribosomal genes

Mammalian and yeast cells repair ribosomal genes that are damaged by chemicals or radiation (x-rays and UV light). However, in most cases, DNA lesions are less efficiently removed from the rDNA locus than from either total genomic DNA or RNA polymerase II-transcribed genes. Thus, repair efficiency of ribosomal genes may vary depending on the type of DNA lesions, the cognate DNA repair mechanism and the cell type.

It has often been reported that the difference in repair efficiency between RNA polymerase I-transcribed rDNA and RNA polymerase II-transcribed genes (mainly the dihydrofolate reductase gene, DHFR, was used as model system) may reflect the different RNA polymerase complexes involved in their transcription. Alternatively, it may reflect the different nuclear localization of these genes.

It is important to remember that the studies summarized below have dealt with repair of *total* rDNA and, therefore, did not distinguish between the transcriptionally active and inactive fractions of these genes (see section 2.1.2). Therefore, these results represent only an average of both rDNA populations and this could explain the lower repair efficiency found in ribosomal genes.

8.4.1.1 Aflatoxin adducts

Aflatoxin B1 (AFB1) is known to be a potent mutagen and exposure to AFB1 is a risk factor for hepatocellular carcinoma. Upon metabolic activation, AFB1 reacts almost exclusively with the N7 of guanine to form guanine DNA adducts [67]. The covalent ABF1 DNA adducts appear to form preferentially within rDNA (~4- to 5-fold greater), which has high GC content, compared to total nuclear DNA. It is noteworthy that AFB1 residues bound to rDNA are also removed ~6-fold more rapidly from rDNA than from bulk DNA [68], and NER is thought to be the repair mechanism involved in the removal of these DNA lesions.

8.4.1.2 Alkylated bases

Stevnsner et al., 1993 [78], treated hamster cells with the bifunctional compound nitrogen mustard (HN2) and the monofunctional compound methyl methanesulphonate (MMS), which produce lesions thought to be repaired by NER and BER, respectively. It was found that intra- and inter- strand adducts induced by HN2 are removed faster from the *DHFR* gene than from rDNA. Conversely, adducts induced by MMS were repaired with equal efficiency in both classes of genes.

In agreement with those results, Wassermann and Damgaard [69] showed that HN2-induced DNA damage can completely inhibit transcription of the *DHFR* gene and reduce (by ~8-fold) transcription of ribosomal genes. These authors also found that during repair, transcription recovers to 34% for rDNA and 76% for the *DHFR* gene. Subsequently, the involvement of RNA polymerase II in the efficient removal of HN2 adducts was studied by treating CHO (Chinese hamster ovary) cells with α-amanitin. These findings suggested that efficient repair of the HN2-induced DNA lesions is dependent on ongoing transcription by RNA polymerase II.

The reduced form of mitomycin C (MMC) reacts with DNA and the degree of crosslinking is proportional to the GC content. Both interstrand and intrastrand crosslinks are formed by MMC and they appear to be repaired by DNA recombination and NER respectively. In a report by Matsumoto and Hanawalt [70], the formation of MMC interstrand crosslinks was analyzed in rDNA, which has high GC content, and repair was followed in normal lymphoblastoid and Fanconi's anemia (FA) cell lines. Cultured cells from FA patients reveal an increased sensitivity and chromosomal instability following exposure to DNA crosslinking agents. In normal cells, most of the crosslinks are removed from the rDNA locus after 48 hours, whereas less efficient repair (50 to 65%) is observed among different

FA cell lines. Thus, in these experiments, repair of crosslinked rDNA was efficient and appears to correlate with cell survival.

Conversely, another group reported that MMC-induced crosslinks in rDNA are stable lesions [71], and are not efficiently repaired. Moreover, these experiments indicated that normal and FA cells do not differ in their capacity to remove crosslinks from rDNA. It was suggested that the repair measured reflects a mean value between repair of active and inactive ribosomal genes, since MMC crosslinks are removed slowly from bulk DNA (see references in [71]) and ~50% of the ribosomal genes resemble bulk chromatin [72].

8.4.1.3 Psoralen photoadducts and DNA crosslinks

Psoralen photo-adducts are often used as a prototype of DNA damage caused by chemical crosslinking agents and are possibly removed by NER, in combination with DNA recombination [3]. In *Tetrahymena* it was found that a dosage of ~1 crosslink per rDNA molecule blocks rRNA synthesis and that crosslinks in rDNA are repaired within 24 h, which restores rDNA transcription [73]. In human cells, psoralen interstrand-crosslinks are efficiently removed from the *DHFR* gene, whereas psoralen monoadducts are removed at a lower rate [74]. Therefore, repair efficiency in this active RNA polymerase II-transcribed gene varies with the type of DNA lesion. In addition, there is a significant difference in the rate of removal of psoralen-induced lesions between the *DHFR* gene and rDNA. In human and hamster cells, repair in the *DHFR* gene is about four times faster than in ribosomal genes [75]. From this study, it was concluded that RNA polymerases I and II might differentially influence the repair rate of active genes.

Interestingly, monoadducts and interstrand crosslinks formed in ribosomal genes are repaired at different rates, depending on the CHO cell lines analyzed [76]. The reason for this repair heterogeneity of the rDNA locus is not known. One possible explanation is that these cell lines age differently and/or that the transcription rate of ribosomal genes is cell lineage dependent.

8.4.1.4 Cisplatin crosslinks

The chemotherapeutic agent cisplatin causes intrastrand- and interstrand- DNA crosslinks [3]. These lesions are believed to be removed by NER and DNA recombination. Normal human cells, FA and XP cell lines were used to measure the formation and repair of cisplatin DNA adducts in the *DHFR* gene and in rDNA [77]. The XPA cells lack NER and FA cells are deficient in repair of DNA interstrand crosslinks. In these experiments it was found that normal cells remove cisplatin adducts from the *DHFR* gene at least twice as fast as FA and XPA cells. Moreover, repair of the rRNA genes was less efficient than repair of the *DHFR* gene. On the other hand, Stevnsner et al., [78] reported that, in hamster cells, repair of cisplatin adducts from the *DFHR* gene and rDNA is equally efficient.

8.4.1.5 Bleomycin strand breaks

The radio-mimetic agent bleomycin (BLM) induces DNA strand breaks [3] that are thought to be repaired by BER. BLM-induced single strand breaks were introduced in the DNA of human cells and repair was followed in both the *DFHR* gene and rDNA [79]. These authors found that single strand breaks are rapidly repaired in both types of genes (~60% removed in 1 hour in the *DFHR* gene and ~80% removed in 1 hour in rDNA). Furthermore, TCR was not observed in either of the two loci analyzed [79]. More recently, it was found that repair of DNA strand breaks induced by ionizing radiation is very fast in the *DFHR* gene, in ribosomal genes and in the genome as a whole [80].

8.4.2 UV-induced DNA damage, global genome repair and transcription-coupled repair in ribosomal genes

At present there is still speculation about the mechanisms by which the DNA repair machinery recognizes and accesses DNA lesions. Of particular interest is the understanding of how NER is coupled to transcription and how it operates in the presence of RNA polymerases. Hence, comparison of NER in genes transcribed by the three classes of RNA polymerases could help to elucidate some of the connections between DNA repair and transcription.

8.4.2.1 UV-induced DNA damage and DNA repair in ribosomal genes

The first study that analyzed repair synthesis of UV-photoproducts in ribosomal genes employed an immunological method in conjunction with a nitrocellulose filter binding technique. In these experiments, it was shown that DNA repair occurs in rDNA of UV-irradiated human diploid fibroblasts [81]. However, since repair synthesis in rDNA was not compared to repair of other DNA sequences in that study, it is not possible to assess the repair efficiency of UV-damaged rDNA from these results.

In a later study, removal of CPDs from ribosomal genes in hamster cells was followed together with removal of DNA lesions induced by cisplatin or alkylating agents (see sections 3.1.4 and 3.1.2). It was found that repair of UV-induced lesions is significantly less efficient in ribosomal genes than in the *DHFR* gene [78]. More recently, after treatment of human cells with either BLM (see section 3.1.5) or UV light, Fritz et al. [79] observed that there is little (or no) repair of UV photodimers in the rDNA locus. From these two reports it was suggested that ribosomal genes could be accessible to BER enzymes but not to the NER machinery. However, the cisplatin-induced crosslinks, thought to be repaired by NER in combination with homologous recombination, were efficiently removed from rDNA, and comparable to crosslinks in the *DHFR* gene [78].

8.4.2.2 Search for transcription-coupled DNA repair in ribosomal genes

The results obtained by Stevnsner et al. [78] imply that there is no TCR in ribosomal genes, since the extent of repair in rDNA is similar to that found in the genome overall [15]. However, the first direct investigation of TCR in ribosomal genes was done in hamster and human cells [82]. These authors found no evidence for TCR in rDNA. Additionally, in CHO cells, CPDs are not removed from rDNA even 24 hours after UV irradiation, whereas TCR of CPDs occurs in the *DHFR* gene. In human cells, CPDs are removed from rDNA but at a slow rate and with no bias for the TS, similar to repair of the genome overall. Moreover, there is no significant difference in the repair rates between growth-arrested and proliferating cells, suggesting that CPD removal is independent of transcription. Subsequently, DNA repair was followed in the rDNA locus of XPC and CS human cells [83]. As for CHO cells, no CPDs were removed from either rDNA strand in XPC cells, suggesting that the slow removal of CPDs in ribosomal genes is due to the transcription-independent GGR pathway. However, in parallel experiments, CS cells also showed lower than normal repair in rDNA. From these results it was concluded that there is no TCR in ribosomal genes and that NER may be impeded by the nucleolar compartmentalization of rDNA.

Balajee et al. [84] further investigated DNA repair of UV-photoproducts in ribosomal genes of hamster cells. The *in situ* sites of DNA repair and transcription were analyzed by light microscopy in combination with immunofluorescent labeling. The results show low DNA repair in the nucleolus despite the presence of abundant RNA polymerase I transcription. Parallel immunofluorescent staining experiments indicate that some NER proteins (the TFIIH subunits p62 and p89, p53, Gadd45 and proliferating cell nuclear antigen) are present in nuclei but are absent in the nucleoli. Therefore, it was suggested that under-representation of NER proteins in nucleoli could explain the inefficient repair of photoproducts in rDNA. However, in disagreement with these findings, it was recently shown that a portion of cellular TFIIH is localized within the nucleolus, at sites where rDNA is actively transcribed [66].

Significantly, the same authors found that, although there is very little removal of CPDs from rDNA in hamster cells, (6,4)s are removed quite efficiently, albeit somewhat slower than from the *DFHR* gene [84]. From these results it was concluded that the differences observed between the processing of CPDs and (6,4)s in the rDNA locus indicate the existence of distinct repair pathways for the removal of these two lesions.

During the search for TCR in rDNA, the existence of two different rDNA populations (one inactive and one active/pre-active) was not considered until the work of Fritz and Smerdon [85]. In these experiments, restriction enzyme digestion was used to selectively release the active rDNA fraction from nuclei. Following fractionation, repair of CPDs was measured in the individual strands of transcriptionally active ribosomal genes and it was found that there is no TCR in the active rDNA. However, repair of ribosomal genes in this mouse cell line was almost nonexistent.

8.4.2.3 Transcription-coupled repair in yeast ribosomal genes

In contrast to what has been found in mammalian cells, CPDs are efficiently removed from the rDNA locus of growing yeast cells [29,86]. Furthermore, since no repair was found in *rad1, 2, 3* and *14* mutants, it was demonstrated that NER is responsible for the removal of CPDs in the rDNA locus [29].

The first evidence for TCR in RNA polymerase I-transcribed genes was reported in yeast by Verhage et al. [29]. Whereas no TCR could be detected in *wt* cells, strand-specific repair was observed in rad*7, 16* and *4* mutants, indicating that TCR probably occurs in yeast rDNA. Nonetheless, strand-specific repair of yeast rDNA is independent of Rad26. This observation is contrary to previous findings in human cells, showing that the Rad26 homologue CSB is involved in removal of CPDs from rDNA [83].

Although the work by Verhage et al. [29] strongly support the existence of TCR in RNA polymerase I-transcribed genes, the existence of this pathway was not directly demonstrated until recently when TCR was measured in *wt* strains [31,64]. As discussed above, it is important to separate the active from the inactive rDNA before analyzing DNA repair. Thus restriction enzyme digestion was used to specifically release the transcriptionally active fraction from yeast nuclei, and differential psoralen crosslinking was used to distinguish between active and inactive rDNA chromatin. Removal of CPDs was followed in both rDNA populations and the data clearly demonstrate that TCR occurs in transcriptionally active rDNA and not in the inactive rDNA fraction. Interestingly, the NTS of active rDNA is repaired faster than either strand of inactive rDNA (Figure 3A). Since the NTS of active genes and both strands of inactive genes are repaired by GGR, these data suggest that active, non-nucleosomal rDNA is more accessible to GGR enzymes than nucleosome-loaded rDNA.

8.4.3 Possible mechanisms involved in TCR of ribosomal genes in yeast

In principle, most of the molecular mechanisms proposed to describe TCR of RNA polymerase II-transcribed genes (see sections 1.2 and 1.3) could also apply to TCR of RNA polymerase I-transcribed genes, providing that repair enzymes have access to rDNA in the nucleolus.

An attractive model suggests that TCR starts with an "obstacle-recognition" step, in which RNA polymerase II is stalled at a lesion. The CS proteins could be involved in removing the arrested RNA polymerase whereas TFIIH and XPG would "assess the nature of the obstruction" and recruit NER enzymes [37]. The same events could also occur for RNA polymerase I-dependent TCR in yeast. It was thought that large protein complexes like TFIIH would have difficulty migrating into the nucleolus, but recent data shown that TFIIH is present at transcription foci in nucleoli of mammalian cells [66].

Another model suggests that the TS is preferentially repaired just upstream of where TFIIH is released from the RNA polymerase II transcription machinery. Beyond that point TCR requires the CS proteins (Rad26) to re-recruit TFIIH to the repair complex when the RNA polymerase is arrested at a lesion. These mechanisms could also apply to TCR by RNA polymerase I in yeast. However, during transcription by RNA polymerase I it appears that TFIIH could play a role in RNA elongation (see section 2.2.2). Therefore, re-recruitment of TFIIH by CS proteins (Rad26) during TCR may differ between RNA polymerase II- and RNA polymerase I- transcribed genes. In fact, TCR in ribosomal genes of yeast is independent of Rad26 [29].

From our current knowledge it seems that the effect of *rad26* mutation, hence the involvement of Rad26 in TCR, is gene-specific and could depend on the level of transcription (reviewed in [23]). It was hypothesized that the requirement for Rad26

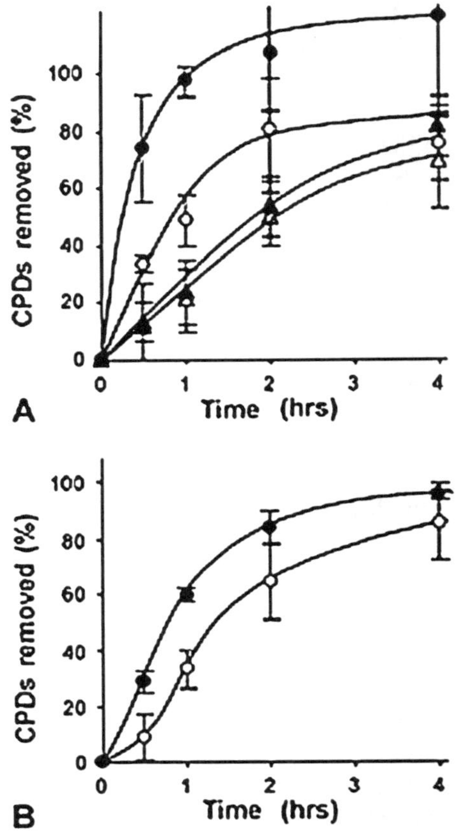

Figure 3. DNA repair of individual strands of active and inactive rDNA. **A.** Repair of individual strands of active and inactive rDNA. Yeast cells were UV-irradiated and allowed to repair for different times. After separation of the active and inactive rDNA populations, DNA was purified and analyzed for the presence of CPDs. DNA repair is expressed as percent of CPDs removed as a function of repair time. Data are from active rDNA (circles) and inactive rDNA (triangles). Solid and open symbols represent data from the TS and NTS, respectively. Data were taken from reference [30]. **B.** Repair of CPDs in *total* rDNA. Yeast cells were UV-irradiated and allowed to repair for different times. Solid and open symbols represent data for the TS and the NTS, respectively. Data were taken from reference [30].

in TCR is inversely related to the transcription rate of the gene and that transcription itself is sufficient to promote TCR. Since ribosomal genes are very actively transcribed, the TCR of these genes would depend on a Rad26-independent TCR pathway, as previously postulated by Verhage et al. [29].

The arrest of RNA polymerase II at a damage site is considered to be the recognition signal for the initiation of TCR (see sections 1.2.3 and 1.2.4). The question arises: do UV photoproducts arrest RNA polymerase I? The observations by Balajee et al. [84] seems to indicate that elongation of RNA polymerase I is relatively unaffected by UV irradiation compared to RNA polymerase II elongation. However, it was previously shown that UV-induced DNA damage prevents elongation of RNA polymerase I in *Xenopus* [87]. Furthermore, human RNA polymerase I is arrested by CPDs present in the transcribed strand, and transcription factor 2 was found to disrupt the RNA polymerase I complex (as well as the RNA polymerase II complex) stalled at a CPD [88].

We have observed that transcription by RNA polymerase I is arrested after UV irradiation of yeast (Conconi et al., submitted). This could be an indirect effect due to cell-cycle arrest and consequent inhibition of cell growth, which closely correlates with decreased rDNA transcription; or a direct effect of UV damage blocking RNA polymerase I elongation. Following a lag period, transcription gradually resumes as CPDs are removed from the rDNA and the cells do not divide during this repair period (Conconi et al., submitted).

In conclusion, it is possible that the initiation and subsequent mechanisms of TCR in RNA polymerase I-transcribed genes are similar to the processes thought to occur in TCR of RNA polymerase II-transcribed genes.

8.4.4 Does TCR of ribosomal genes occur in mammalian cells?

At present it remains to be determined whether TCR of RNA polymerase I genes is a unique feature of yeast cells. Some possible reasons have been put forward for the current lack of observable TCR in mammalian cells.

One hypothesis is that the unique structural environment of the nucleolus may limit accessibility of repair proteins. Based on DNA repair studies of ribosomal genes in mammalian cells, where lesions are repaired by BER more rapidly than lesions repaired by NER, it was proposed that the nucleolus may be a barrier to the NER complex but not to BER. Though this cannot be ruled out for all components of NER, at least the protein complex TFIIH has been observed in the nucleolus. Furthermore, (6-4)PPs in ribosomal genes are efficiently repaired by NER in hamster cells [84]. Thus, at least in this case, excision repair proteins need to migrate into the nucleolus, whether the NER complex involves a repairosome [89] or results from sequential assembly of NER factors [44]. Clearly, the yeast nucleolus does not inhibit migration of NER proteins since NER of ribosomal genes is very efficient (e.g. see [30]).

Another plausible explanation for the low efficiency of NER found in nucleoli of mammalian cells is based on evidence that only GGR is involved in repair of rDNA. Hence rodent cells, which have low levels of GGR, have little or no repair of rDNA. On the other hand, human cells have higher levels of GGR and rDNA is moderately repaired. In agreement with this hypothesis, rDNA is not repaired in human XPC cells, which lack GGR. Thus it was proposed that rDNA is repaired by GGR [82]. However, as discussed in Section 1.2.1, ongoing transcription is necessary for TCR and the above study did not measure transcription of rRNA.

An explanation for the low efficiency of CPD removal from RNA polymerase I-transcribed genes in mammalian cells was that recombinational repair (rather than NER) could be responsible for maintaining sequence integrity in the highly conserved rDNA sequences. Moreover, amplification and unequal sister chromosome exchange, along with gene conversion, could also play a role to ensure the elimination of mutations in rDNA. Once again, the proportion of active ribosomal genes was not determined, which is important since the ratios of active and inactive genes and transcription levels can vary markedly from cell line to cell line and during cell growth. Indeed, TCR in yeast was not observed when repair of total rDNA was monitored (Figure 3B). Consequently, the existence of TCR in ribosomal genes of mammalian cells is still an open question, which may be addressed by measuring repair of separated active and inactive ribosomal genes in cells having high levels of rRNA transcription.

Acknowledgements

Work on this manuscript was supported by NIH grants ES02614 and ES04106 from the National Institute of Environmental Health Sciences.

Abbreviations

(6-4)PPs, pyrimidine-(6-4)-pyrimidone photoproducts; AFB1, aflatoxin B1; BER, base excision repair; BLM, bleomycin; CHO, Chinese hamster ovary; CPDs, cyclobutane pyrimidine dimers; CS, Cockayne syndrome; CSA, CS group A; CSB, CS group B; FA, Fanconi's anemia; HN2, nitrogen mustard; GGR, global genomic repair; DFHR, dihydrofolate reductase; MMC, mitomycin C; MMS, methylmethanesulphonate; NER, nucleotide excision repair; NTS, non-transcribed DNA strand; rDNA, ribosomal genes; TCR, transcription-coupled repair; TFIIH, transcription factor H of RNA polymerase II; TIF, transcription initiation factor; TS, transcribed DNA strand; XP, *Xeroderma pigmentosum*; XPA, XP group A; XPB, XP group B; XPC, XP group C; XPD, XP group D; XPG, XP group G.

References

1. T. Lindahl (1993). Instability and decay of the primary structure of DNA. *Nature*, **362**, 709–715.
2. J.H. Hoeijmakers (2001). Genome maintenance mechanisms for preventing cancer. *Nature*, **411**, 366–374.
3. E.C. Freidberg, G.C. Walker, W. Siede (1990). *DNA Repair and Mutagenesis*, Am. Soc. Microbiol. Press, Washington, DC.
4. S.H. Reed, R. Waters (2000). DNA repair. In: D. Cooper (Ed.), *Encyclopedia of the Human Genome*. Nature Publishing Group.
5. T. Lindahl, R.D. Wood (1999). Quality control by DNA repair. *Science*, **286**, 1897–1905.
6. M.J. Smerdon, A. Conconi (1999). Modulation of DNA damage and DNA repair in chromatin. *Prog. Nucleic Acid Res. Mol. Biol.*, **62**, 227–255.
7. F. Thoma (1999). Light and dark in chromatin repair: repair of UV-induced DNA lesions by photolyase and nucleotide excision repair. *EMBO J.*, **18**, 6585–6598.

8. K.E. Van Holde (1989). *Chromatin*, Springer-Verlag, Berlin, Germany.
9. A.P. Wolffe, J.J. Hayes (1999). Chromatin disruption and modification. *Nucleic Acids Res.*, **27**, 711–720.
10. K. Ura, J.J. Hayes (2002). Nucleotide excision repair and chromatin remodeling. *Eur. J. Biochem.*, **269**, 2288–2293.
11. D.P. Batty, R.D. Wood (2000). Damage recognition in nucleotide excision repair of DNA. *Gene*, **241**, 193–204.
12. P.C. Hanawalt (2001). Controlling the efficiency of excision repair. *Mutat. Res.*, **485**, 3–13.
13. S. Tornaletti, P.C. Hanawalt (1999). Effect of DNA lesions on transcription elongation. *Biochimie*, **81**, 139–146.
14. E. Citterio, W. Vermeulen, J.H. Hoeijmakers (2000). Transcriptional healing. *Cell*, **101**, 447–450.
15. V.A. Bohr, C.A. Smith, D.S. Okumoto, P.C. Hanawalt (1985). DNA repair in an active gene: removal of pyrimidine dimers from the DHFR gene of CHO cells is much more efficient than in the genome overall. *Cell*, **40**, 359–369.
16. I. Mellon, G. Spivak, P.C. Hanawalt (1987). Selective removal of transcription-blocking DNA damage from the transcribed strand of the mammalian DHFR gene. *Cell*, **51**, 241–229.
17. I. Mellon, P.C. Hanawalt (1989). Induction of the *Escherichia coli* lactose operon selectively increases repair of its transcribed DNA strand. *Nature*, **342**, 95–98.
18. L. Lommel, C. Carswell-Crumpton, P.C. Hanawalt (1995). Preferential repair of the transcribed DNA strand in the dihydrofolate reductase gene throughout the cell cycle in UV-irradiated human cells. *Mutat. Res.*, **336**, 181–92.
19. H.J. Ruven, R.J. Berg, C.M. Seelen, J.A. Dekkers, P.H. Lohman, L.H. Mullenders, A.A. van Zeeland (1993). Ultraviolet-induced cyclobutane pyrimidine dimers are selectively removed from transcriptionally active genes in the epidermis of the hairless mouse. *Cancer Res.*, **53**, 1642–1645.
20. W.G. McGregor, R.-H. Chen, L.L. Lukash, V.M. Maher, J.J. McCormick (1991). Cell cycle-dependent strand bias for UV-induced mutations in the transcribed strand of excision repair-proficient human fibroblasts but not in repair-deficient cells. *Mol. Cell. Biol.*, **11**, 1927–1934.
21. P. Menichini, H. Vrieling, A.A. van Zeeland (1991). Strand-specific mutation spectra in repair-proficient and repair-deficient hamster cells. *Mutat. Res.*, **251**, 143–155.
22. L.H. Mullenders, H. Vrieling, J. Venema, A.A. van Zeeland (1991). Hierarchies of DNA repair in mammalian cells: biological consequences. *Mutat. Res.*, **250**, 223–228.
23. R.A. Verhage, M. Tijsterman, P. van de Putte, J. Brouwer (1998). Transcription-coupled and global genome nucleotide excision repair. In: F. Eckstein, D.M.J. Lilley (Eds), *Nucleic Acids and Molecular Biology*, Vol. 12, (pp. 157–172). Springer-Verlag, Berlin, Germany.
24. M. Carreau, D. Hunting (1992). Transcription-dependent and independent DNA excision repair pathways in human cells. *Mutat. Res.*, **274**, 57–64.
25. S.A. Leadon, D.A. Lawrence (1991). Preferential repair of DNA damage on the transcribed strand of the human metallothionein genes requires RNA polymerase II. *Mutat. Res.*, **255**, 67–78.
26. F.C. Christians, P.C. Hanawalt (1992). Inhibition of transcription and strand-specific DNA repair by α-amanitin in Chinese hamster ovary cells. *Mutat. Res.*, **274**, 93–101.
27. K.S. Sweder, P.C. Hanawalt (1992). Preferential repair of cyclobutane pyrimidine dimers in the transcribed strand of a gene in yeast chromosomes and plasmids is dependent on transcription. *Proc. Natl. Acad. Sci. USA*, **89**, 10696–10700.
28. S.A. Leadon, D.A. Lawrence (1992). Strand-selective repair of DNA damage in the yeast GAL7 gene requires RNA polymerase II. *J. Biol. Chem.*, **267**, 23175–23182.

29. R.A. Verhage, P. Van de Putte, J. Brouwer (1996). Repair of rDNA in *Saccharomyces cerevisiae:* RAD4-independent strand-specific nucleotide excision repair of RNA polymerase I transcribed genes. *Nucleic Acids Res.*, **24**, 1020–1025.
30. A. Conconi, V.A. Bespalov, M.J. Smerdon (2002). Transcription-coupled repair in RNA polymerase I-transcribed genes of yeast. *Proc. Natl. Acad. Sci. USA*, **99**, 649–654.
31. A. Meier, M. Livingstone-Zatchej, F. Thoma (2002). Repair of active and silenced rDNA in yeast: the contributions of photolyase and transcription-couples nucleotide excision repair. *J. Biol. Chem.*, **277**, 11845–11852.
32. A. Dvir, J.W. Conaway, R.C. Conaway (2001). Mechanism of transcription initiation and promoter escape by RNA polymerase II. *Curr. Opin. Genet. Dev.*, **11**, 209–214.
33. V. Moncollin, P. Vichi, J.-M. Egly (1998). A transcription factor involved in DNA repair and cell-cycle regulation. In: J.A. Nickoloff, M.F. Hoekstra (Eds.), *DNA Repair in Higher Eucaryotes* (pp. 143–159). Humana Press Inc., Totowa, New Jersey.
34. L. Zawel, K.P. Kumar, D. Reinberg (1995). Recycling of the general transcription factors during RNA polymerase II transcription. *Genes Dev.*, **9**, 1479–1490.
35. P.K. Cooper, T. Nouspikel, S.G. Clarkson, S.A. Leadon (1997). Defective transcription-coupled repair of oxidative base damage in Cockayne syndrome patients from XP group G. *Science*, **275**, 990–993.
36. F. Le Page, A. Klungland, D.E. Barnes, A. Sarasin, S. Boiteux (2000). Transcription coupled repair of 8-oxoguanine in murine cells: the Ogg1 protein is required for repair in nontranscribed sequences but not in transcribed sequences. *Proc. Natl. Acad. Sci. USA*, **97**, 8397–8402.
37. P.C. Hanawalt (2000). DNA repair, the bases for Cockayne syndrome. *Nature*, **405**, 415–416.
38. C.P. Selby, A. Sancar (1991). Gene- and strand-specific repair in vitro: partial purification of a transcription-repair coupling factor. *Proc. Natl. Acad. Sci. USA*, **88**, 8232–8236.
39. C.P. Selby, A. Sancar (1993). Molecular mechanism of transcription-repair coupling. *Science*, **260**, 53–58.
40. C. Troelstra, A. van Gool, J. de Wit, W. Vermeulen, D. Bootsma, J.H. Hoeijmakers (1992). ERCC6, a member of a subfamily of putative helicases, is involved in Cockayne's syndrome and preferential repair of active genes. *Cell*, **71**, 939–953.
41. D.B. Bregman, R. Halaban, A.J. van Gool, K.A. Henning, E.C. Friedberg, S.L. Warren (1996). UV-induced ubiquitination of RNA polymerase II: a novel modification deficient in Cockayne syndrome cells. *Proc. Natl. Acad. Sci. USA*, **93**, 11586–11590.
42. J.N. Ratner, B. Balasubramanian, J. Corden, S.L. Warren, D.B. Bregman (1998). Ultraviolet radiation-induced ubiquitination and proteasomal degradation of the large subunit of RNA polymerase II. Implications for transcription-coupled DNA repair. *J. Biol. Chem.*, **273**, 5184–5189.
43. E.C. Woudstra, C. Gilbert, J. Fellows, L. Jansen, J. Brouwer, H. Erdjument-Bromage, P. Tempst, J.Q. Svejstrup (2002). A Rad26-Def1 complex coordinates repair and RNA pol II proteolysis in response to DNA damage. *Nature*, **415**, 929–933.
44. S. Prakash, L. Prakash (2000). Nucleotide excision repair in yeast. *Mutat. Res.*, **451**, 13–24.
45. S.N. Guzder, Y. Habraken, P. Sung, L. Prakash, S. Prakash (1995). Reconstitution of yeast nucleotide excision repair with purified Rad proteins, replication protein A, and transcription factor TFIIH. *J. Biol. Chem.*, **270**, 12973–12976.
46. M.J. Smerdon, F. Thoma (1990). Site-specific DNA repair at the nucleosome level in a yeast minichromosome. *Cell*, **61**, 675–684.

47. C.P. Selby, E.M. Witkin, A. Sancar (1991). *Escherichia coli mfd* mutant deficient in "mutation frequency decline" lacks strand-specific repair: in vitro complementation with purified coupling factor. *Proc. Natl. Acad. Sci. USA*, **88**, 11574–11578.
48. C.M. Green, G. Almouzni (2002). When repair meets chromatin: First in series on chromatin dynamics. *EMBO Rep.*, **3**, 28–33.
49. M. Tijsterman, R.A. Verhage, P. van de Putte, J.G. Tasseron-de Jong, J. Brouwer (1997). Transitions in the coupling of transcription and nucleotide excision repair within RNA polymerase II-transcribed genes of *Saccharomyces cerevisiae*. *Proc. Natl. Acad. Sci. USA*, **94**, 8027–8032.
50. Y. Tu, S. Bates, G.P. Pfeifer (1997). Sequence-specific and domain-specific DNA repair in *Xeroderma pigmentosum* and Cockayne syndrome cells. *J. Biol. Chem.*, **272**, 20747–20755.
51. L.E. Jansen, H. den Dulk, R.M. Brouns, M. de Ruijter, J.A. Brandsma, J. Brouwer (2000). Spt4 modulates Rad26 requirement in transcription-coupled nucleotide excision repair. *EMBO J.*, **19**, 6498–6507.
52. S. Li, M.J. Smerdon (2002). Rpb4 and Rpb9 mediate subpathways of transcription-coupled DNA repair in *Saccharomyces cerevisiae*. *EMBO J.*, **21**, 5921–5929.
53. S.K. Lee, S.L. Yu, L. Prakash, S. Prakash (2002). Yeast RAD26, a homolog of the human CSB gene, functions independently of nucleotide excision repair and base excision repair in promoting transcription through damaged bases. *Mol. Cell Biol.*, 4383–4389.
54. I. Grummt (1999). Regulation of mammalian ribosomal gene transcription by RNA polymerase I. *Prog. Nucleic Acid Res. Mol. Biol.*, **62**, 109–154.
55. R.H. Reeder (1999). Regulation of RNA polymerase I transcription in yeast and vertebrates. *Prog. Nucleic Acid Res. Mol. Biol.*, **62**, 293–327.
56. B. Sollner-Webb (1991). RNA editing. *Curr. Opin. Cell Biol.*, **3**, 1056–1061.
57. R. Visintin, A. Amon (2000). The nucleolus: the magician's hat for cell cycle tricks. *Curr. Opin. Cell Biol.*, **12**, 752.
58. F. Thoma (1991). Structural changes in nucleosomes during transcription: strip, split or flip?. *Trends Genet.*, **7**, 175–177.
59. A.P. Wolffe (1998). *Chromatin: Structure and Function*, 3rd ed., Academic Press, New York.
60. R. Lucchini, J.M. Sogo (1998). The dynamic structure of ribosomal RNA gene chromatin, In: M.R. Paule (Ed.), *Transcription of Ribosomal RNA Genes by Eukaryotic RNA Polymerase I*, (pp. 254–276). Landes Bioscience, Austin, Texas.
61. J.M. Sogo, A. Conconi, R. Widmer (1989). The use of psoralen-crosslinking for the analysis of the chromatin structure during transcription. In: P.E. Nielsen (Ed.), *Photochemical Probes in Biochemistry*, (pp. 179–194). Kluwer Academic Publishers.
62. A. Conconi, J.M. Sogo, C.A. Ryan (1992). Ribosomal gene clusters are uniquely proportioned between open and closed chromatin structures in both tomato leaf cells and exponentially growing suspension cultures. *Proc. Natl. Acad. Sci. USA*, **89**, 5256–5260.
63. R. Dammann, R. Lucchini, T. Koller, J.M. Sogo (1993). Chromatin structures and transcription of rDNA in yeast *Saccharomyces cerevisiae*. *Nucleic Acids Res.*, **21**, 2331–2338.
64. R. Dammann, R. Lucchini, T. Koller, J M. Sogo (1995). Transcription in the yeast rRNA gene locus: distribution of the active gene copies and chromatin structure of their flanking regulatory sequences. *Mol. Cell Biol.*, **15**, 5294–5303.
65. P. Seither, S. Iben, I. Grummt (1998). Mammalian RNA polymerase I exists as a holoenzyme with associated basal transcription factors. *J. Mol. Biol.*, **275**, 43–53.

66. S. Iben, H. Tschochner, M. Bier, D. Hoogstraten, P. Hozak, J.M. Egly, I. Grummt (2002). TFIIH plays an essential role in RNA polymerase I transcription. *Cell*, **109**, 297–306.
67. R.G. Croy, J.M. Essigmann, V.N. Reinhold, G.N. Wogan (1978). Identification of the principal aflatoxin B1-DNA adduct formed *in vivo* in rat liver. *Proc. Natl. Acad. Sci. USA*, **75**, 1745–1749.
68. T.R. Irvin, G.N. Wogan (1985). Quantitative and qualitative characterization of aflatoxin B1 adducts formed in vivo within the ribosomal RNA genes of rat liver DNA. *Cancer Res.*, **45**, 3497–3502.
69. K. Wassermann, J. Damgaard (1994). Ongoing activity of RNA polymerase II confers preferential repair of nitrogen mustard-induced N-alkylpurines in the hamster dihydrofolate reductase gene. *Cancer Res.*, **54**, 175–181.
70. A. Matsumoto, J.M. Vos, P.C. Hanawalt (1989). Repair analysis of mitomycin C-induced DNA crosslinking in ribosomal RNA genes in lymphoblastoid cells from Fanconi's anemia patients. *Mutat. Res.*, **217**, 185–192.
71. J.P. Rey, R. Scott, H. Muller (1993). Induction and removal of interstrand crosslinks in the ribosomal RNA genes of lymphoblastoid cell lines from patients with Fanconi anemia. *Mutat. Res.*, **289**, 171–180.
72. A. Conconi, R.M. Widmer, T. Koller, J.M. Sogo (1989). Two different chromatin structures coexist in ribosomal RNA genes throughout the cell cycle. *Cell*, **57**, 753–761.
73. X. Fengquin, H. Nielsen, W. Zhen, P.E. Nielsen (1993). 8-Methoxypsoralen DNA interstrand cross-linking of the ribosomal RNA genes in *Tetrahymena thermophila*. Distribution, repair and effect on rRNA synthesis. *Photochem. Photobiol.*, **58**, 238–245.
74. J.M. Vos, P.C. Hanawalt (1987). Processing of psoralen adducts in an active human gene: repair and replication of DNA containing monoadducts and interstrand cross-links. *Cell*, **50**, 789–799.
75. J.M. Vos, E.L. Wauthier (1991). Differential introduction of DNA damage and repair in mammalian genes transcribed by RNA polymerases I and II. *Mol. Cell. Biol.*, **11**, 2245–2252.
76. E.L. Wauthier, P.C. Hanawalt, J.M. Vos (1990). Differential repair and replication of damaged DNA in ribosomal RNA genes in different CHO cell lines. *J. Cell Biochem.*, **43**, 173–183.
77. W. Zhen, M.K. Evans, C.M. Haggerty, V.A. Bohr (1993). Deficient gene specific repair of cisplatin-induced lesions in *Xeroderma pigmentosum* and Fanconi's anemia cell lines. *Carcinogenesis*, **14**, 919–924.
78. T. Stevnsner, A. May, L.N. Petersen, F. Larminat, M. Pirsel, V.A. Bohr (1993). Repair of ribosomal RNA genes in hamster cells after UV irradiation, or treatment with cisplatin or alkylating agents. *Carcinogenesis*, **14**, 1591–1596.
79. L.K. Fritz, C. Suquet, M.J. Smerdon (1996). Strand breaks are repaired efficiently in human ribosomal genes. *J. Biol. Chem.*, **271**, 12972–12976.
80. M. Ljungman (1999). Repair of radiation-induced DNA strand breaks does not occur preferentially in transcriptionally active DNA. *Radiat. Res.*, **152**, 444–449.
81. S.M. Cohn, M.W. Lieberman (1984). The distribution of DNA excision-repair sites in human diploid fibroblasts following ultraviolet irradiation. *J. Biol. Chem.*, **259**, 12463–12469.
82. F.C. Christians, P.C. Hanawalt (1993). Lack of transcription-coupled repair in mammalian ribosomal RNA genes. *Biochemistry*, **32**, 10512–10518.
83. F.C. Christians, P.C. Hanawalt (1994). Repair in ribosomal RNA genes is deficient in xeroderma pigmentosum group C and in Cockayne's syndrome cells. *Mutat. Res.*, **323**, 179–187.

84. A.S. Balajee, I. Dianova, V.A. Bohr (1999). Oxidative damage-induced PCNA complex formation is efficient in xeroderma pigmentosum group A but reduced in Cockayne syndrome group B cells. *Nucleic Acids Res.*, **27**, 4476–4482.
85. L.K. Fritz, M.J. Smerdon (1995). Repair of UV damage in actively transcribed ribosomal genes. *Biochemistry*, **34**, 13117–13124.
86. A. Conconi, P. Jager-Vottero, X. Zhang, B.C. Beard, M.J. Smerdon (2000). Mitotic viability and metabolic competence in UV-irradiated yeast cells. *Mutat. Res.*, **459**, 55–64.
87. P. Labhart, R.H. Reeder (1989). High initiation rates at the ribosomal gene promoter do not depend upon spacer transcription. *Proc. Natl. Acad. Sci. USA*, **86**, 3155–3158.
88. R. Hara, C.P. Selby, M. Liu, D.H. Price, A. Sancar (1999). Human transcription release factor 2 dissociates RNA polymerase I and II stalled at a cyclobutane thymine dimer. *J. Biol. Chem.*, **274**, 24779–24786.
89. J.Q. Svejstrup, Z. Wang, W.J. Feaver, X. Wu, D.A. Bushnell, T.F. Donahue, E.C. Friedberg, R.D. Kornberg (1995). Different forms of TFIIH for transcription and DNA repair: Holo-TFIIH and a nucleotide excision repairosome. *Cell*, **80**, 21–28.

Chapter 9

Global genome nucleotide excision repair: key players and their functions

Marcel Volker and Leon H.F. Mullenders

Table of contents

Abstract

Nucleotide excision repair (NER) is among the best-studied DNA repair pathways. Initially the various steps, the proteins involved and their mechanism of action were characterised in vitro, but recently developed techniques have enabled the detailed study of NER in vivo. These in vivo studies have shed light on some long-standing questions regarding NER, in particular the order and mutual dependency of binding of the various components to the complex, their intranuclear associations and dynamics. Furthermore, proteins have been identified that are not required in vitro. Here, we present an overview of recent findings regarding NER in vivo.

9.1 Introduction

Of all exogenous agents that damage genomic DNA, and hence present a threat to the integrity of genetic information, the ultraviolet (UV)-component of sunlight is the one with the highest clinical relevance because of its abundance. Short-wave UV light (UV-B and -C) causes predominantly two types of lesions: cyclobutane pyrimidine dimers (CPD) and pyrimidine 6-4 pyrimidone photoproducts (6-4PP) in a 3:1 ratio. Many organisms, including bacteria, yeasts, plants and animals possess specialised pathways for repair of UV-induced photolesions. One major group of specialised proteins consists of photoreactivating enzymes that directly revert photolesions to the original bases using the far-blue component of visible light [for a review, see [1]]. A second group is comprised of endonucleases that specifically recognise UV-lesions and incise DNA adjacent to the lesion leading to rapid repair, e.g. CPD-specific DNA glycosylase/AP lyase or the *Schizosaccharomyces pombe* ultraviolet DNA endonuclease UVDE.

Uniquely, placental mammals rely solely on the nucleotide excision repair (NER) pathway to repair photolesions. In striking contrast to UV-lesion specific repair pathways, NER is capable of removing a very wide range of lesions from genomic DNA. The evolutionarily highly conserved NER pathway is responsible for the repair of bulky, helix-distorting lesions, apart from UV-induced photolesions encompassing numerous chemical adducts such as those caused by cisplatin, benzo(a)pyrene, and N-acetoxy-2-acetylaminofluorene (NA-AAF). NER is a complex process, in humans involving over 30 polypeptides, that uses a cut-and-paste mechanism to excise an oligonucleotide of $\sim$30 nt containing the DNA lesion, followed by gap-filling DNA synthesis using the undamaged strand as a template. Two subpathways can be distinguished: global genome repair (GGR), a general pathway that removes lesions from the entire genome, and transcription-coupled repair (TCR), a specialized pathway that removes lesions from the transcribed strand of active genes only and strictly depends on ongoing transcription [2–4]. To a large extent the two subpathways use the same set of proteins. The major difference between them lies in the initial step, i.e. recognition of the DNA lesion. In GGR, specialised damage recognition proteins execute this step, whereas in TCR the stalling of a transcribing RNA polymerase onto a lesion is thought to trigger repair.

The importance of the NER process, especially for the removal of UV-induced lesions, at the organismal level is underlined by the rare human disorder xeroderma pigmentosum (XP). XP is a severe autosomal recessive disorder caused by defects in genes encoding NER proteins. Patients suffering from XP display extreme sun-sensitivity, hyperpigmentation of the skin and a highly (>1000-fold) increased incidence of cancer in sun-exposed parts of the skin. Seven complementation groups are distinguished (XP-A through G), and with the exception of XP-C and XP-E, in which only GGR is compromised [5,6], cells from all complementation groups are compromised in both subpathways of NER.

The most striking difference between the GGR pathway of NER and other DNA repair processes is the substrate specificity. Processes like base excision repair (BER) deal with a defined, narrow set of lesions and repair is initiated by lesion-specific proteins such as the abovementioned CPD-specific DNA glycosylase. In strong contrast, NER is capable of processing a wide variety of bulky, helix-distorting lesions. One attractive model to explain the low fastidiousness of NER is the 'bipartite' model put forward by Hess et al. [7]. This model suggests that NER is capable of processing lesions that to some extent both distort the stacking of the DNA helix and alter the chemical composition of the DNA. For proper biological functioning of a repair system it is essential that its specificity limits repair exclusively to valid DNA lesions. Gratuitous repair not only is a waste of cellular resources; unnecessary generation of DNA strand breaks may even endanger genome stability. On the other hand, too high a stringency will lead to lesions being undetected, also posing a threat to several DNA-metabolic processes. How global genome NER has struck a balance between those two 'evils' is subject of this discourse.

9.2 NER *in vitro*

The NER reaction has been studied extensively. In vitro repair systems have been established using cell-free extracts or purified proteins and non-chromatinised DNA templates with defined lesions [8–10]. Based on results obtained from the in vitro assays, NER can be divided into the stages: 1) damage recognition 2) helix unwinding 3) dual incision 4) excision and release of damage-containing oligomer 5) gap-filling DNA resynthesis, and 6) ligation of the nicked strand [11]. The minimal set of essential proteins to perform NER in vitro up to and including the dual incision step consists of the following factors [8–10]: the XPC-hHR23B heterodimer and XPA, both of which show affinity for damaged DNA; transcription factor IIH (TFIIH), which is essential for promoter opening in transcription by RNA polymerase I [12] and RNA polymerase II [13]; the mammalian single-stand DNA binding protein replication protein A (RPA); and the two structure-specific endonucleases XPG and the ERCC1-XPF heterodimer. DNA resynthesis and ligation also require RPA, plus an additional 4 proteins: PCNA, which acts as a sliding clamp for proteins active in DNA synthesis and is loaded onto the DNA by RF-C; DNA polymerase delta or epsilon are required to resynthesize across the gap; finally DNA ligase I serves to ligate the nick.

9.3 NER *in vivo*

9.3.1 Sequential incorporation of factors into the GGR complex

An important question regarding NER in vivo is its organisation. NER proteins have to constantly search the genome for damages and after their detection, damages have to be rapidly removed to prevent the obstruction of transcription or replication. GGR can be organised in at least two principally different ways.

Firstly, most or all proteins might form a pre-assembled holocomplex that moves through the nucleus as a single entity. Evidence for this so-called 'repairosome' in yeast cells was presented by Svestrup et al. [14]; however, other researchers did not find evidence for such a complex [15]. The major advantage of a holocomplex is that a lesion can be processed immediately after it is detected, but it also presents specific problems such as low access to lesions inside heterochromatin, low diffusion rates (i.e., slow kinetics of lesion recognition) and inflexibility in allowing multiple roles for repair proteins.

A fundamentally different way of organisation is that every protein is incorporated sequentially into an 'emerging' NER complex. Thus, most problems associated with an NER holocomplex are circumvented: single proteins and small subcomplexes can easily access the DNA in tightly packed chromatin and diffuse through the nucleus significantly faster than a holocomplex, ensuring rapid screening of the entire genome. Also, since proteins only interact strongly at the site of a lesion, they can freely function in other pathways than NER when no DNA damages are present. Recently, much evidence has been put forward supporting the latter model [16–21] and the idea of a repairosome in human cells has been virtually abandoned.

The following section will represent the order of incorporation of proteins and their actions as they have emerged from recent investigations. Apart from the more classical in vitro methods, a newly developed method has greatly contributed to the understanding of NER in the living cell: the so-called 'local irradiation' technique [22,23]. This technique allows partial UV-irradiation of a nucleus, resulting in UV-damaged areas with a diameter of 3 to 12 μm. Subsequently, the relocalization of proteins to the irradiated area can be assayed either in fixed cells using immunofluorescent labeling [18,22,23] or in living cells with fluorescently tagged proteins [19,20]. A schematic representation of the method is shown in Figure 1.

9.3.2 Damage recognition

Several proteins and protein complexes have been proposed to play a role in the damage recognition step of GGR: the UV-damaged DNA binding protein UV-DDB [6,24,25], XPC-hHR23B [26,27], XPA [28,29], RPA [30] and the XPA-RPA heterodimer [31–33].

UV-DDB is the heterodimer of the p48 and p127 proteins and was originally identified as a factor in HeLa whole-cell extracts that readily bound to UV-damaged

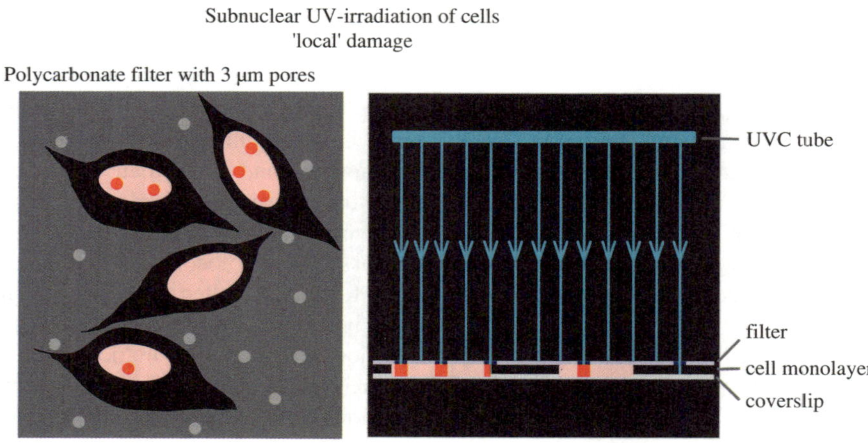

Figure 1. Schematic representation of the local UV-irradiation method. Monolayers of cells are grown on glass coverslips in a petridish. Prior to irradiation, cells are covered by a polycarbonate filter with pores of a single diameter ranging from 3–12 μm. Cells are then irradiated with UVC light. The filter blocks all the UV-light, resulting in exposure of the cells only at the position of pores.

DNA in electrophoretic mobility shift assays. Indeed, UV-DDB displays the highest affinity for UV-induced photolesions of all putative damage recognition factors, plus a moderate affinity for several other types of lesions. Nevertheless UV-DDB is not required for NER in vitro, which led to the suggestion that it played a role specifically in the repair of lesions within chromatin [34]. Contradictory findings exist reporting either stimulation [35], no effect [36] or even inhibition [8] by UV-DDB of NER in vitro. Moreover, human XP group E and rodent cells, which are deficient in UV-DDB show only a limited repair defect [6], in agreement with the very mild clinical phenotype of XP-E. Interestingly, rodent cells despite their lack of CPD repair show no reduced clonal survival following UV-irradiation compared to normal human cells; this apparent discrepancy has been dubbed the 'rodent repairadox' [reviewed in [37]]. These observations have lead to the hypothesis that UV-DDB is not required for damage recognition of all lesions but is necessary only for repair of those lesions that are otherwise poorly recognised by XPC-hHR23B, such as CPD [6]. Apart from playing a role in damage recognition, UV-DDB might also participate in repair-associated chromatin remodeling, a process that will be discussed in section 4.1.

Initial evidence lead to the hypothesis that the XPA-RPA complex was the damage recognition factor in both GGR and TCR. XP-C cells only show compromised GGR, whereas in XP-A cells also TCR is affected. Furthermore, initial biochemical characterisations showed that XPA has a high affinity for damaged over undamaged DNA [29,28]. Shortly hereafter, it was found that XPA interacts with RPA and the resulting complex displayed a higher affinity for damaged DNA than XPA alone [31]. The XPA-RPA complex therefore seemed a likely candidate for general damage recognition, and a series of experiments from Wakasugi and

Sancar supported this notion [33]. XPC was thought to be involved in a GGR-specific role, such as stabilisation of the open intermediate formed after the lesion has been recognised [26,32,38]. However, recent in vitro and in vivo evidence has now put XPC (or more accurately the XPC-hHR23B complex) forward as being the initial damage sensor in GGR [18,21,27]. For instance, using the local irradiation method we were able to show that in the absence of XPA, XPC rapidly relocalizes to UV-damages, whereas XPA does not accumulate in local UV-spots when no XPC is present (Figure 2A). Furthermore, damage recognition in TCR is now thought to occur by stalling of the transcribing RNA polymerase onto a lesion. Together this has led to the notion that XPA is not a damage recognition factor, but instead a 'damage verification protein' in a later stage of NER to ascertain that

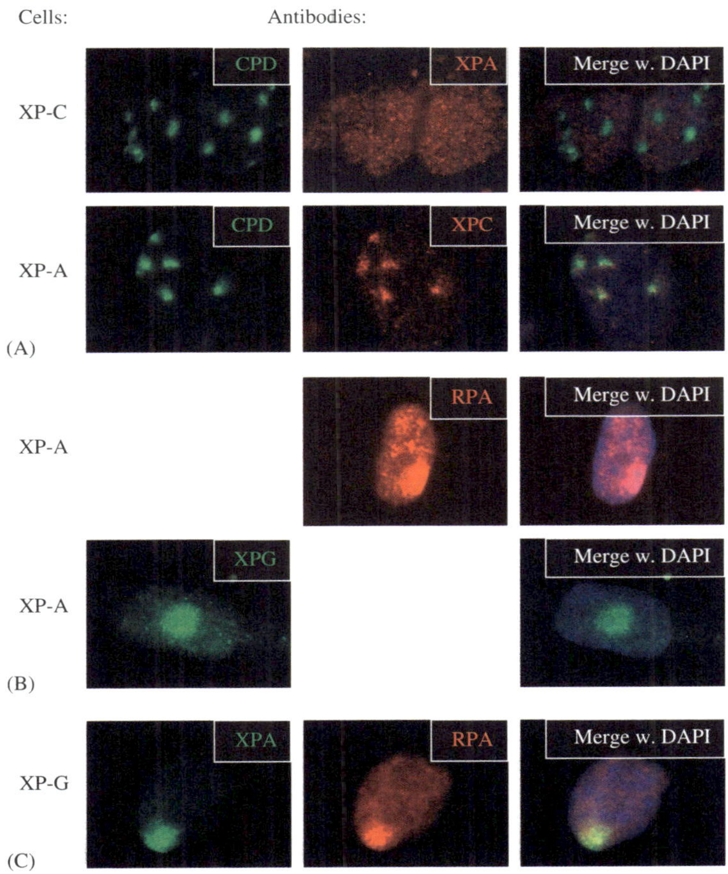

Figure 2. Typical results from local irradiation experiments. All cells shown were locally UV-irradiated followed by fixation 30 minutes after irradiation. (A) XPC, not XPA, is the initiating protein in global genome NER. (B) XPA is not required for the recruitment of XPG and RPA to the NER complex. (C) Functional XPG is not required for the recruitment of XPA and RPA to the NER complex.

the DNA covered by the NER complex indeed harbors a lesion. For an extensive review on the evidence supporting the respective roles of XPA and XPC-hHR23B the reader is referred to [39].

In the cell, XPC is found complexed with the hHR23B protein [40] which stabilises XPC by protecting it from proteasomal degradation [41]. The resulting heterodimer displays an affinity for a broad spectrum of DNA lesions, e.g. 6-4PP, CPD and cisplatin- and NA-AAF adducts [26,27,42]. Upon binding to DNA, XPC-hHR23B bends the DNA [43] and slightly opens up the double helix [27]. The resulting 'bubble' structure presumably facilitates the recruitment of subsequent NER factors, in combination with protein-protein interactions of XPC-hHR23B.

9.3.3 Recruitment of TFIIH, partial helix unwinding

The second protein that is incorporated into the NER complex is TFIIH [44]. In the transcription process and presumably in the NER reaction, TFIIH opens up the DNA by virtue of its two ATP-dependent DNA helicase subunits with opposite polarity, thus creating short stretches of single-stranded DNA (ssDNA) encompassing the promoter (in transcription) or the lesion (in NER). The importance of the role of TFIIH in NER is underlined by the fact that the XP complementation groups B and D are comprised of mutations in the two helicases p89 and p80 respectively, also known as XPB and XPD. Apart from helix unwinding, the helicases of TFIIH might perform another function. It is speculated that obstruction of the helicase activities by the lesion helps to discriminate between the damaged and the non-damaged strand [45,46]. The structure formed after helix unwinding by TFIIH might stabilise the binding between TFIIH and the DNA, and thus facilitate the further recruitment of NER factors to the GGR complex via their interactions with TFIIH.

9.3.4 Damage demarcation and full helix unwinding

After the initial complex of XPC-hHR23B and TFIIH is formed, XPG, XPA and RPA are recruited to the NER complex [18,21,20].

XPG is the endonuclease that cuts at the 3′ side of the lesion. XPG interacts fairly strongly with TFIIH [17,47] and relocalizes to local UV-induced damages in the absence of functional XPA and XPF [18], strongly suggesting that XPG is recruited to the NER complex predominantly by its interactions with TFIIH.

XPA as discussed above serves as a damage verification protein. The damage recognition factor XPC-hHR23B possesses an intrinsic affinity for undamaged DNA [26,42], and is known to interact relatively strongly with TFIIH even in the absence of DNA damage [10,48]. Hence, formation of an NER complex on undamaged DNA can be expected to initiate. As mentioned above, it is of great importance that such an erroneous NER complex does not reach the incision stage. Therefore, it is highly beneficial for the cell to include a damage verification step in the NER reaction. XPA might execute this damage verification in several different ways. First, binding of XPA to a lesion could stabilise a bona fide NER complex. If no lesion is present the complex is not stabilised by XPA and is able to disassemble. Second, XPA indirectly stimulates cutting of XPG and ERCC1-XPF through its DNA damage-dependent stimulation of ATP-hydrolysis by TFIIH, which in turn alleviates the inhibition of the endonucleases [49]. Moreover, the presence of XPA is required to allow recruitment of ERCC1-XPF [31,18]. In short, dual incision will not take place if XPA is not present in the NER complex. Although initial observations suggested that XPA binds to lesions as a heterodimer complexed with RPA [31] recent in vivo observations have established that XPA interacts with RPA only within the context of the NER complex, and that XPA moves through the nucleus and binds to the NER complex as a single, separate entity.

RPA likely is incorporated into the NER complex in the so-called 8–10 nt binding mode, followed by stretching to the 30 nt binding mode, creating two stretches of ssDNA of essentially the same length as the released oligomer. Apart from protecting the ssDNA from endonuclease attacks and resolving secondary structures, RPA also plays a crucial role in correctly positioning the endonucleases XPG and ERCC1-XPF along the DNA.

It is unclear how RPA is attracted to the NER complex. Investigators assaying interactions between various NER proteins found RPA to interact only very weakly with other proteins in the complex [10], suggesting that RPA is not recruited to the complex via protein-protein interactions. On the other hand, the findings by Winkler et al. imply that the presence of XPA is required to allow full opening of the helix. Nevertheless, in XP-A cells RPA is readily detectable at sites of local UV-damage [20]. Furthermore, relocalization of RPA to locally induced UV-damage was assayed in a severe XP-B/CS mutant in which TFIIH presumably is inactive in NER helix unwinding, resulting in incomplete opening of the helix. Also in this situation, RPA efficiently relocalizes to sites of local UV-damage in these cells [20]. Both these observations suggest that full opening around the lesion is not required to allow incorporation of RPA into the NER complex, and that the small stretches of ssDNA created by XPC-hHR23B and TFIIH, or alternatively weak protein-protein interactions are sufficient for RPA to join the NER complex.

In this stage of NER the order of binding of proteins, i.e. RPA, XPG and XPA is unclear. Based on in vitro experiments in which XPA was necessary to allow binding of both RPA and XPG to the complex, and incorporation of RPA stimulated binding of XPG, Riedl et al. suggested the order XPA -> RPA -> XPG [21]. On the other hand, experiments using local irradiation of various XP-mutant cells show that recruitment of both XPG and RPA to the NER complex occurs independent of the presence of XPA [18,20] and Figure 2B, indicating that XPA is not necessarily the first protein of these three to be incorporated. Conversely, we found functional XPG to be dispensable for the incorporation of XPA into the NER complex, although we could not exclude that truncated XPG present in the cells we used is responsible for the observed recruitment of XPA [18] and Figure 2C. A possible explanation for this apparent inconsistency is that these factors bind in no specific order, and the idea of a fixed and strictly sequential order of binding in this phase of complex formation is incorrect. Nevertheless, these factors could still mutually stabilise binding of one another to the complex.

9.3.5 Dual incision

Like XPG, ERCC1-XPF is a structure-specific endonuclease that cuts in duplex DNA at junctions with ssDNA. Whereas XPG cuts at the 3′ side, ERCC1-XPF incises at the 5′ side of the lesion. The recruitment of the ERCC1-XPF heterodimer completes the formation of the preincision complex and as mentioned above depends strictly on functional XPA both in vitro and in vivo [18,31]. Furthermore the combined presence of DNA damage and XPA is required to fully stimulate incision activity by ERCC1-XPF [49]. In the absence of ERCC1-XPF, a complex can be formed that includes all above-mentioned proteins and that is capable of executing uncoupled 3′ incision via the XPG endonuclease [32,50].

Several regulatory mechanisms prevent the endonucleases from untargeted cutting in the undamaged DNA strand or even in random DNA. First, ATP hydrolysis activity of the TFIIH helicase subunits is required to stimulate dual incision [32,51]. Second, as mentioned above, the presence of XPA together with a DNA lesion stimulates this ATP hydrolysis [49]. Finally, the polarity of RPA binding to DNA is responsible for the proper positioning of the endonucleases [52], ensuring that only the damage-containing strand is incised and no double-strand breaks are introduced. When all these conditions are met, incisions are made at the 5′ and 3′ side of the lesion and the oligonucleotide containing the lesion is removed.

It is conceivable that one or more proteins are released from the complex before repair is completed. XPC-hHR23B has been reported to leave the NER complex upon recruitment of XPG [21,53], whereas TFIIH, XPA and both endonucleases would only dissociate following dual incision [21]. The advantage of this 'early release' would be that proteins can quickly be re-used in other repair events after having executed their function in the current repair event, leading to more effective repair.

9.3.6 DNA resynthesis and ligation

Dual incision is an absolute requirement for the next step in the NER process: recruitment of PCNA. It was found that in XP-cells that are unable to perform dual incision, PCNA did not associate with chromatin, i.e. participate in NER, after UV-irradiation [54]. PCNA is involved in many cellular processes that involve DNA synthesis, serving as a sliding clamp for DNA polymerase. PCNA is loaded onto the DNA by RF-C in an ATP-dependent manner. Binding of PCNA to the DNA is stabilised by an activation step of the RF-C::PCNA complex at ds- to ssDNA transitions with a 3′-OH end. These structures are characteristically formed after 5′ incision by ERCC1-XPF and subsequent removal of the lesion-containing oligomer, possibly providing a clue as to how PCNA is associated with the post-incision NER events. All proteins involved in the post-incision steps of NER interact with PCNA, underlining its crucial role in this stage of NER. First, either DNA polymerase delta or epsilon can be utilised to resynthesize the excised strand. It is currently not known if there is discrimination between the two polymerases and if so, what is the basis for such discrimination. Subsequently, DNA ligase I is involved in ligation of the gap. Finally, the transition to an even later step in the repair process is heralded by the interaction of PCNA with CAF-1, a chromatin assembly factor responsible for the restoration of the chromatin structure following the completion of repair. The consequences of this interaction are discussed in section 4.3.

9.4 Challenges put forward by a chromatin environment

The above-mentioned core NER reaction, up to and including ligation by DNA ligase I, has been reconstituted in vitro, using purified proteins and naked DNA. In vivo however, NER has to function in tightly packed (hetero)chromatin and at different stages of the cell cycle. Several researchers have reported a negative effect of chromatin on the rate of NER in vitro [55–57]. Nevertheless, NER has a higher rate in vivo than in any of the in vitro repair systems, indicating the existence of in vivo accessory systems that enhance the rate of repair despite the inhibitory effect of chromatin. To account for the in principle inhibitory effect of chromatin on NER, the 'access-repair-restore' model has been put forward [for a review see [58]]. In this model, before core NER occurs, the chromatin in the vicinity of the lesion is relaxed or in another way remodeled to improve accessibility for NER proteins. Accordingly, after repair is completed the higher-order chromatin structure is restored. A model for global genome nucleotide excision repair and its interplay with chromatin remodeling is shown in Figure 3.

9.4.1 Access: UV-DDB

One important complication for repair in vivo is the paradox between lesion accessibility and lesion detection. Before chromatin relaxation proteins can do their work, a lesion needs to be detected, while on the other hand, a lesion can only be detected if the chromatin is accessible to damage-recognising proteins. Recent findings indicate that UV-DDB functions both in lesion detection and chromatin remodeling. Thus, UV-DDB might solve the paradox between lesion detection and accessibility, being the protein complex that detects lesions in their native chromatin environment, as was suggested by Rapic-Otrin and co-workers [34].

UV-DDB has been a subject of extensive study in the last few years. First, it has become increasingly clear that UV-DDB indeed plays a direct role in NER damage recognition. For long, it was found that cells lacking UV-DDB activity (not only human XP-E cells but also rodent cells) are selectively deficient in repair of CPD. Oddly, although UV-DDB displays a higher affinity for 6-4PP than CPD, UV-DDB is not strictly required for repair of these lesions [6,25], instead it performs a stimulatory role in their repair [35,59,74]. UV-DDB seems to be consumed following its binding to DNA lesions, possibly by proteasome-mediated degradation [60], limiting its stimulatory effect on repair, and we therefore propose that UV-DDB can participate in the repair of only one lesion before being degraded [74]. Effectively, this limitation results in accelerated repair of a limited number of 6-4PP, followed by repair at a lower rate of the remaining lesions [74]. Remarkably,

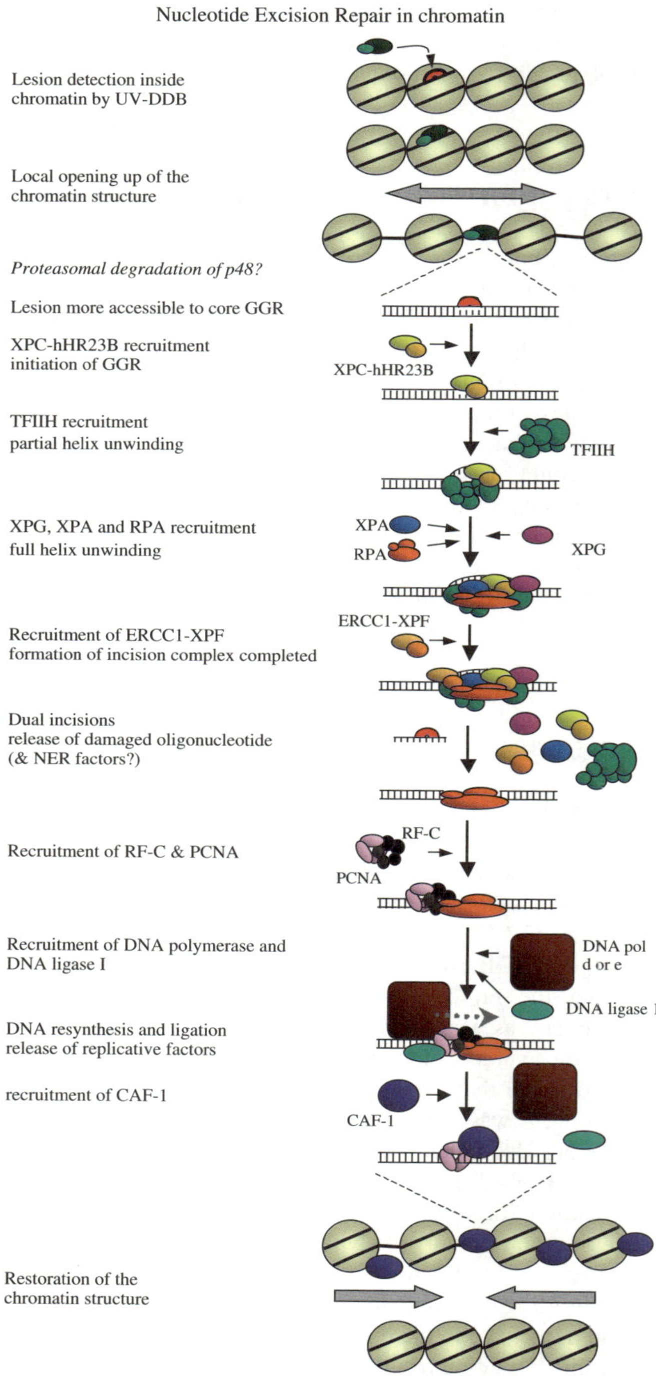

Figure 3. Model for nucleotide excision repair within the chromatin context.

degradation of the p48 subunit of UV-DDB is observed in repair-deficient cells as well [60], suggesting that it is not repair that consumes UV-DDB, but perhaps pre-repair chromatin remodeling. Several recent findings suggest a role for UV-DDB in chromatin reorganisation. Firstly, p48 bears homology to certain chromatin-remodeling proteins [61]. Secondly, both p48 and p127 are found to interact with the histone acetylating CBP/p300 protein [60,62]. Hyperacetylation of histones is known to accelerate NER in vivo [63]. Finally, both subunits of UV-DDB can be found in a large complex that also contains the COP9 signalosome, cullin 4A and Roc1, and that displays ubiquitin ligase activity [64]. Interestingly, two observations link this complex to the UV-induced breakdown of p48. Not only does p48 associate with chromatin following UV-irradiation, supposedly induced by its binding to UV-damaged DNA [34], moreover cullin 4A is capable of specifically targeting p48 for proteasomal degradation [65].

9.4.2 Access: other accessory proteins

Several other proteins that are dispensable for NER in vitro have been found to influence NER specifically in a chromatin environment: Gadd45a [66,67], HMGN1 [68], the STAGA complex [69], SWI/SNF [56]. Also p53 was recently reported to play a role in global chromatin relaxation following UV [70], but an extensive treatment of these proteins falls beyond the scope of this essay.

9.4.3 Restore: CAF-1

Little is known yet about the events that take place after repair is completed. The only protein to be firmly implicated in post-repair events in the cell is CAF-1. In vitro CAF-1 is capable of loading histones onto DNA [71]. The observation by Green and Almouzni that CAF-1 relocalizes to locally induced UV-damages only in cells that are capable of executing NER [72] suggests that CAF-1 indeed functions in restoring the correct chromatin structure after processes such as repair that obviously disrupt chromatin structure. The interaction of CAF-1 with PCNA is responsible for implicating CAF-1 into NER [73]. As PCNA also attracts DNA polymerases and DNA ligase I towards sites of NER, CAF-1 would be present at the place it is needed at the time it is needed there.

9.5 Conclusion

NER is a complex repair process, in vivo involving considerably more proteins than in the core in vitro reaction, which already comprises over 30 polypeptides. The obvious reason for this added complexity is the chromatin environment of DNA inside the nucleus. The process of repair inside chromatin and the interactions between repair and chromatin remodeling are an emerging field of research that will produce many exciting new findings and unexpected links in the years to come. To study this field, the same approach can be utilised that has recently also greatly benefited the study of the core NER reaction in the live cell: confocal studies using GFP-tagged proteins or immunofluorescent labeling using specific antibodies, combined with local UV-irradiation. Using these techniques NER has come forward as a highly dynamic process in vivo. The way of organisation of NER, i.e. separate factors that associate at the site of a DNA lesion, functions as a two-edged sword. It provides the system speed and flexibility to ensure that lesions are repaired in varying chromatin surroundings and through all phases of the cell cycle. Additionally it provides a step-by-step increasing accuracy enabling NER to distinguish with high precision between undamaged and damaged DNA. UV-DDB, XPC-hHR23B and XPA all possess affinity for damaged DNA; additionally the 'kinetic proofreading' by the TFIIH helicases may add another step of discrimination. Finally, the absence of strong interactions between NER proteins outside the context of a DNA lesion enables them to function in other cellular processes. Proteins such as TFIIH (transcription) and RPA (replication and other DNA repair pathways), would be hampered to execute their multiple roles if NER was to be organised in a tightly bound holocomplex.

Acknowledgements

This work was supported by the Netherlands Organization for Scientific Research (NWO) program "Nucleotide excision repair in its nuclear context" and by European Union project. nr. QRLT-1999-00181. The authors thank Dr. H. Vrieling for helpful discussions on the manuscript.

References

1. T. Carell, L.T. Burgdorf, L.M. Kundu, M. Cichon (2001). The mechanism of action of DNA photolyases, *Curr. Opin. Chem. Biol.*, **5**, 491–498.
2. I. Mellon, V.A. Bohr, C.A. Smith, P.C. Hanawalt (1986). Preferential DNA repair of an active gene in human cells, *Proc. Natl. Acad. Sci. USA*, **83**, 8878–8882.
3. S.A. Leadon, D.A. Lawrence (1991). Preferential repair of DNA damage on the transcribed strand of the human metallothionein genes requires RNA polymerase II, *Mutat. Res.*, **255**, 67–78.
4. J. Venema, Z. Bartosova, A.T. Natarajan, A.A. van Zeeland, L.H. Mullenders (1992). Transcription affects the rate but not the extent of repair of cyclobutane pyrimidine dimers in the human adenosine deaminase gene, *J. Biol. Chem.*, **267**, 8852–8856.
5. J. Venema, A. van Hoffen, V. Karcagi, A.T. Natarajan, A.A. van Zeeland, L.H. Mullenders (1991). Xeroderma pigmentosum complementation group C cells remove pyrimidine dimers selectively from the transcribed strand of active genes, *Mol. Cell Biol.*, **11**, 4128–4134.
6. B.J. Hwang, J.M. Ford, P.C. Hanawalt, G. Chu (1999). Expression of the p48 xeroderma pigmentosum gene is p53-dependent and is involved in global genomic repair, *Proc. Natl. Acad. Sci. USA*, **96**, 424–428.
7. M.T. Hess, U. Schwitter, M. Petretta, B. Giese, H. Naegeli (1997). Bipartite substrate discrimination by human nucleotide excision-repair, *PNAS*, **94**, 6664–6669.

8. A. Aboussekhra, M. Biggerstaff, M.K. Shivji, J.A. Vilpo, V. Moncollin, V.N. Podust, M. Protic, U. Hubscher, J.M. Egly, R.D. Wood (1995). Mammalian DNA nucleotide excision repair reconstituted with purified protein components, *Cell*, **80**, 859–868.
9. T. Bessho, A. Sancar, L.H. Thompson, M.P. Thelen (1997). Reconstitution of human excision nuclease with recombinant XPF-ERCC1 complex, *J. Biol. Chem.*, **272**, 3833–3837.
10. S.J. Araujo, F. Tirode, F. Coin, H. Pospiech, J.E. Syvaoja, M. Stucki, U. Hubscher, J.M. Egly, R.D. Wood (2000). Nucleotide excision repair of DNA with recombinant human proteins: definition of the minimal set of factors, active forms of TFIIH, and modulation by CAK, *Genes Dev.*, **14**, 349–359.
11. W.L. de Laat, N.G. Jaspers, J.H. Hoeijmakers (1999). Molecular mechanism of nucleotide excision repair, *Genes Dev.*, **13**, 768–785.
12. S. Iben, H. Tschochner, M. Bier, D. Hoogstraten, P. Hozak, J.M. Egly, I. Grummt (2002). TFIIH plays an essential role in RNA polymerase I transcription, *Cell*, **109**, 297–306.
13. O. Flores, H. Lu, D. Reinberg (1992). Factors involved in specific transcription by mammalian RNA polymerase II. Identification and characterization of factor IIH, *J Biol. Chem.*, **267**, 2786–2793.
14. J.Q. Svejstrup, Z. Wang, W.J. Feaver, X. Wu, D.A. Bushnell, T.F. Donahue, E.C. Friedberg, R.D. Kornberg (1995). Different forms of TFIIH for transcription and DNA repair: holo-TFIIH and a nucleotide excision repairosome, *Cell*, **80**, 21–28.
15. S.N. Guzder, P. Sung, L. Prakash, S. Prakash (1996). Nucleotide excision repair in yeast is mediated by sequential assembly of repair factors and not by a pre-assembled repairosome, *J. Biol. Chem.*, **271**, 8903–8910.
16. A.B. Houtsmuller, S. Rademakers, A.L. Nigg, D. Hoogstraten, J.H. Hoeijmakers, W. Vermeulen (1999). Action of DNA repair endonuclease ERCC1/XPF in living cells, *Science*, **284**, 958–961.
17. S.J. Araujo, E.A. Nigg, R.D. Wood (2001). Strong functional interactions of TFIIH with XPC and XPG in human DNA nucleotide excision repair, without a preassembled repairosome, *Mol. Cell Biol.*, **21**, 2281–2291.
18. M. Volker, M.J. Mone, P. Karmakar, A. van Hoffen, W. Schul, W. Vermeulen, J.H. Hoeijmakers, R. van Driel, A.A. van Zeeland, L.H. Mullenders (2001). Sequential assembly of the nucleotide excision repair factors in vivo, *Mol. Cell*, **8**, 213–224.
19. D. Hoogstraten, A.L. Nigg, H. Heath, L.H. Mullenders, R. van Driel, J.H. Hoeijmakers, W. Vermeulen, A.B. Houtsmuller (2002). Rapid switching of TFIIH between RNA polymerase I and II transcription and DNA repair in vivo, *Mol. Cell*, **10**, 1163–1174.
20. S. Rademakers, M. Volker, D. Hoogstraten, A.L. Nigg, M.J. Mone, A.A. van Zeeland, J.H.J. Hoeijmakers, A.B. Houtsmuller, W. Vermeulen (2003). Xeroderma pigmentosum Group A protein loads as a separate factor onto DNA lesions, *Mol. Cell. Biol.*, **23**, 5755–5767.
21. T. Riedl, F. Hanaoka, J.M. Egly (2003). The comings and goings of nucleotide excision repair factors on damaged DNA, *The EMBO Journal*, **22**, 5293–5303.
22. M.J. Mone, M. Volker, O. Nikaido, L.H. Mullenders, A.A. van Zeeland, P.J. Verschure, E.M. Manders, R. van Driel (2001). Local UV-induced DNA damage in cell nuclei results in local transcription inhibition, *EMBO Rep.*, **2**, 1013–1017.
23. S. Katsumi, N. Kobayashi, K. Imoto, A. Nakagawa, Y. Yamashina, T. Muramatsu, T. Shirai, S. Miyagawa, S. Sugiura, F. Hanaoka, T. Matsunaga, O. Nikaido, T. Mori (2001). In situ visualization of ultraviolet-light-induced DNA damage repair in locally irradiated human fibroblasts, *Journal of Investigative Dermatology*, **117**, 1156–1161.
24. G. Chu, E. Chang (1988). Xeroderma pigmentosum group E cells lack a nuclear factor that binds to damaged DNA, *Science*, **242**, 564–567.

25. J.Y. Tang, B.J. Hwang, J.M. Ford, P.C. Hanawalt, G. Chu (2000). Xeroderma pigmentosum p48 gene enhances global genomic repair and suppresses UV-induced mutagenesis, *Mol. Cell*, **5**, 737–744.
26. J.T. Reardon, D. Mu, A. Sancar, Overproduction, purification, and characterization of the XPC subunit of the human DNA repair excision nuclease, *J. Biol. Chem.*, **271**, (1996) 19451–19456.
27. K. Sugasawa, J.M. Ng, C. Masutani, S. Iwai, P.J. van der Spek, A.P. Eker, F. Hanaoka, D. Bootsma, J.H. Hoeijmakers (1998). Xeroderma pigmentosum group C protein complex is the initiator of global genome nucleotide excision repair, *Mol. Cell*, **2**, 223–232.
28. P. Robins, C.J. Jones, M. Biggerstaff, T. Lindahl, R.D. Wood (1991). Complementation of DNA repair in xeroderma pigmentosum group A cell extracts by a protein with affinity for damaged DNA, *EMBO J.*, **10**, 3913–3921.
29. C.J. Jones, R.D. Wood (1993). Preferential binding of the xeroderma pigmentosum group A complementing protein to damaged DNA, *Biochemistry*, **32**, 12096–12104.
30. J.T. Reardon, A. Sancar (2002). Molecular anatomy of the human excision nuclease assembled at sites of DNA damage, *Mol. Cell Biol.*, **22**, 5938–5945.
31. L. Li, X. Lu, C.A. Peterson, R.J. Legerski (1995). An interaction between the DNA repair factor XPA and replication protein A appears essential for nucleotide excision repair, *Mol. Cell Biol.*, **15**, 5396–5402.
32. D. Mu, D.S. Hsu, A. Sancar (1996). Reaction mechanism of human DNA repair excision nuclease, *J. Biol. Chem.*, **271**, 8285–8294.
33. M. Wakasugi, A. Sancar (1999). Order of assembly of human DNA repair excision nuclease, *J. Biol. Chem.*, **274**, 18759–18768.
34. V.R. Otrin, M. McLenigan, M. Takao, A.S. Levine, M. Protic, Translocation of a UV-damaged DNA binding protein into a tight association with chromatin after treatment of mammalian cells with UV light, *J. Cell Sci.*, 110 (Pt 10) (1997) 1159–1168.
35. M. Wakasugi, M. Shimizu, H. Morioka, S. Linn, O. Nikaido, T. Matsunaga (2001). Damaged DNA-binding protein DDB stimulates the excision of cyclobutane pyrimidine dimers in vitro in concert with XPA and replication protein A, *J. Biol. Chem.*, **276**, 15434–15440.
36. V. Rapic, I. Kuraoka, T. Nardo, M. McLenigan, A.P. Eker, M. Stefanini, A.S. Levine, R.D. Wood (1998). Relationship of the xeroderma pigmentosum group E DNA repair defect to the chromatin and DNA binding proteins UV-DDB and replication protein A, *Mol. Cell Biol.*, **18**, 3182–3190.
37. P.C. Hanawalt (2001). Revisiting the rodent repairadox, *Environ. Mol. Mutagen.*, **38**, 89–96.
38. P.J. van der Spek, A. Eker, S. Rademakers, C. Visser, K. Sugasawa, C. Masutani, F. Hanaoka, D. Bootsma, J.H. Hoeijmakers (1996). XPC and human homologs of RAD23: intracellular localization and relationship to other nucleotide excision repair complexes, *Nucleic Acids Res.*, **24**, 2551–2559.
39. B.S. Thoma, K.M. Vasquez (2003). Critical DNA damage recognition functions of XPC-hHR23B and XPA-RPA in nucleotide excision repair, *Mol. Carcinog.*, **38**, 1–13.
40. C. Masutani, K. Sugasawa, J. Yanagisawa, T. Sonoyama, M. Ui, T. Enomoto, K. Takio, K. Tanaka, P.J. van der Spek, D. Bootsma (1994). Purification and cloning of a nucleotide excision repair complex involving the xeroderma pigmentosum group C protein and a human homologue of yeast RAD23, *EMBO J.*, **13**, 1831–1843.
41. J.M. Ng, W. Vermeulen, G.T. Van Der Horst, S. Bergink, K. Sugasawa, H. Vrieling, J.H. Hoeijmakers (2003). A novel regulation mechanism of DNA repair by damage-induced and RAD23-dependent stabilization of xeroderma pigmentosum group C protein, *Genes Dev.*, **17**, 1630–1645.

42. D. Batty, V. Rapic'-Otrin, A.S. Levine, R.D. Wood (2000). Stable binding of human XPC complex to irradiated DNA confers strong discrimination for damaged sites, *J. Mol. Biol.*, **300**, 275–290.
43. Janicijevic, K. Sugasawa, Y. Shimizu, F. Hanaoka, N. Wijgers, M. Djurica, J.H. Hoeijmakers, C. Wyman (2003). DNA bending by the human damage recognition complex XPC-HR23B, *DNA Repair (Amst)*, **2**, 325–336.
44. M. Yokoi, C. Masutani, T. Maekawa, K. Sugasawa, Y. Ohkuma, F. Hanaoka (2000). The xeroderma pigmentosum group C protein complex XPC-HR23B plays an important role in the recruitment of transcription factor IIH to damaged DNA, *J. Biol. Chem.*, **275**, 9870–9875.
45. H. Naegeli, L. Bardwell, E.C. Friedberg (1992). The DNA helicase and adenosine triphosphatase activities of yeast Rad3 protein are inhibited by DNA damage. A potential mechanism for damage-specific recognition, *J. Biol. Chem.*, **267**, 392–398.
46. H. Naegeli, L. Bardwell, E.C. Friedberg (1993). Inhibition of Rad3 DNA helicase activity by DNA adducts and abasic sites: implications for the role of a DNA helicase in damage-specific incision of DNA, *Biochemistry*, **32**, 613–621.
47. N. Iyer, M.S. Reagan, K.J. Wu, B. Canagarajah, E.C. Friedberg (1996). Interactions involving the human RNA polymerase II transcription/nucleotide excision repair complex TFIIH, the nucleotide excision repair protein XPG, and Cockayne syndrome group B (CSB) protein, *Biochemistry*, **35**, 2157–2167.
48. R. Drapkin, J.T. Reardon, A. Ansari, J.C. Huang, L. Zawel, K. Ahn, A. Sancar, D. Reinberg (1994). Dual role of TFIIH in DNA excision repair and in transcription by RNA polymerase II, *Nature*, **368**, 769–772.
49. G.S. Winkler, K. Sugasawa, A.P. Eker, W.L. de Laat, J.H. Hoeijmakers (2001). Novel functional interactions between nucleotide excision DNA repair proteins influencing the enzymatic activities of TFIIH, XPG, and ERCC1- XPF, *Biochemistry*, **40**, 160–165.
50. E. Evans, J.G. Moggs, J.R. Hwang, J.M. Egly, R.D. Wood (1997). Mechanism of open complex and dual incision formation by human nucleotide excision repair factors, *EMBO J.*, **16**, 6559–6573.
51. D.L. Svoboda, J.S. Taylor, J.E. Hearst, A. Sancar (1993). DNA repair by eukaryotic nucleotide excision nuclease. Removal of thymine dimer and psoralen monoadduct by HeLa cell-free extract and of thymine dimer by Xenopus laevis oocytes, *J Biol. Chem.*, **268**, 1931–1936.
52. W.L. de Laat, E. Appeldoorn, K. Sugasawa, E. Weterings, N.G. Jaspers, J.H. Hoeijmakers (1998). DNA-binding polarity of human replication protein A positions nucleases in nucleotide excision repair, *Genes Dev.*, **12**, 2598–2609.
53. M. Wakasugi, A. Sancar (1998). Assembly, subunit composition, and footprint of human DNA repair excision nuclease, *Proc. Natl. Acad. Sci. USA*, **95**, (1998) 6669–6674.
54. A. Aboussekhra, R.D. Wood (1995). Detection of nucleotide excision repair incisions in human fibroblasts by immunostaining for PCNA, *Exp. Cell Res.*, **221**, 326–332.
55. M. Missura, T. Buterin, R. Hindges, U. Hubscher, J. Kasparkova, V. Brabec, H. Naegeli, Double-check probing of DNA bending and unwinding by XPA-RPA: an architectural function in DNA repair, *EMBO J.*, **20**, (2001) 3554–3564.
56. R. Hara, A. Sancar (2002). The SWI/SNF Chromatin-Remodeling Factor Stimulates Repair by Human Excision Nuclease in the Mononucleosome Core Particle, *Mol. Cell Biol.*, **22**, 6779–6787.
57. D. Wang, R. Hara, G. Singh, A. Sancar, S.J. Lippard (2003). Nucleotide excision repair from site-specifically platinum-modified nucleosomes, *Biochemistry*, **42**, 6747–6753.
58. C.M. Green, G. Almouzni (2002). When repair meets chromatin. First in series on chromatin dynamics, *EMBO Rep.*, **3**, 28–33.

59. M.E. Fitch, I.V. Cross, S.J. Turner, S. Adimoolam, C.X. Lin, K.G. Williams, J.M. Ford (2003). The DDB2 nucleotide excision repair gene product p48 enhances global genomic repair in p53 deficient human fibroblasts, *DNA Repair (Amst)*, **2**, 819–826.
60. V. Rapic-Otrin, M.P. McLenigan, D.C. Bisi, M. Gonzalez, A.S. Levine (2002). Sequential binding of UV DNA damage binding factor and degradation of the p48 subunit as early events after UV irradiation, *Nucleic Acids Res.*, **30**, 2588–2598.
61. B.J. Hwang, S. Toering, U. Francke, G. Chu (1998). p48 Activates a UV-damaged-DNA binding factor and is defective in xeroderma pigmentosum group E cells that lack binding activity, *Mol. Cell Biol.*, **18**, 4391–4399.
62. Datta, S. Bagchi, A. Nag, P. Shiyanov, G.R. Adami, T. Yoon, P. Raychaudhuri (2001). The p48 subunit of the damaged-DNA binding protein DDB associates with the CBP/p300 family of histone acetyltransferase, *Mutat. Res.*, **486**, 89–97.
63. M.J. Smerdon, S.Y. Lan, R.E. Calza, R. Reeves (1982). Sodium butyrate stimulates DNA repair in UV-irradiated normal and xeroderma pigmentosum human fibroblasts, *J Biol. Chem.*, **257**, 13441–13447.
64. R. Groisman, J. Polanowska, I. Kuraoka, J. Sawada, M. Saijo, R. Drapkin, A.F. Kisselev, K. Tanaka, Y. Nakatani (2003). The ubiquitin ligase activity in the DDB2 and CSA complexes is differentially regulated by the COP9 signalosome in response to DNA damage, *Cell*, **113**, 357–367.
65. Nag, T. Bondar, S. Shiv, P. Raychaudhuri (2001). The xeroderma pigmentosum group E gene product DDB2 is a specific target of cullin 4A in mammalian cells, *Mol. Cell Biol.*, **21**, 6738–6747.
66. M.L. Smith, J.M. Ford, M.C. Hollander, R.A. Bortnick, S.A. Amundson, Y.R. Seo, C.X. Deng, P.C. Hanawalt, A.J. Fornace, Jr. (2000). p53-mediated DNA repair responses to UV radiation: studies of mouse cells lacking p53, p21, and/or gadd45 genes, *Mol. Cell Biol.*, **20**, 3705–3714.
67. T. Maeda, A.N. Hanna, A.B. Sim, P.P. Chua, M.T. Chong, V.A. Tron (2002). GADD45 regulates G2/M arrest, DNA repair, and cell death in keratinocytes following ultraviolet exposure, *J. Invest. Dermatol.*, **119**, 22–26.
68. Y. Birger, K.L. West, Y.V. Postnikov, J.H. Lim, T. Furusawa, J.P. Wagner, C.S. Laufer, K.H. Kraemer, M. Bustin (2003). Chromosomal protein HMGN1 enhances the rate of DNA repair in chromatin, *EMBO J.*, **22**, 1665–1675.
69. E. Martinez, V.B. Palhan, A. Tjernberg, E.S. Lymar, A.M. Gamper, T.K. Kundu, B.T. Chait, R.G. Roeder (2001). Human STAGA complex is a chromatin-acetylating transcription coactivator that interacts with pre-mRNA splicing and DNA damage-binding factors in vivo, *Mol. Cell Biol.*, **21**, 6782–6795.
70. C.P. Rubbi, J. Milner (2003). p53 is a chromatin accessibility factor for nucleotide excision repair of DNA damage, *EMBO J.*, **22**, 975–986.
71. P.H. Gaillard, E.M. Martini, P.D. Kaufman, B. Stillman, E. Moustacchi, G. Almouzni (1996). Chromatin assembly coupled to DNA repair: a new role for chromatin assembly factor I, *Cell*, **86**, 887–896.
72. C.M. Green, G. Almouzni (2003). Local action of the chromatin assembly factor CAF-1 at sites of nucleotide excision repair in vivo, *EMBO J.*, **22**, 5163–5174.
73. J.G. Moggs, P. Grandi, J.P. Quivy, Z.O. Jonsson, U. Hubscher, P.B. Becker, G. Almouzni (2000). A CAF-1-PCNA-mediated chromatin assembly pathway triggered by sensing DNA damage, *Mol. Cell Biol.*, **20**, 1206–1218.
74. J. Moser, M. Volker, H. Kool, S. Alekseev, H. Vrieling, A. Yasui, A. A. van Zeeland, L.H. Mullenders (2005). The UV-damaged DNA binding protein mediates efficient targeting of the nucleotide excision the repair complex to UV-induced photoresins. *DNA repair (Amst.)* **4**, 571–582.

Chapter 10

Efficient repair of UV-induced DNA damage in terminally differentiated human keratinocytes

Dennis H. Oh and Kelvin Yeh

Table of contents

Abstract

Terminally differentiated cells do not generally repair their bulk DNA as efficiently as their undifferentiated precursors, presumably because mutagenic replication of damaged DNA is no longer an issue and because transcription-coupled DNA repair suffices in active genes. However, keratinocytes in the skin's epidermis are unusual in that they must terminally differentiate in a highly organized, multi-step sequence in both time and space despite being subject to repetitive damage from external genotoxic agents such as ultraviolet radiation (UVR). We hypothesized that keratinocytes must maintain repair of their global genome, even in the terminally differentiated state, to ensure successful epidermal maturation. Using an immunoblot assay, we measured global genomic repair of UVR-induced cyclobutane pyrimidine dimers (CPD) and pyrimidine(6-4)pyrimidone photoproducts (6-4 PP) in undifferentiated and differentiated cultured, normal human keratinocytes. Undifferentiated proliferating cells were grown in media with <0.07 mM Ca^{2+}, while terminal differentiation was induced by several methods, including placing cells in media containing 1.2 mM Ca^{2+} with or without phorbol ester. Terminal differentiation was confirmed by the lack of genomic incorporation of [^{3}H]-labeled thymidine and expression of characteristic differentiation markers. Following damage by 10 J/m^2 of 254 nm UVR, undifferentiated cells repaired both 6-4 PP and CPD in their global genomes over 24 hours as quickly and completely as observed in normal human fibroblasts. However, in contrast to some other terminally differentiated cell types, differentiated keratinocytes also repaired both 6-4 PP and CPD as efficiently as their undifferentiated precursors; only after prolonged periods under differentiating conditions did global repair eventually diminish. p53, a key regulator of global nucleotide excision repair of CPD, was induced following UVR in undifferentiated cells, but p53 levels appeared to be significantly decreased in the differentiated state. These results demonstrate that, unlike some other terminally differentiated cell types, differentiated keratinocytes efficiently repair UVR-induced DNA damage in their global genome, and they support a model in which normal keratinocyte differentiation requires maintenance of global genome repair of UVR-induced DNA damage.

10.1 Introduction

The epidermis of human skin has a variety of structural and physiological functions. One of its most important roles is to provide a competent barrier, both to maintain normal internal homeostasis by preventing excessive loss of fluid and nutrients and to protect the body from deleterious external physical, chemical and infectious agents [1,2]. The epidermal barrier forms as a result of a highly coordinated temporal and spatial program of terminal differentiation in keratinocytes [3]. Starting as proliferative cells in the stratum basale, keratinocytes stop dividing and migrate upwards, expressing new keratins as well as other structural proteins and enzymes, and undergo profound changes in shape as they flatten, ultimately becoming stratum corneum, a complex and compact mixture of acellular protein and lipids that forms the epidermal permeability barrier. Successful differentiation and barrier formation must occur despite repetitive stresses from the external environment, including genotoxic agents such as ultraviolet radiation (UVR). Although a great deal has been learned about the general process of DNA repair in mammalian cells, relatively little attention has been directed to repair and its regulation in differentiated keratinocytes. Yet these cells comprise the predominant cell type of the epidermis and bear the primary burden of UVR-induced injury. Determining how differentiating keratinocytes respond to ultraviolet injury may yield insights into how DNA repair is regulated in a dynamically changing cell and more generally will be important in understanding the overall photobiology of human skin.

Cultured keratinocytes can largely recapitulate their *in vivo* epidermal differentiation program in culture if the medium in which they are grown is supplemented with higher concentrations of Ca^{2+} [4,5]. This calcium switch is most dramatic in murine keratinocytes, but it is also present in human cells. Cultured human keratinocytes placed in higher concentrations of Ca^{2+} undergo major changes in gene expression, leading to production of markers characteristic of terminally differentiated epithelium, including early markers such as involucrin [5], as well as later markers such as transglutaminase [5], and loricrin [6]. Reminiscent of epidermal differentiation, changes in shape as well as intercellular bridges reflecting formation of desmosomes also occur in the presence of Ca^{2+}. Addition of other agents such as phorbol esters [7], vitamin D [8] and vitamin C [9] can further promote differentiation.

UVR-induced DNA damage, predominantly cyclobutane pyrimidine dimers (CPD) and pyrimidine(6-4)pyrimidone photoproducts (6-4 PP), is repaired by nucleotide excision repair (NER). NER is mediated by two distinct pathways – global genomic repair (GGR) operates on the bulk of the genome while transcription-coupled repair (TCR) operates on transcribed strands of active genes [10,11]. These pathways, with their separate mechanisms for detecting DNA damage, permit cells to use TCR to repair transcribed genes necessary for crucial functions following DNA damage, while relying on GGR to maintain the integrity of the entire genome so that subsequent progeny do not inherit mutations during replication. The necessity of GGR in differentiated cells that no longer replicate is therefore unclear, and a reasonable expectation is that such cells could dispense with GGR altogether while relying on TCR to repair genes that are necessary for the differentiated function. There is experimental evidence from several model differentiating systems that GGR is indeed reduced upon differentiation [12].

Published reports of DNA repair in cultured proliferating human keratinocytes have generally indicated that their repair efficiency and capacity are comparable to those generally observed in fibroblasts [13–15]. However, the few studies that have assessed DNA repair in differentiated human keratinocytes have come to variable conclusions [16–19]. Liu et al. separated cultured keratinocytes into basal and differentiating cells and observed that repair replication in the first two hours following UVR was decreased in differentiated cells [16]. These experiments did not distinguish between CPD and the 6-4 PP but probably predominantly detected

6-4 PP repair during the early post-damage period; later times more relevant to CPD repair were not examined. Mitchell et al., using a radioimmunoassay, observed that cultured keratinocytes removed 6-4 PP more efficiently than cells derived from whole mouse epidermis. These authors inferred that epidermis consists of a mixed population of undifferentiated and differentiated keratinocytes, with differentiated cells possessing a reduced repair capacity, while cultured, proliferating keratinocytes were skewed toward undifferentiated populations with normal repair activity. More recently, Li et al. have employed host cell reactivation of an exogenously damaged reporter plasmid, finding that NER of both CPD and 6-4 PP in proliferating murine keratinocytes is dependent on p53, but not in cells differentiated in high Ca^{2+} concentrations. They have also observed that human keratinocytes treated with 1 mM Ca^{2+} for 48 hours possessed significantly less host cell reactivation than those grown in low Ca^{2+} concentrations [19].

In contrast to reports that NER is diminished in differentiated keratinocytes, some workers have found differentiated cells repair as well or even better than their undifferentiated precursors. D'Ambrosio et al., employing a micrococcal endonuclease assay, determined the level of CPD in epidermis shaved from irradiated buttocks of individuals and observed almost complete removal in 24 hours [20]. Eggset et al. used antibodies raised against CPD and observed loss of antibody sites throughout the epidermis in sections taken from biopsies of irradiated human skin [17]. Though it appeared that cells in the upper zones of the epidermis lost antibody sites more slowly than those in the lower epidermis, the larger doses received by the upper epidermis as well as the non-quantitative nature of the method prohibited definitive conclusion, and it was reported that all sites appeared to be repaired by 24 hours. In another study, using cultured mouse keratinocytes induced to differentiate in high Ca^{2+} concentrations, Tokura et al. surprisingly found that psoralen adducts were actually removed more efficiently in differentiated murine keratinocytes [21]. Moreover, retinoids which maintain the undifferentiated state in cultured keratinocytes impeded the ability of keratinocytes to remove psoralen photoadducts. Most recently, Bernerd et al. have observed that skin equivalents reconstituted from normal keratinocytes displayed significant repair of CPD throughout the epidermis by 24–96 hours following UVR [22].

The variability in these results may be due to several factors, including the use of cultured cells versus intact epidermis, the use of broadband ultraviolet B versus germicidal lamp irradiation sources, and differences in assays for DNA damage and NER activity. Other differences may also exist in differentiated keratinocytes derived from rodents and humans. Hypothesizing that maintenance of GGR is important even in differentiated keratinocytes that must survive repetitive DNA damage, and in an attempt to clarify the response of these cells to UVR-induced DNA damage, we have measured NER activity in cultured undifferentiated and differentiated human keratinocytes using an immunoblot assay. Our results indicate that, while GGR does eventually diminish, it persists for significant periods following the appearance of the terminally differentiated keratinocyte phenotype.

10.2 Methods

Cells: Normal human keratinocytes were isolated from neonatal foreskins using standard techniques [23], and cultured in serum-free medium (Medium 154CF with Human Keratinocyte Growth Supplement, Cascade Biologics, Portland, OR) containing 0.07 mM Ca^{2+}. Keratinocytes were differentiated by either 1) allowing cells to grow to confluence and switching to medium containing 1.2 mM Ca^{2+} for at least 6 days, or 2) treating subconfluent cells with 1.2 mM Ca^{2+} and 50 nM 12-*O*-tetradecanoylphorbol-13-acetate (TPA, Sigma, St. Louis, MO) for at least 6 days.

Tritiated thymidine incorporation: To determine replication activity, undifferentiated or differentiated cells were incubated in the presence of 0.07 mM Ca^{2+} medium in the presence of [Methyl-^{3}H]-thymidine (Amersham, Piscataway, NJ) [24]. At various times, cells were lysed, and DNA was purified by phenol extraction, quantified, and fixed onto nitrocellulose, and activity was quantified by scintillation counting. Specific activity was then computed.

Western immunoblotting: Cells were harvested by trypsinization and gently pelleted by centrifugation. Protein was extracted by incubation with lysis buffer (50 mM HEPES, pH 7.4, 1% triton X-100, 0.1% SDS, 150 mM NaCl, 1 mM EDTA) supplemented with 1 μg/ml each of leupeptin, pepstatin, and aprotinin, and 20 μg/ml PMSF. Protein was quantified using the bicinchoninic acid method (Pierce Biotechnology, Rockford, IL). Extracts were separated on 10% SDS-PAGE gels, and electroeluted onto nitrocellulose membranes. Membrane blots were incubated with primary antibodies for 2 hours, followed by secondary antibodies conjugated to horseradish peroxidase for 2 hours and detected by enhanced chemiluminescence (ECL, Amersham, Piscataway, NJ). A monoclonal antibody (B.C1) specific for keratinocyte transglutaminase was a gift from Professor R. Rice (University of California, Davis [25]). Monoclonal antibodies were obtained commercially for p53 (DO-1, Santa Cruz Biotechnology, Santa Cruz, CA) and involucrin (SY5, Neomarkers, Union City, CA).

Global genomic repair immunoassay: Cells were initially prelabeled with [Methyl-^{3}H]-thymidine, washed with phosphate-buffered saline, and irradiated from above with a germicidal lamp emitting predominantly at 254 nm with fluence of 0.017 mW/cm^{2}. DNA was prepared either by trypsinization and gentle centrifugation of cells followed by DNA purification using spin columns (DNeasy, Qiagen, Union City, CA), or by direct lysis of cells in 0.5% SDS in TE buffer supplemented with 100 μg/ml Proteinase K, followed by phenol extraction and ethanol precipitation. DNA was quantified fluorometrically using Hoechst 33258 [26] and damage was detected with an immunoblot assay using monoclonal antibodies against CPD (TDM-2) and 6-4 PP (6-4M-2) that were gifts from Professor Toshio Mori (Nara Medical University, Nara, Japan [27]), as previously described [28]. The resulting blots were quantitated using a phosphorimager. The linearity of the assay for both 6-4 PP and CPD was confirmed over a range of DNA damage produced by 0–10 J/m^{2} 254 nm UVR (data not shown).

10.3 Results

10.3.1 Human keratinocytes terminally differentiate in culture

Human keratinocytes grown in 0.07 mM Ca^{2+} continue to proliferate, as demonstrated by their exponential incorporation of tritiated thymidine (Figure 1). In contrast, keratinocytes differentiated for 6 days in either 1.2 mM Ca^{2+} at confluence or 1.2 mM Ca^{2+} supplemented with TPA in subconfluent culture stop dividing and do not incorporate significant amounts of tritiated thymidine. Under these differentiating conditions, cells expressed the differentiation markers, involucrin and epidermal transglutaminase, whereas undifferentiated precursors did not (Figure 2). In comparison to 1.2 mM Ca^{2+} alone, the addition of TPA arrested replication of keratinocytes more quickly (data not shown) as well as enhanced the expression of involucrin and transglutaminase. These results confirm that terminal differentiation can be induced in human keratinocytes under these culture conditions.

10.3.2 Global repair persists in differentiated human keratinocytes

Using a slot blot immunoassay with monoclonal antibodies to CPD and 6-4 PP, we measured GGR in cultured human keratinocytes that were either proliferating and undifferentiated or that were terminal and differentiated (Figures 3 and 4). Following damage by 10 J/m^2 254 nm irradiation, undifferentiated cells repaired almost all 6-4 PP within the first 2 hours and approximately 50% of CPD by

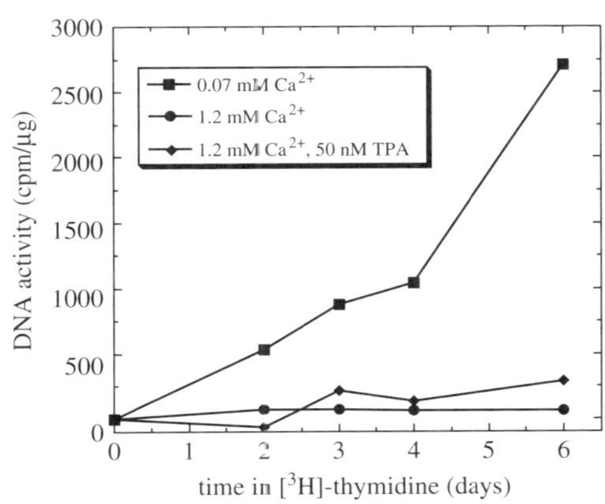

Figure 1. [^{3}H]-thymidine incorporation in human keratinocytes. [Methyl-^{3}H]-thymidine was added to cells that had been incubated in 0.07 mM Ca^{2+}, or differentiated in 1.2 mM Ca^{2+} with or without TPA. Genomic DNA was purified, quantitated, slot-blotted onto nitrocellulose membranes and specific activity was computed following liquid scintigraphy.

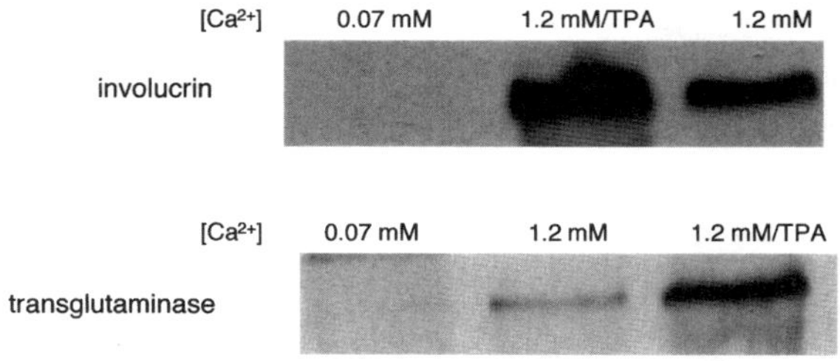

Figure 2. Induction of terminal differentiation markers in human keratinocytes. Cells were allowed to proliferate undifferentiated in 0.07 mM Ca^{2+}, or differentiated for 6 days in 1.2 mM Ca^{2+} with or without the phorbol ester, TPA. Whole cell protein extracts were then prepared, and protein was quantified. 75 μg/lane of total protein was separated by denaturing SDS-polyacrylamide gel electrophoresis and analyzed by Western blotting, using monoclonal antibodies to involucrin or keratinocyte transglutaminase.

24 hours. Interestingly, differentiated cells also continued to repair both 6-4 PP and CPD comparably. Only after significant times – 20 days for keratinocytes in 1.2 mM Ca^{2+} and 12 days for cells in 1.2 mM Ca^{2+} and TPA – after terminal differentiation appeared did GGR significantly decrease. At these times, repair of 6–4 PP was reduced to approximately 50% of undifferentiated cells, while repair of CPD was more severely affected, with little repair by 24 hours.

10.3.3 p53 levels are reduced in differentiated keratinocytes

To investigate whether p53 levels correlated with repair activity, whole cell extracts were analyzed by Western immunoblotting (Figure 5). In undifferentiated

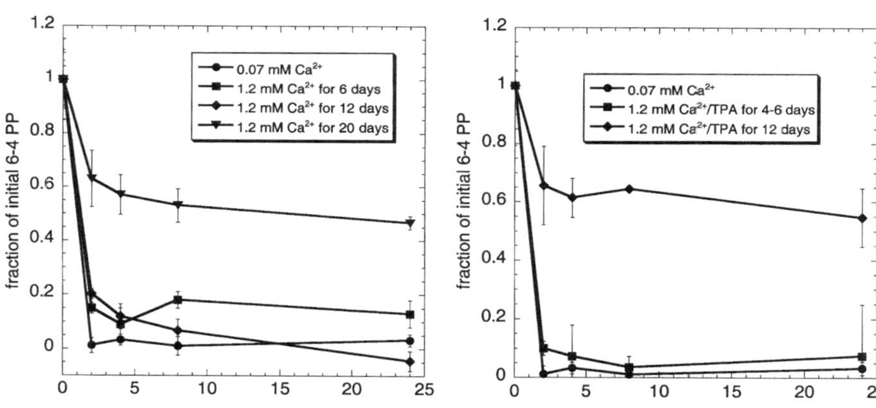

Figure 3. Global genomic repair of pyrimidine(6-4)pyrimidone photoproducts. Human keratinocytes were allowed to grow undifferentiated, or were differentiated for varying lengths of time in either 1.2 mM Ca^{2+} with or without TPA. Following 10 J/m^2 irradiation at 254 nm, genomic DNA was purified and analyzed for 6-4 PP using an immunoblot assay.

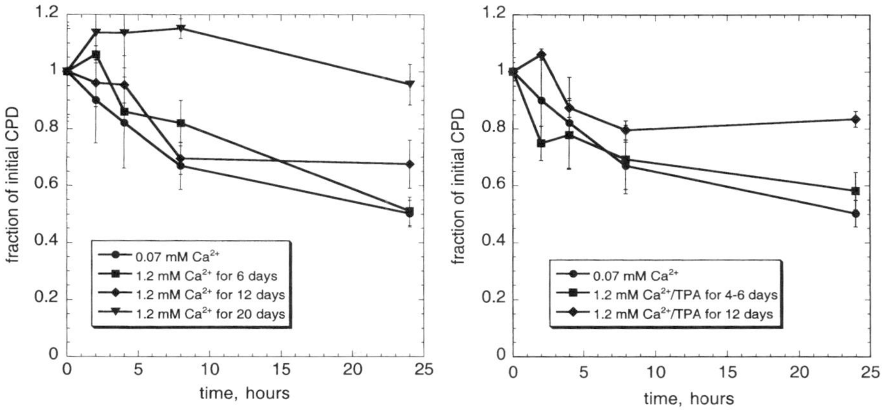

Figure 4. Global genomic repair of cyclobutane pyrimidine dimers. Human keratinocytes were allowed to grow undifferentiated, or were differentiated for varying lengths of time in either 1.2 mM Ca^{2+} with or without TPA. Following 10 J/m^2 irradiation at 254 nm, genomic DNA was purified and analyzed for CPD using an immunoblot assay.

cells, p53 was induced four-fold by 6 hours after UV and these levels persisted at 24 hours. Cells differentiated in 1.2 mM Ca^{2+} and TPA for 6 days showed similar basal levels but a blunted induction of p53. In contrast, cells differentiated in 1.2 mM Ca^{2+} for 6 days showed nearly undetectable basal p53 levels and no induction following UVR. Because p53 levels might artifactually appear reduced

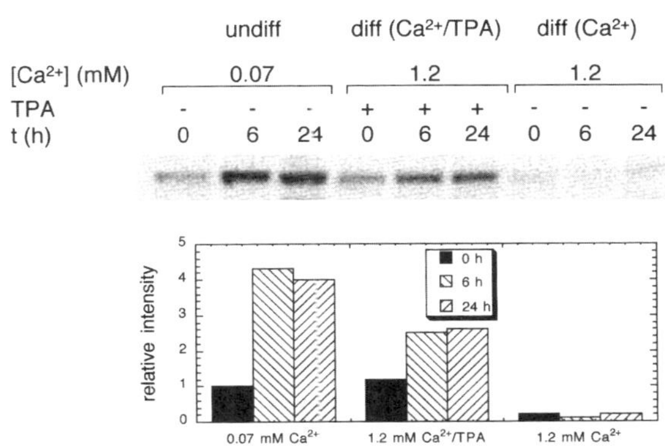

Figure 5. p53 expression in keratinocytes following ultraviolet radiation. Human keratinocytes were allowed to grow undifferentiated, or were differentiated for 6 days in either 1.2 mM Ca^{2+} with or without TPA. Following 10 J/m^2 irradiation at 254 nm, cells were harvested in parallel for both protein and DNA. Protein extracts were analyzed for p53 by Western blotting. To reduce the likelihood that differences between undifferentiated and differentiated cells could be attributed to changes in total protein, the amount of protein loaded in each lane was normalized to the same quantity of DNA, but was approximately 70–80 μg.

due to an increase in total cellular protein accompanying the differentiation process, these experiments were performed both by loading equivalent amounts of total protein (data not shown) and by loading protein normalized to equivalent amounts of cellular DNA (data shown in Figure 5). Both approaches produced similar results.

10.4 Discussion

Human skin possesses a number of acute responses to UVR-induced injury, such as DNA repair, apoptosis, increased pigmentation, cytokine release and immune modulation, as well as epidermal thickening. The contribution of different cells of the skin to each of these responses is unclear, although by their sheer numbers, the response of terminally differentiated keratinocytes would seem to be important. We have shown that human keratinocytes terminally differentiated in culture continue to possess the global NER activity of their undifferentiated precursors well after they have stopped replicating and expressed markers of differentiation characteristic of keratinocytes in the spinous layer. Only after significant periods of time in differentiating conditions did keratinocytes display reduced GGR. This reduced GGR activity was most pronounced for CPD, as even 6-4 PP were repaired to 50% within several hours in cells differentiated for prolonged times. We employed two different methods of inducing differentiation, and while both produced qualitatively comparable results, addition of the phorbol ester TPA to 1.2 mM Ca^{2+}-containing media accentuated both differentiation as well as the eventual loss of repair. However, the absence of an immediate effect of TPA on the kinetics of repair of either 6-4 PP or CPD suggests that the protein kinase C pathway or AP-1 modulated genes are unlikely to play a role in directly governing GGR in differentiated keratinocytes.

To the extent that cultured keratinocytes differentiated in culture recapitulate the epidermal differentiation program, these results may serve to reconcile prior experiments that have provided evidence for either diminished [16,18,29] or efficient [17,20] NER in differentiated keratinocytes. Keratinocytes are the predominant cell of the epidermis, a stratified squamous epithelium, and must change their gene expression profiles even in the post-mitotic state as they migrate from the lower basal to upper granular layer over approximately two weeks before becoming anucleate in the stratum corneum [3]. Under repetitive UVR, a differentiating keratinocyte of the epidermis may need to repair not only genes that are necessary for its current layer-specific state, but also genes that will be necessary for its future roles in the spinous and granular layers. It is unclear if TCR functions in differentiated keratinocytes, although there is evidence that NER does selectively operate in expressed genes in other types of differentiated cells [30,31]. If present in differentiated keratinocytes, TCR may be able to repair newly activated genes that have accumulated mild to intermediate quantities of damage while inactive in lower layers of epidermis. However, if excessive damage has accumulated due to prior frequent or intense UVR, TCR may be insufficient, and apoptosis may be triggered.

While apoptosis appears to be an important mechanism for dealing with extensively sun-damaged keratinocytes in skin [32], it may be detrimental to epidermal barrier formation for a significant number of keratinocytes at any level of the epidermis to be routed to that path. For example, the disease xeroderma pigmentosum in which GGR is defective is characterized in part by aberrant epidermal differentiation and barrier formation. Indeed, the original description of this disease focused not on the early predisposition to skin cancers for which it has become so well known, but rather on the "parchment-like dryness" and "thinness" of sun-exposed skin, and the disease was actually classified as an atrophic disorder [33]. Therefore, keratinocytes may need to maintain GGR even when differentiated until the most of the genes necessary for epidermal barrier formation have been expressed. Consistent with this model, recent elegant experiments with skin equivalents reconstituted with xeroderma pigmentosum complementation group C keratinocytes have shown reductions and aberrant distributions in several markers of differentiation in the epidermis in the absence of exogenous DNA damage [22], suggesting that an even more fundamental link may exist between NER and keratinocyte differentiation.

It is currently unclear if persistent NER activity in the differentiated keratinocyte is passively acquired from the undifferentiated precursor cell or is actively maintained. One possible explanation for the presence of repair in the differentiated state is that transcription of NER genes ceases upon terminal differentiation, but that NER mRNA and proteins synthesized in undifferentiated keratinocytes persist long enough to function in the differentiated state. However, it is notable that XPA protein injected into repair-deficient xeroderma pigmentosum fibroblasts lacking XPA had a half-life of only 14 hours, and would be unlikely to function after days or weeks [34]. If NER is actively maintained in differentiated keratinocytes, however, it is equally unclear what cellular mechanisms are responsible. In fibroblasts as well as proliferating, undifferentiated keratinocytes, GGR of CPD and other bulky adducts appears to depend on p53 levels [28,29,35,36]. Interestingly, it has been reported that NER of ultraviolet B radiation-induced damage in murine keratinocytes, while dependent on p53 activity in the undifferentiated state, decreased during differentiation independent of p53 [29]. These results may not be directly comparable to those of the current study due to potential differences between murine and human keratinocytes, between the use of host-cell reactivation and immunoblot assays for repair, and possibly between ultraviolet B and C radiation. Both our current results as well as those published previously for murine keratinocytes indicate that p53 expression is decreased, to varying degrees, sometimes substantially in differentiated keratinocytes [37]. The decreased levels of p53 in the current study were detected while concurrent GGR activity was essentially as efficient as undifferentiated cells. In murine keratinocytes, it was observed that p53, while present in lower levels, did possess an enhanced ability to transactivate transcription of a reporter plasmid transfected into differentiated cells [37], and if true in human cells could possibly compensate for the apparent dissociation between levels of p53 and GGR activity. Alternatively, although the functional significance is unclear, it has been reported that p53 exists as alternatively spliced forms in undifferentiated and differentiated keratinocytes [38].

Acknowledgments

We thank Professors Toshio Mori and Robert Rice for generously providing monoclonal antibodies, Sally Pennypacker for expert cell culture support, and Professors Philip Hanawalt, Corey Largman, Peter Elias, and Walter Holleran and Dr. Thierry Nouspikel for early helpful discussions. This work was supported by grants from the Dermatology Foundation, the Hampton Medical Research Fund of the University of California at San Francisco, and the Merit Review Entry Program of the Medical Research Service, Department of Veterans Affairs.

References

1. P.M. Elias, K.R. Feingold (1999). Skin as an organ of protection. In: I.M. Freedberg, et al. (Eds), *Dermatology in General Medicine* (pp. 164–174). McGraw-Hill, New York.
2. Z. Nemes, P.M. Steinert (1999). Bricks and mortar of the epidermal barrier. *Exp. Mol. Med.*, **31**, 5–19.
3. E. Fuchs, S. Raghavan (2002). Getting under the skin of epidermal morphogenesis. *Nat. Rev. Genet.*, **3**, 199–209.
4. H. Hennings, et al. (1980). Calcium regulation of growth and differentiation of mouse epidermal cells in culture. *Cell*, **19**, 245–254.
5. S. Pillai, et al. (1988). Biochemical and morphological characterization of growth and differentiation of normal human neonatal keratinocytes in a serum-free medium. *J. Cell. Physiol.*, **134**, 229–237.
6. D. Hohl, et al. (1991). Transcription of the human loricrin gene *in vitro* is induced by calcium and cell density and suppressed by retinoic acid. *J. Invest. Dermatol.*, **96**, 414–418.
7. D.C. Ng, et al. (2000). Requirement of an AP-1 site in the calcium response regin of the involucrin promoter. *J. Biol. Chem.*, **275**, 24080–24088.
8. D.D. Bikle, et al. (2001). Calcium- and vitamin D-regulated keratinocyte differentiation. *Mol. and Cell. Endocrinology*, **177**, 161–171.
9. Y. Uchida, et al. (1999). Vitamin C stimulates sphingolipid synthesis in submerged human keratinocyte cultures. *J. Invest. Dermatol.*, **112**, 578.
10. V.A. Bohr, et al. (1985). DNA repair in an active gene: Removal of pyrimidine dimers from the DHFR gene of CHO cells is much more efficient than in the genome overall. *Cell*, **40**, 359–369.
11. I. Mellon, G. Spivak, P.C. Hanawalt (1987). Selective removal of transcription-blocking DNA damage from the transcribed strand of the mammalian DHFR gene. *Cell*, **51**, 241–249.
12. T. Nouspikel, P.C. Hanawalt (2002). DNA repair in terminally differentiated cells. *DNA Repair*, **1**, 59–75.
13. L.B. Taichman, R.B. Setlow (1979). Repair of ultraviolet light damage to the DNA of cultured human epidermal keratinocytes and fibroblasts. *J. Invest. Dermatol.*, **73**, 217–219.
14. M.K. Reusch, et al. (1988). Comparative removal of pyrimidine dimers from human epidermal keratinocytes *in vivo* and *in vitro*. *J. Invest. Dermatol.*, **91**, 349–352.

15. A.A. Schothorst, et al. (1991). Pyrimidine dimer induction and repair in cultured human skin keratinocytes or melanocytes after irradiation with monochromatic ultraviolet radiation. *J. Invest. Dermatol.*, **96**, 916–920.
16. S.-C. Liu, S. Parsons, P.C. Hanawalt (1983). DNA repair in cultured keratinocytes. *J. Invest. Dermatol.*, **81**, 179s-183s.
17. G. Eggset, G. Volden, H. Krokan (1983). U.V.-induced DNA damage and its repair in human skin *in vivo* studied by sensitive immunohistochemical methods. *Carcinogenesis*, **4**, 745–750.
18. D.L. Mitchell, J.E. Cleaver, J.H. Epstein (1990). Repair of pyrimidine(6-4)pyrimidone photoproducts in mouse skin. *J. Invest. Dermatol.*, **95**, 55–59.
19. G. Li, et al. (1996). Decreased DNA repair but normal apoptosis in ultraviolet-irradiated skin of p53 transgenic mice. *Am. J. Pathol.*, **148**, 1113–1123.
20. S.M. D'Ambrosio, et al. (1981). Excision repair of UV-induced pyrimidine dimers in human skin *in vivo. J. Invest. Dermatol.*, **77**, 311–313.
21. Y. Tokura, R.L. Edelson, F.P. Gasparro (1991). Formation and removal of 8-MOP-DNA photoadducts in keratinocytes: Effects of calcium concentration and retinoids. *J. Invest. Dermatol.*, **96**, 942–949.
22. F. Bernerd, et al. (2001). Clues to epidermal cancer proneness revealed by reconstruction of DNA repair-deficient xeroderma pigmentosum skin *in vitro. Proc. Natl. Acad. Sci.USA*, **98**, 7817–7822.
23. M.R. Pittelkow, R.E. Scott (1986). New techniques for the *in vitro* culture of human skin keratinocytes and perspectives on their use for grafting of patients with extensive burns. *J. Invest. Dermatol.*, **61**, 771–777.
24. G. Spivak, P.C. Hanawalt (1995). Determination of damage and repair in specific DNA sequences. *Methods: A companion to Methods in Enzymology*, **7**, 147–161.
25. S.M. Thacher, R.H. Rice (1985). Keratinocyte-specific transglutaminase of cultured human epidermal cells: Relation to cross-linked envelope formation and terminal differentiation. *Cell*, **40**, 685–695.
26. J. Sambrook, D.W. Russel (2001). Fluorometric quantitation of DNA using Hoechst 33258. In: *Molecular Cloning: A Laboratory Manual* (pp. A8.22–23). Cold Spring Harbor Laboratory Press, Cold Spring Harbor.
27. T. Mori, et al. (1991). Simultaneous establishment of monoclonal antibodies specific for either cyclobutane pyrimidine dimer or (6-4) photoproduct from the same mouse immunized with ultraviolet-irradiated DNA. *Photochem. Photobiol.*, **54**, 225–232.
28. J.M. Ford, P.C. Hanawalt (1997). Expression of wild-type p53 is required for efficient global genomic nucleotide excision repair in UV-irradiated human fibroblasts. *J. Biol. Chem.*, **272**, 28078–28080.
29. G. Li, et al. (1997). Differentiation-dependent p53 regulation of nucleotide excision repair in keratinocytes. *Am. J. Pathol.*, **150**, 1457–1764.
30. A.L. Islas, P.C. Hanawalt (1995). DNA repair in the MYC and FMS proto-oncogenes in ultraviolet light-irradiated human HL60 promyelocytic cells during differentiation. *Cancer Res.*, **55**, 336–341.
31. T. Nouspikel, P.C. Hanawalt (2000). Terminally differentiated human neurons repair transcribed genes but display attenuated global DNA repair and modulation of repair gene expression. *Mol. Cell. Biol.*, **20**, 1562–1570.
32. L.L. Hill, et al. (1999). Fas ligand: a sensor for DNA damage critical in skin cancer etiology. *Science*, **285**, 898–900.
33. F. Hebra, M. Kaposi (1874). On diseases of the skin, including exanthemata. *The New Sydenham Society*, **61**, 252–258.

34. M. Yamaizumi, et al. (1986). Microinjection of partially purified protein factor restores DNA damage specifically in group A of xeroderma pigmentosum cells. *Proc. Natl. Acad. Sci.*, **83**, 1476–1479.
35. D.R. Lloyd, P.C. Hanawalt (2000). p53-dependent global genomic repair of benzo[a]pyrene-7,8-diol-9,10-epoxide adducts in human cells. *Cancer Res.*, **60**, 517–521.
36. T. Maeda, et al. (1999). Enhanced repair of benzo(a)pyrene-induced DNA damage in human cells treated with thymidine dinucleotides. *Mutat. Res.*, **433**, 137–145.
37. W.C. Weinberg, et al. (1995). p53-mediated transcriptional activity increases in differentiating epidermal keratinocytes in association with decreased p53 protein. *Oncogene*, **10**, 2271–2279.
38. P.A. Rehberger, et al. (1997). Differential expression of the regularly spliced wild-type p53 and its COOH-terminal alternatively spliced form during epidermal differentiation. *Cell Growth Differ.*, **8**, 851–860.

Chapter 11

Reactivation of UV-damaged viruses and reporter genes in mammalian cells

Andrew J. Rainbow, Photini Pitsikas, Colleen Caney, Ihor P. Boszko, Bruce C. McKay and Murray A. Francis

Table of contents

Abstract

UV-damaged viruses and reporter genes have been used to examine the DNA repair capacity of mammalian cells. In particular, the non-replicating recombinant adenovirus, Ad5HCMVspllacZ, which expresses the β-galactosidase reporter gene, has been used to examine both constitutive and inducible repair of UV-damaged DNA in human cells. Prior UV-irradiation of normal human fibroblasts with low UV fluences resulted in a transient enhancement of host cell reactivation (HCR) for expression of the UV-damaged reporter gene. Using a quantitative PCR technique we have reported that similar prior low UV fluences to normal fibroblasts results in an enhanced rate of removal of photoproducts from the UV-damaged reporter gene indicating the enhanced HCR for reporter gene expression results from UV-induced DNA repair. The UV-enhanced HCR in normal human fibroblasts involves p53, one or more of the pRb family of proteins and TCR. We propose a model for UV-inducible DNA repair that includes a p53 dependent transcriptional activation of one or more genes involved in nucleotide excision repair (NER). Since a UV-induced p53-dependent transcriptional upregulation requires the removal of UV-induced transcription terminating lesions from the upregulated cellular gene, our model is consistent with a dependence of UV-inducible DNA repair on the TCR pathway of NER.

11.1 Introduction

There are many early reports showing that pretreatment of a variety of different mammalian cells with chemical or physical DNA damaging agents results in an increased survival (or enhanced host cell reactivation (HCR)) for several nuclear replicating double-stranded DNA viruses damaged by UV or ionising radiation (for a review see references [1,2]). In analogy with the phenomenon of Weigle reactivation for bacteriophage and the SOS response in bacteria, it has been suggested that the enhanced reactivation of DNA-damaged viruses results, in part at least, from an induced DNA repair pathway in mammalian cells [3–7]. More recently several investigators have used the expression of UV-irradiated reporter genes to examine HCR of damaged DNA in mammalian cells [8–15].

In this chapter we will describe results obtained using the host cell reactivation of UV-damaged reporter genes to examine both constitutive and inducible repair of UV-damaged DNA in human cells.

11.2 Nucleotide excision repair of damaged DNA

The integrity of the human genome is constantly being compromised by alterations induced by a wide variety of exogenous physical and chemical agents, as well as by products of cellular metabolism. Several highly conserved repair pathways have evolved to remove damage from cellular DNA and disruption of each of these DNA repair pathways is associated with carcinogenesis (as reviewed in reference [16]).

The nucleotide excision repair (NER) pathway repairs a wide range of bulky DNA adducts induced by numerous carcinogenic and antineoplastic compounds, including ultraviolet (UV) irradiation from the sun. NER can be divided into two interrelated subpathways: (1) transcription coupled repair (TCR) which preferentially removes DNA damage at a faster rate from the transcribed strand of actively transcribed genes, and (2) global genomic repair (GGR) which removes damage from throughout the entire genome and from the non-transcribed strand of active genes [17,18] For a review on the genomic heterogeneity of NER see [19]. Individuals with the genetic diseases xeroderma pigmentosum (XP) have some deficiency in NER and show an increased incidence of a variety of skin cancers [20,21]. Several additional links between NER and carcinogenesis have been reported. Mutations in the p53 tumour suppressor gene, the most commonly altered gene in malignancy [22], have been shown to result in reduced NER [12,23–27]. Decreased NER has also been observed in a variety of tumour and transformed cell lines [28–30]. These reports suggest that DNA repair mechanisms are disrupted in tumour cells and that DNA repair contributes to resistance to neoplasia.

XP is composed of a minimum of seven complementation groups (XP-A through XP-G), each displaying a general deficiency in NER which compromises at least the GGR subpathway and usually TCR as well. The exception is XP-C, which retains viable TCR in spite of a severe deficiency in GGR [31,32]. The genetic disease Cockayne syndrome (CS) is also associated with a deficiency in NER, although unlike XP patients, CS patients do not display an increased risk of skin cancer. Two complementation groups of CS (CS-A and CS-B) have been identified, each of which exhibits a deficiency in TCR, while the GGR pathway appears to function normally [33,34]. The XPB and XPD proteins are components of the transcription factor TFIIH [35,36] which plays a role in both NER and transcription by RNA polymerase II (RNAPII) [37]. The CSA and CSB proteins co-immunoprecipitate [38] and are required for TCR. CSA interacts with TFIIH directly [38] whereas CSB does so via XPG [39]. TFIIH, XPG and the RPA/XPA/XPF/ERCC1 complex [40,41] are required for both subpathways of NER, whereas the XPC/HHR23B complex [42] appears to be required only for GGR [32,33,43].

11.3 Host cell reactivation of DNA-damaged reporter genes in mammalian cells

Host cell reactivation (HCR) of reporter gene activity has been assessed in mammalian cells using either recombinant adenoviruses or plasmid constructs by a number of different laboratories.

11.3.1 Recombinant adenovirus-based reporter genes

Recombinant non-replicating adenovirus constructs have been used to introduce UV-damaged [11–14] or cisplatin-damaged [44] reporter genes into non-treated human and rodent cells in order to assess the repair of damaged DNA in the absence of cellular stress using host cell reactivation (HCR) of reporter gene activity as an endpoint. UV-induced lesions in the template strand of active genes inhibit progression of RNA polymerase II [45] and a single UV-induced cyclobutane pyrimidine dimer (CPD) is thought to be sufficient to inhibit reporter gene expression [13,46]. Thus HCR of reporter gene activity is thought to require the repair of transcription blocking DNA lesions and reflect repair of DNA lesions in the transcribed strand.

The level of reporter gene expression in cells infected by adenovirus vectors is greatly influenced by the promoter controlling expression of the transgene. The CMV immediate early (IE) promoter is one of the most commonly used promoters in eukaryotic expression vectors, due primarily to its ability to yield high expression levels in many different mammalian cell types. Xu and colleagues [47] have demonstrated that the human CMV immediate early promoter directs the highest level of expression in the widest variety of mammalian cell types *in vitro* when compared to that directed by the human β-actin, Ad major late, and SV40 early and late promoters. In addition, Addison and colleagues [48] showed that the murine CMV immediate early promoter [49] is also an extremely strong promoter and that it rivals the human CMV promoter for in *vitro* expression in both murine and human cells. *In vivo*, the highest level of expression reported to date occurred following intravenous delivery to the mouse of an Ad vector under the control of the human CMV promoter [50].

Ad5HCMVsp1*lacZ* is a recombinant non-replicating adenovirus containing the *lacZ* reporter gene under the control of the human cytomegalovirus (HCMV) immediate early promoter inserted into the deleted E1 region which can efficiently infect and express the reporter gene in human fibroblasts [51]. A typical protocol to examine HCR of reporter gene activity for UV-irradiated recombinant Ad expressing the *lacZ* gene is shown in Figure 1. Using this approach we have shown that HCR of reporter gene activity for UV-damaged DNA is reduced in several XP (groups A to G) and CS (groups A and B) cell strains, relative to that in normal diploid human fibroblasts [11,13] indicating that both TCR and GGR contribute to expression of the UV-damaged reporter gene in human cells (Figure 2 and Table 1). Fibroblasts from patients with xeroderma pigmentosum from complementation group C (XP-C) showed HCR levels ranging from 0.25–0.75 of the average HCR value obtained for the NER-proficient normal primary human fibroblasts used in the experiments (Figure 2 and Table 1 [11,13]). The result for XP-C is surprising since XP-C cells are reported to be proficient in transcription coupled repair (TCR) following UV exposure and thus would be expected to reactivate UV-induced lesions in the transcribed strand of the reporter gene to levels obtained in normal fibroblasts. Blockage of RNA polymerase II by UV-induced DNA lesions in the reporter gene does not appear to be sufficient to promote the preferential repair of these transcription blocking lesions from the reporter gene in non-UV-treated XP-C cells [11,13]. Since the adenovirus genome is approximately five orders of magnitude smaller than the human genome, the number of lesions introduced into cells with a UV-irradiated virus in our experiments is minimal compared to the number introduced into the host cell genome following UV treatments used to examine

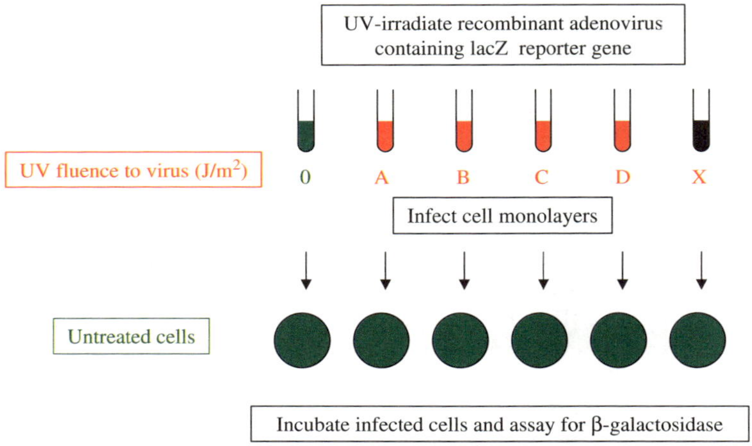

Figure 1. Host cell reactivation of a UV-damaged recombinant adenovirus based reporter gene in human cells. Cells were seeded in 96 well microtitre plates at a density of 2×10^4 cells/well and 18–24 h later infected for 90 min at 37°C with unirradiated or UV-irradiated Ad5HCMVsp1*lacZ* in a total volume of 40 μl. Ad5HCMVsp1*lacZ* was UV-irradiated with a range of fluences (A to D) using a germicidal lamp (General Electric model G8T5) emitting predominantly at 254 nm at an incident fluence rate of 2 $J/m^2/s$. 12–48 h later, infected cells were harvested and scored for β-galactosidase activity as reported previously [13]). Lysates from wells infected with heavily irradiated Ad5HCMVsp1*lacZ* (10,000 J/m^2), X, served as a measure of background levels for β-galactosidase activity. [Adapted from reference 92.]

repair in cellular DNA. Also of interest was the finding that several of the CS fibroblast strains tested retained a considerable ability to repair the UV-damaged reporter gene in non-treated cells (Table 1) in spite of their being characterized as deficient in repair of the transcribed strand of active genes following UV irradiation of the cell [13]. We interpret these results to indicate that GGR contributes more strongly than TCR to the constitutive repair of the transcribed strand of the reporter gene in unirradiated mammalian cells.

11.3.2 Plasmid-based reporter genes

Several studies using plasmid-borne reporter genes have reported decreased HCR for expression of the UV-damaged reporter gene in NER-deficient XP cells [46,52–54] and CS cells [52,54] compared to repair-proficient human cells. It is assumed that the HCR for the expression of plasmid-based and recombinant adenovirus-based reporter genes reflects, in part at least, the NER pathways of the host cell which remove UV lesions from cellular DNA. However, since the DNA conformation and associated chromatin structure differ for plasmid, viral and cellular DNA, there may be some subtle differences in the removal of UV lesions from cellular DNA compared to plasmid and viral DNA. For plasmid constructs,

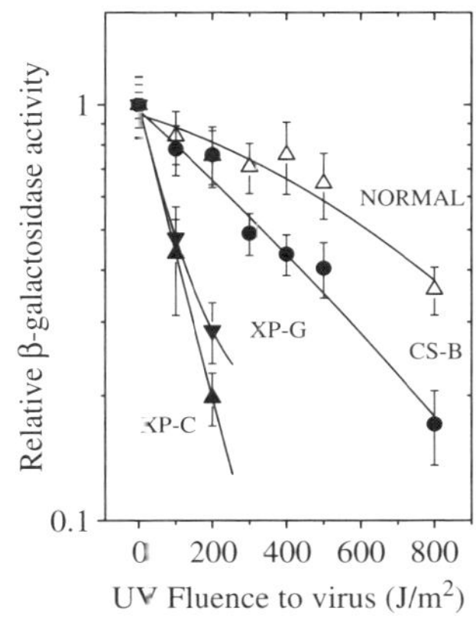

Figure 2. Nucleotide excision repair deficient cell strains show reduced HCR of UV-irradiated reporter activity. Shows results of typical experiments for repair proficient normal GM 3440 (Δ); and repair deficient XP-C (XP3BE (▲)), XP-G (XP2BI (▼)) and CS-B (CS1AN (●)) primary human fibroblasts. Using the protocol described in Figure 1, cells were infected with unirradiated or UV-irradiated virus at 10–20 plaque forming units per cell and scored for β-galactosidase activity 40–44 h later. Each point is the average of 4 replicates; error bars represent one standard error. D_{37} values were determined for each strain using the equation $\ln(SF) = -(\alpha X + \beta X^2 + \gamma)$, where X corresponds to the UV fluence in J/m^2. Average relative D_{37} values from multiple experiments are summarized in Table 1. [Adapted from reference [13] by permission of Oxford University Press.]

HCR studies have typically involved the transfection of a DNA-damaged plasmid carrying a reporter gene into the cells of interest [8,24,55–57]. Since primary human fibroblasts take up exogenous DNA 10–100 times less efficiently than either rodent or human cell lines derived from tumour tissue or transformed by viral antigens [58–60], most experiments examining the reactivation of plasmid-borne UV-damaged reporter genes have been performed in SV40-transformed repair-deficient XP and CS cell lines and repair proficient normal cell lines derived from human tumours rather than primary fibroblast strains. However, SV40-transformed cells and many human tumours have alterations in the p53 tumour suppressor, the pRb tumour suppressor and other stress-activated pathways, which have been shown to affect the GGR and/or TCR pathway of NER [12,25–27,61,62]. Furthermore, the cellular response to DNA damage is stimulated by at least some transfection procedures [63,64], leading to cell cycle arrest [63], suggesting that the transfection procedure itself may affect the outcome of DNA repair experiments. In contrast, recombinant non-replicating adenovirus reporter constructs have the ability to infect and express high levels of recombinant gene products in most cell types including primary human fibroblasts. Furthermore, non-replicating adenovirus reporter constructs do not appear to elicit the DNA damage response, or inhibit host DNA synthesis following infection [65]. The use of transformed cells and tumour cells in plasmid-based HCR assays compared to the use of primary fibroblasts in the recombinant adenovirus-based asssays, together with the differences in cellular response

Table 1. HCR of β-gal activity for UV-irradiated Ad5HCMVsp1*lacZ* in unirradiated cells

NIGMS designation	Cell Strain	Group	D_{37} (J/m^2)	Relative HCR[a] (no. determinants)
	Average	normal	1190±192	1.00
	423	normal	958±128	0.89±0.08 (10)
GM38		normal	1427±163	1.05±0.02 (3)
GM969		normal	1221±206	1.02±0.003 (19)
GM8399		normal	1139±114	1.02±0.03 (10)
GM3440		normal	1449±330	1.16±0.10 (5)
GM13026	XPCS2BA	XP-B	697±353	0.52±0.30 (2)
GM30a	XP3BE	XP-C	211±47	0.25±0.05[b] (12)
GM677	XP2BE	XP-C	293±87	0.41±0.07[b] (5)
GM10881	XP1BE	XP-C	1021±422	0.75±0.13[c] (3)
GM3615	XP1BR	XP-D	737±410	0.33±0.11[b] (8)
GM10430	XP6BE	XP-D	149±39	0.09±0.03[d] (2)
GM4313	XP2YO	XP-F	644±190	0.56±0.07[b] (4)
GM3021	XP2BI	XP-G	249±43	0.30±0.04[b] (10)
GM1856	CS3BE	CS-A	900±199	0.57±0.08[b] (7)
GM2965		CS-A	1121±193	0.90±0.15 (2)
GM739	CS1AN	CS-B	1090±249	0.79±0.11[d] (9)
GM1629	CS1BE	CS-B	1107±155	0.79±0.07[d] (5)

[a] Calculated by D_{37}(experimental line)/D_{37} (average normal controls) within individual experiments. The number of determinants for each cell strain is included in parenthesis.
[b] Significantly less than 1 by one-tailed t-test ($p<0.005$);
[c] $0.05<P<0.1$;
[d] Significantly less than 1 by one-tailed t-test ($p<0.05$). Adapted from reference [13] with permission from Oxford University Press.

to plasmid transfection compared to viral infection may lead to differences in the host mediated repair of plasmid compared to viral DNA. Therefore, a direct comparison of the relative contribution of TCR and GGR to repair in the transcribed strand of plasmid-based reporter genes using SV40-transformed cells and tumour cells with that obtained using an adenovirus-based reporter in primary human fibroblasts may not be appropriate.

11.3.3 Host mediated removal of photolesions from UV-damaged reporter genes

UV-induced DNA lesions are removed from both plasmid-borne [56,57] and recombinant adenovirus-borne [15] reporter genes when introduced into repair proficient human cells and the removal is reduced when the same reporter genes are introduced into NER-deficient cells. Using a quantitative PCR technique we have examined the host mediated removal of photolesions from an adenovirus based reporter gene. We show a significant removal of UV photoproducts from the *lacZ* reporter gene after infection of normal human fibroblasts with UV irradiated Ad5HCMVsp1*lacZ*, and a significantly diminished lesion removal following infection of NER deficient XP-C, XP-A and CS-B cells (Figure 3 [15]). This supports a

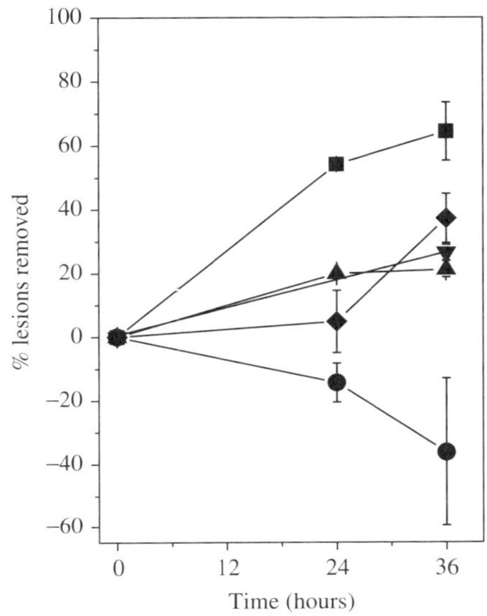

Figure 3. Removal of UV-induced lesions from the *lacZ* region of Ad5HCMVsp1*lacZ*. Cells were seeded in 24-well plates at 1×10^5 cells per well and 24 h later infected with either non-irradiated or UV-irradiated virus in a total volume of 200 μl at a multiplicity of infection of 50 plaque-forming units (PFU) per cell. Following viral adsorption for 60 min at 37°C, the viral suspension was aspirated and replaced with 1 ml of supplemented growth medium until cell lysis and DNA extraction. Samples of infected cells were collected at various times after infection. Primers flanking a 2.6 kb region of the *lacZ* reporter gene were added to equal amounts of DNA extracted from Ad5HCMVsp1lacZ infected cells and amplified by PCR using radiolabelled nucleotides as substrates. PCR product was measured by quantitation of incorporated radiolabelled deoxyribonucleotide and survival curves were obtained for the amount of PCR product following infection of normal, XP-C, XP-A and CS-B human fibroblast strains. By applying the principles of the Poisson distribution, the number of lesions induced or remaining was calculated for each UV dose to the virus at each time point [average # lesions/fragment = −ln (SF)]. The average lesion removal from the *lacZ* fragment was then plotted as a function of time for four (pooled) normal fibroblast lines (■), GM00030A (XP-C) (▲), GM00510 (XP-C) (▼), GM00082 (XP-A) (◆), and GM01428 (CS-B) (●). Data points represent the mean (± one standard error) of relative lesion removal for three doses (two PCR reactions/determination at each dose of 100, 200 & 300 J/m^2) from two or more experiments). [Adapted from reference [15] with permission from Kluwer Academic/Plenum Publishers, New York, NY.]

model in which the basal levels of lesion removal from this vector in unirradiated cells involve both the GGR and TCR pathways. The removal of UV-induced DNA lesions from a plasmid born reporter gene has also been reported for NER-proficient HT-1080 cells and at a reduced level for NER deficient SV40-transformed XP group A, XP group C and CS group B cells [56,57]. Thus HCR of reporter gene activity is thought to reflect the repair of transcription-blocking DNA lesions from the transcribed strand of the reporter gene.

11.4 Enhanced host cell reactivation of a UV-damaged reporter gene following pretreatment of mammalian cells with DNA damaging agents

Examination of DNA repair in mammalian cells generally requires that the cells are treated with a DNA damaging agent in some manner, which makes it difficult to determine if the repair pathways are constitutively active or induced by the DNA damaging agent. In contrast, the use of viral probes for DNA repair allows the virus and cell to be treated with a DNA damaging agent independently and thus allows an examination of both constitutive and inducible pathways affecting survival of the virus or expression of the reporter gene.

We and others have examined HCR of a UV-damaged reporter gene in pre-treated compared to non-treated cells [10,12,13,15]. A typical protocol for recombinant adenovirus-based reporter gene enhanced HCR experiments is shown in Figure 4. Using this approach we have reported that pre-treatment of normal human fibroblasts with low UV fluences (Figure 5a [12,13]) as well as heat shock [11,12] results in enhanced HCR of the UV-damaged reporter suggesting the presence of inducible DNA repair in human cells. More recent results from our laboratory showed that the UV-enhanced HCR in normal cells was transient reaching a maximum when cells were infected between 12 and 24 h after UV (and scored at 24 h after infection) and disappearing when cells were infected at 72 h after UV (Figure 6, [66]). Using a quantitative PCR technique we have reported that prior UV-irradiation of normal human fibroblasts results in an enhanced rate of removal of UV-induced photoproducts from the *lacZ* reporter gene in AdHCMVsp1*lacZ* following infection of human fibroblasts (Figure 7, [15]). We therefore suggest that the enhanced HCR for expression of UV-damaged reporter genes reported by us and others reflects the presence of one or more inducible DNA repair pathways in mammalian cells. Since both TCR and GGR contribute to expression of the UV-damaged reporter gene in untreated normal human fibroblasts, the enhanced HCR in pre-UV-irradiated cells could result from an upregulation of TCR and/or GGR in the transcribed strand of the reporter gene.

A UV-enhanced HCR response was also detected in transcription coupled repair (TCR) proficient xeroderma pigmentosum (XP) group C cells, but was absent or delayed in TCR deficient XP and CS cells (Figures 5 and 8). These results suggest that pre-UV treatment of cells results in an induced repair of UV-damaged DNA in the transcribed strand of the reporter gene in normal cells through an enhancement of TCR and/or through a mechanism that involves the TCR pathway. The UV-enhanced HCR response was also absent in Li-Fraumeni syndrome (LFS) cells (Figure 9, [12]), SV40-transformed human fibroblasts and normal human fibroblasts expressing the human papilloma virus (HPV) E6 or E7 gene [66]. Since SV40 transformation and HPV E7 expression abrogate the retinoblastoma (pRb) proteins, whereas HP V E6 abrogates p53 and LFS cells express mutant p53, these results indicate that the UV-induced DNA repair in normal human fibroblasts involves p53, one or more of the pRb family of proteins and TCR.

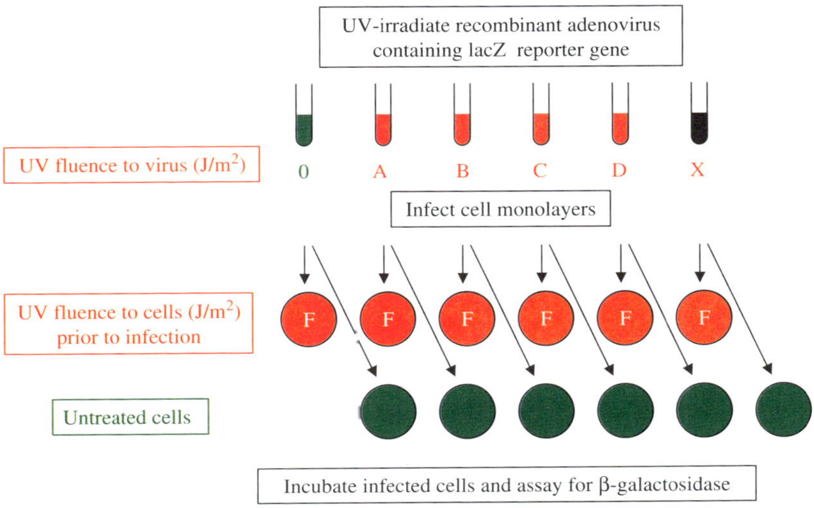

Figure 4. UV-enhanced host cell reactivation of a UV-damaged recombinant adenovirus based reporter gene in human cells. Cells were seeded in 96 well microtitre plates at a density of 2×10^4 cells/well, 18–24 h prior to UV-treatment of cells. The growth medium was then aspirated, replaced with 40 μl phosphate buffered saline (PBS) and sets of cell monolayers were either UV-irradiated with a fluence F or were mock-irradiated and received no UV treatment. At an appropriate time after treatment, both UV-irradiated and non-irradiated sets of cells were infected for 90 min at 37°C in a total volume of 40 μl with unirradiated Ad5HCMVsp1*lacZ* or Ad5HCMVsp1*lacZ* which had been UV-irradiated with a range of fluences (A to D). Infected cells were incubated for a period of time (usually 12–48 h) before harvesting and scoring for β-galactosidase activity as reported previously [13]. Lysates from wells infected with heavily irradiated Ad5HCMVsp1*lacZ* (10,000 J/m^2), X, served as a measure of background levels for β-galactosidase activity. UV irradiation of virus was as for Figure 1 and as described previously [13]. UV irradiation of cells was performed using a germicidal lamp (General Electric model G8T5) emitting predominantly at 254 nm at an incident fluence rate of 1 J/m^2/s. Enhanced HCR in UV-treated compared to non-treated cells suggests inducible repair of the UV-damaged reporter gene. [Adapted from reference 92.]

A p53-dependent enhancement in HCR for a CMV driven plasmid based and UV-damaged reporter gene has been reported in UVB-treated murine fibroblasts [10]. In addition, thymine dinucleotides have been shown to induce the reactivation of a UV-damaged reporter gene under control of the SV40 early promoter, by a process that may also involve p53 [9]. Furthermore, Huang and colleagues [67] report that transcription from a p53-driven promoter in the presence of wild-type p53 results in up regulation of both transcription and repair of a UV-damaged reporter gene, and that the enhanced DNA repair of the reporter gene is a separate and distinct activity of p53, but is dependent on p53 driven transcription.

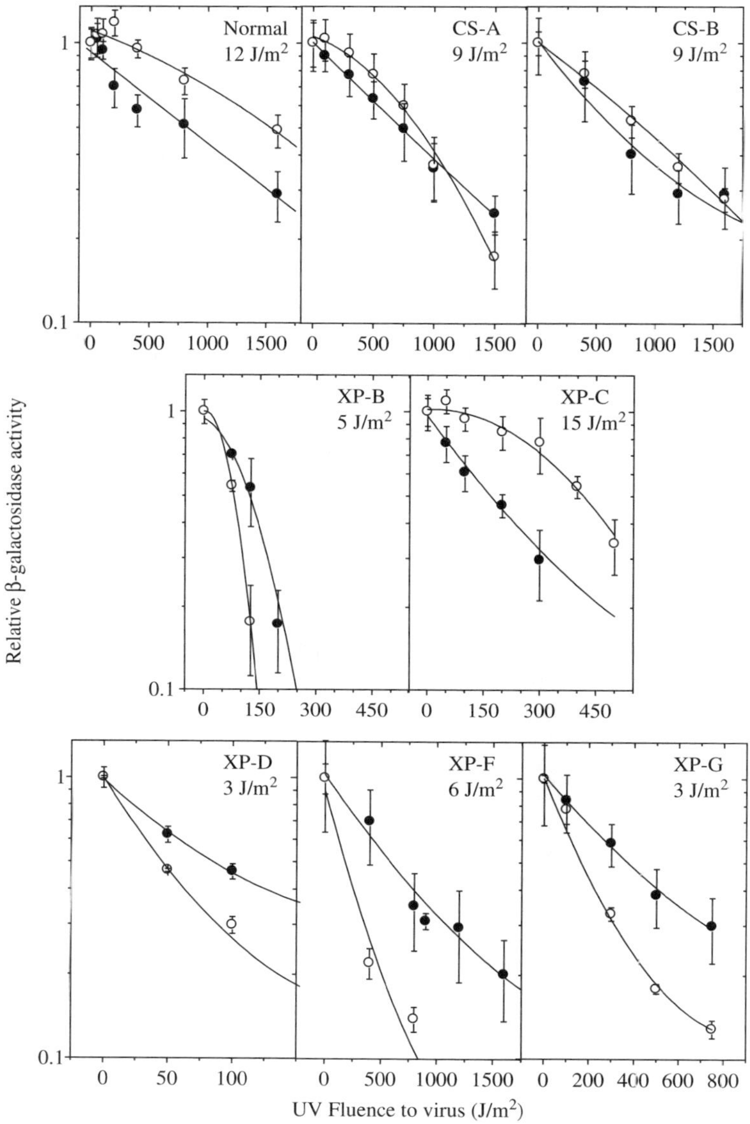

Figure 5. Pre-UV irradiation of cells results in enhanced HCR of UV-irradiated reporter activity in normal and XP-C but not in other TCR deficient XP and CS cells. Results of typical experiments representing unirradiated (●) and UV-irradiated (○) primary human fibroblasts. The experimental protocol was as described in Figure 4. Immediately following UV exposure, cells were infected with unirradiated or UV-irradiated virus at 10–20 plaque-forming units per cell and scored for β-galactosidase activity 40–44 h later. UV exposures to cells are indicated on the figure and cell strains presented are GM 3440 (normal), CS3BE (CS-A), CS1BE (CS-B), XPCS2BA (XP-B), XP3BE (XP-C), XP1BR (XP-D), XP2YO (XP-F), and XP2BI (XP-G). Each point is the average of 4–6 replicates; error bars represent one standard error. [Adapted from reference [13] with permission from Oxford University Press.]

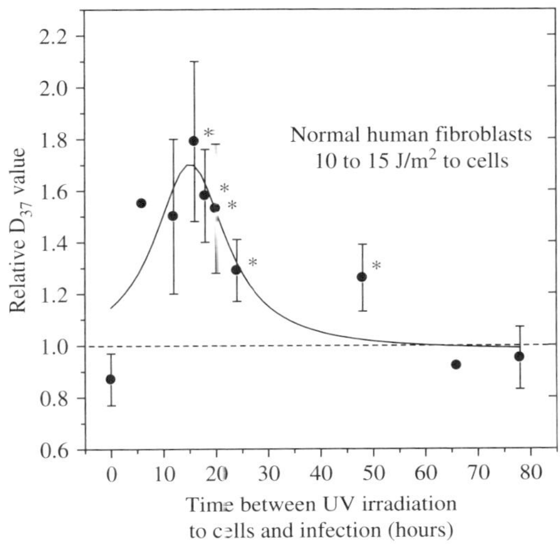

Figure 6. Pre-UV irradiation of cells results in a transient enhancement of HCR for the UV-damaged reporter gene in normal human fibroblasts. Effect of 10–15 J/m^2 UV treatment of normal human fibroblasts at various times prior to infection with UV-irradiated Ad5HCMVsp1*lacZ*. The experimental protocol was as described in Figure 4. Normal human fibroblasts were UV-irradiated or left unirradiated and subsequently infected with unirradiated or UV-irradiated Ad5HCMVsp1*lacZ* at various times after UV treatment of cells. Infected cells were scored for β-galactosidase activity 24 h later. The relative D_{37} value represents the D_{37} value for survival of β-galactosidase activity for UV-irradiated Ad5HCMVsp1*lacZ* in pre-UV-treated cells normalized to the values obtained in untreated cells. Each point is the average of multiple independent experiments; error bars represent one standard error. * indicates the relative D_{37} value is significantly greater than 1, $P<0.05$.

11.5 Evidence for inducible DNA repair

Other evidence for damage-induced DNA repair pathways in mammalian cells comes from a number of studies including the enhanced DNA repair capacity of mammalian cells following carcinogen treatment [8], the p53-mediated enhancement of NER by the DNA damaged induced GADD45 gene [68,69] and the identification of a novel DNA repair response which is induced by irradiation of cells at the G1/S border [70]. Pretreatment of normal human lung fibroblasts with the drug emodin enhances NER of UV and cisplatin-damaged DNA [71] and pretreatment of cells with dinucleotides prior to UV irradiation increased the repair of UV-induced DNA damage as assessed by unscheduled DNA synthesis [9]. Pretreatment of normal human fibroblasts with low doses of quinacrine mustard resulted in an enhanced rate of removal of CPD by NER [72], also giving evidence for an inducible NER response in human cells. Although most of these studies

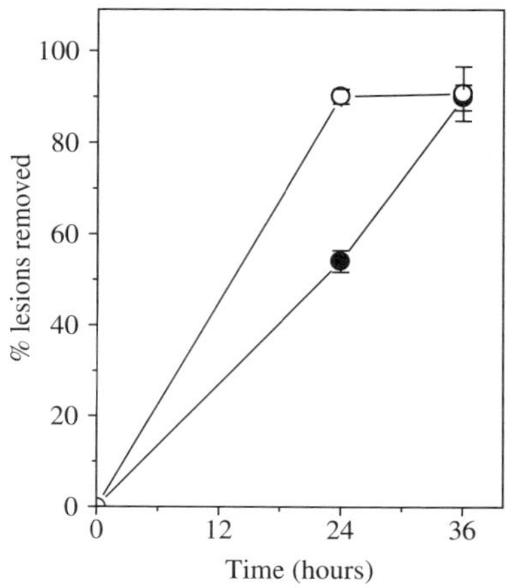

Figure 7. Enhanced lesion removal from the UV-damaged reporter gene in pre-UV-treated normal lung fibroblasts. Effect of 10 J/m^2 UV treatment of human fibroblast cells (IMR90) 24 hours prior to infection of with UV-irradiated Ad5HCMVsp1*lacZ*. The experimental protocol was similar to that described in Figure 3. Lesion removal is shown for the *lacZ* region following infection of untreated IMR90 cells (●) and cells pre-treated with 10 J/m^2 (○). Data points represent the mean (± one standard error) of relative lesion removal for three UV doses (two PCR reactions/determination at each dose of 100, 200 & 300 J/m^2) from two independent experiments. [Adapted from reference [15] with permission from Kluwer Academic/Plenum Publishers, New York, NY.]

provide evidence for an induction of the GGR rather than the TCR pathway of NER in human cells (as reviewed in reference [26]) there is recent evidence for an inducible TCR response in hamster cells. Germanier and colleagues [73] report an increase in the rate of lesion removal from the actively transcribed cellular dihydrofolate reductase (DHFR) gene following pre-treatment of Chinese hamster ovary (CHO) cells with a low UV dose (2 J/m^2) 3 h prior to a higher dose of 20 J/m^2. The enhancement in rate of lesion removal was confined to the transcribed strand of the DHFR gene indicating that TCR is inducible in CHO cells. Since CHO cells express mutant p53 these results suggest the presence of a p53 independent inducible TCR in mammalian cells.

Some mammalian cells exhibit a hypersensitivity to low doses of x-rays or cisplatin, but increased resistance following higher doses of these agents [74–76]. The increased radioresistance at higher x-ray doses is absent in some DNA repair deficient cell lines [74] and hyper-sensitivity at low doses of x-rays or cisplatin can be removed by pretreatment of cells with priming doses of a DNA damaging agent [75,76], suggesting an inducible DNA repair response in mammalian cells. Pre-exposure of human cells with low priming doses of ionising radiation leads also to an enhanced removal of thymine glycols after higher doses [77] providing evidence for an inducible repair of base damage in human cells.

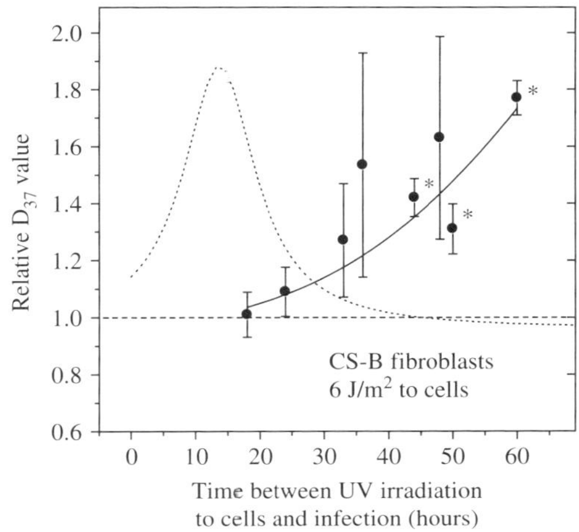

Figure 8. UV enhancement of HCR for the UV-damaged reporter gene is delayed in CS cells. Effect of 6 J/m^2 UV treatment of CSB fibroblasts at various times prior to infection with UV-irradiated Ad5HCMVsp1*lacZ*. The experimental protocol was as described in Figure 4. Cells were UV-irradiated or left unirradiated and subsequently infected with unirradiated or UV-irradiated Ad5HCMVsp1*lacZ* at various times after UV treatment of cells. Infected cells were scored for β-galactosidase activity 24 h later. The relative D_{37} value represents the D_{37} value for survival of β-galactosidase activity for UV-irradiated Ad5HCMVsp1*lacZ* in pre-UV-treated cells normalized to the values obtained in untreated cells. Each point is the average of multiple independent experiments; error bars represent one standard error. * indicates the relative D_{37} value is significantly greater than 1, $P<0.05$. The dotted line shows the data for normal fibroblasts from Figure 6. UV-enhanced HCR was not detected in CS-B cells when cells were infected within 24 h after pre-UV-treatment of cells (see also Figure 5) but was detected in CS-B cells when cells were infected at later times (36–60 h) after UV-treatment of cells.

11.6 Evidence for the involvement of p53 in NER

Over the past few years it has become clear that p53 and/or p53-regulated gene products contribute to NER of UV-induced DNA damage in mammalian cells [12,24,25,61,62,68,69,78]. For recent reviews on the involvement of p53 in NER see references [26,79]. There is evidence that some of the p53-regulated gene products are involved in inducible NER including the p53-mediated and DNA damage-induced GADD45 gene [24,69], the p48(XPE) gene [80,81] and the XPC gene [82,83]. In particular, transcription from the p48 gene, which is mutated in GGR-deficient, damage-specific DNA binding (DDB) protein deficient, XP-E cells [81] is up regulated (in a p53-dependent manner) in response to UV treatment in human cells.

UV-induced DNA lesions efficiently block transcript elongation and induce the p53 response [84–86]. Although p53 contributes to the transcriptional activation of

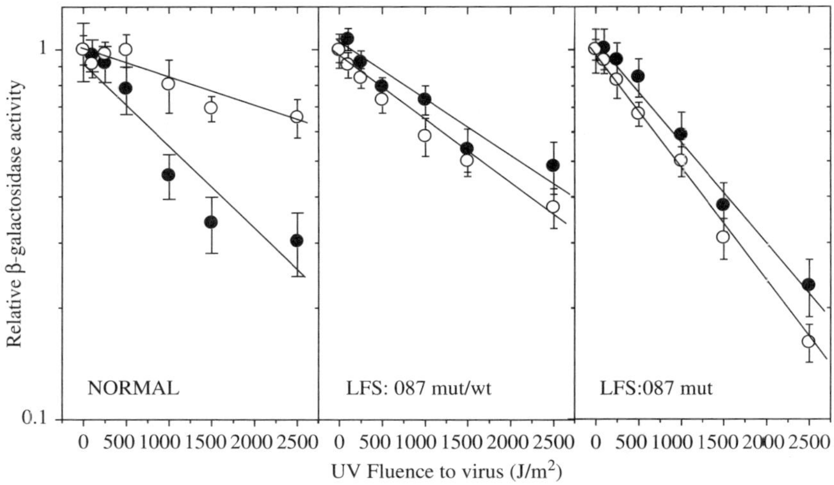

Figure 9. Pre-UV irradiation of cells results in enhanced HCR of UV-irradiated reporter activity in normal but not in Li-Fraumeni syndrome cells. Results of typical experiments representing unirradiated (●) and UV-irradiated with 15 J/m^2 (○) normal human fibroblasts and Li-Fraumeni syndrome (LFS) cells. The experimental protocol was as described in Figure 4. Cell strains and cell lines presented are GM 9503 (normal), LFS: 087 mut/wt (heterozygous for a mutation in p53), LFS 087 mut (expressing only mutant p53). Immediately following UV exposure cells were infected with unirradiated or UV-irradiated virus at 10–20 plaque forming units per cell and scored for β-galactosidase activity 40–44 h later. Each point is the average of 3–6 replicates, error bars represent one standard error. [Adapted from references 12,92.]

many p53 responsive genes (reviewed in reference [87]), accumulation of the protein products from these p53 responsive genes requires that these genes are free of UV-induced pyrimidine dimers [88]. For example, accumulation of the p53 regulated p21 protein following UV exposure is delayed in TCR deficient cells (Figure 10, [88]). Thus, TCR-deficient fibroblasts do not recover RNA synthesis following UV as efficiently as repair-proficient cells [84–86,89] and do not accumulate the protein products of p53-responsive genes as efficiently as repair-proficient cells [88]. The UV-induced increase in p48 transcription which leads to enhanced GGR would require removal of any UV-induced lesions from the p48 gene and therefore be delayed in TCR-deficient cells compared to repair proficient cells. It is thus possible that the p53 and TCR dependent UV-enhanced repair of the UV-damaged reporter gene reported previously for normal human fibroblasts [12,13], results from an up regulation of GGR in the transcribed strand of the reporter gene mediated by a UV-induced up regulation of the p48 gene product. Previous reports have also suggested that the DDB protein is responsible for the enhanced repair of UV-damaged expression vectors [90]. This provides a model for a UV-inducible GGR response in human cells that is dependent on p48 (and possibly XPC) transcription and therefore dependent also on p53 and TCR

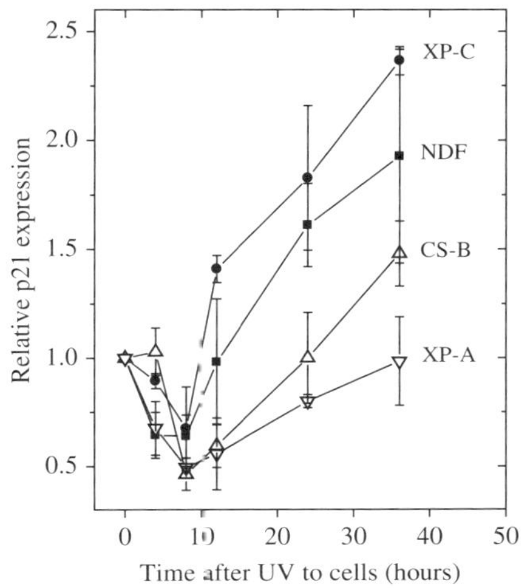

Figure 10. Expression of the p53 dependent p21 gene is decreased at early times following 30 J/m^2 UV in human fibroblasts and the time of recovery is dependent on TCR. A quantitative ELISA assay was used to quantify p21 expression at several times following 30 J/m^2 UV irradiation in normal human fibroblasts, GM38A (■); XP-C (XP2BE) (●); CS-A (CS3BE) (Δ) and XP-A (XP12BE) (▽) cells. p21 levels in UV irradiated samples were normalized to unirradiated controls collected at the same time. Each point represents the meanstandard error of at least 3 independent experiments. [Adapted from reference 88.]

(Figure 11, [26]). The UV-enhanced HCR of the UV-damaged reporter gene in normal fibroblasts and the delayed expression of UV-enhanced HCR in GGR+/TCR- CS-B cells are consistent with the above model. However, the UV-enhanced HCR of the UV-damaged reporter gene in GGR−/TCR+, XP-C cells is not consistent with an upregulation of GGR and suggests the UV-enhanced HCR in XP-C cells results from some other mechanism.

Although there is compelling evidence for an involvement of p53 in GGR, the role of p53 in TCR is more controversial. Some [27,91] but not all [25,61,62] reports suggest that TCR also may be regulated by a p53-dependent mechanism in human cells. Therrien and colleagues [27] showed that the rate of repair of UV-induced CPD was reduced along both the transcribed and the non-transcribed strands of the p53 and/or c-jun loci in Li-Fraumeni syndrome (LFS) cells expressing mutant p53 and human fibroblasts expressing the human papilloma virus (HPV) E6 oncoprotein that functionally inactivates p53. The reduction in the rate of CPD repair for the LFS cells compared to normal cells was considerably greater in the transcribed (6-fold) compared to the non-transcribed strand (3-fold) providing evidence that both TCR and GGR are dependent on wild-type p53 in UV-irradiated human cells. It is therefore possible that TCR can be upregulated in mammalian cells through both p53 dependent and p53 independent mechanisms [72]. Our results for UV-enhanced HCR of a UV-damaged reporter gene in XP-C cells are consistent with a model in which pre-treatment of cells with UV results in

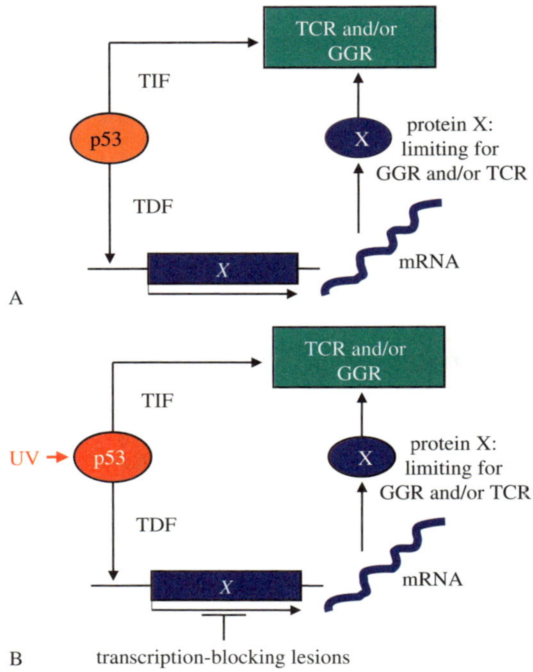

Figure 11. Potential roles for p53 in NER. p53 may participate in NER through either transactivation-dependent functions (TDF) or transactivation-independent functions (TIF) of p53. TIF of p53 may involve protein-protein interactions between p53 and proteins involved in DNA repair. Whereas TDF of p53 may be required to regulate the expression of one or more proteins limiting for NER, p53 could potentially contribute to these processes even in unstressed cells (A). Following UV irradiation (B), p53 accumulates, which may stimulate the direct interaction with DNA repair proteins. However, the TDF of p53 will be impeded by the presence of transcription blocking lesions in the p53 target genes (B). [Adapted from reference [26] by permission of Oxford University Press.]

an up regulation of TCR. It appears likely that up regulation of both GGR and TCR can contribute to UV-enhanced HCR of a UV-damaged reporter gene in mammalian cells.

References

1. A.J. Rainbow (1981). Reactivation of viruses. In: H.F. Stich, RHC San (Eds), *Short Term Tests for Chemical Carcinogens* (pp. 20–35). Springer-Verlag, New York.
2. M.J. Defais, P.C. Hanawalt, A. Sarasin (1983). Viral probes for DNA repair. *Adv. Radiat. Biol.*, **268**, 1–37.
3. W.P. Jeeves, A.J. Rainbow (1983). Gamma-ray-enhanced reactivation of irradiated adenovirus in xeroderma pigmentosum and cockayne syndrome fibroblasts. *Radiat. Res.*, **94**, 480–498.

4. W.P. Jeeves, A.J. Rainbow (1983). UV-enhanced reactivation of UV- and gamma-irradiated adenovirus in normal human fibroblasts. *Int. J. Radiat. Biol.*, **43**, 599–623.
5. W.P. Jeeves, A.J. Rainbow (1983). UV-enhanced reactivation of UV- and gamma-irradiated adenovirus in Cockayne syndrome and xeroderma pigmentosum fibroblasts. *Int. J. Radiat. Biol.*, **43**, 625–647.
6. C.B. Bennett, A.J. Rainbow (1988). Enhanced reactivation and mutagenesis of UV-irradiated adenovirus in normal human fibroblasts. *Mutagenesis*, **3**, 157–164.
7. T.C. Brown, P.A. Cerutti (1989). UV-enhanced reactivation of UV-damaged SV40 is due to restoration of viral early gene functions. *Mutat. Res.*, **218**, 211–217.
8. M. Protic, E. Roilides, A.S. Levine, K. Dixon (1988). Enhancement of DNA repair capacity of mammalian cells by carcinogen treatment. *Somat. Cell. Mo.l Genet.*, **14**, 351–357.
9. M.S. Eller, T. Maeda, C. Magnoni, D. Atwal, B.A. Gilchrest (1997). Enhancement of DNA repair in human skin cells by thymidine dinucleotides: evidence for a p53-mediated mammalian SOS response. *Proc. Nat.l Acad. Sci. USA*, **94**, 12627–12632.
10. G. Li, V.C. Ho (1998). p53-dependent DNA repair and apoptosis respond differently to high and low-dose ultraviolet radiation. *Br. J. Dermatol.*, **139**, 3–10.
11. B.C. McKay, A.J. Rainbow (1996). Heat-shock enhanced reactivation of a UV-damaged reporter gene in human cells involves the transcription coupled DNA repair pathway. *Mutat. Res.*, **363**, 125–135.
12. B.C. McKay, M.A. Francis, A.J. Rainbow (1997). Wildtype p53 is required for heat shock and ultraviolet light enhanced repair of a UV-damaged reporter gene. *Carcinogenesis*, **18**, 245–249.
13. M.A. Francis, A.J. Rainbow (1999). UV-enhanced reactivation of a UV-damaged reporter gene suggests transcription-coupled repair is UV-inducible in human cells. *Carcinogenesis*, **20**, 19–26.
14. K. Valerie, A. Singhal (1995). Host cell reactivation of reporter genes introduced into cells by adenovirus as a convenient way to measure cellular DNA repair. *Mutat. Res.*, **336**, 91–100.
15. I.P. Boszko, A.J. Rainbow (2002). Removal of UV photoproducts from an adenovirus-encoded reporter gene following infection of unirradiated and UV-irradiated human fibroblasts. *Somatic Cell Mol. Genet.*, **25**, 303–315.
16. E.C. Friedberg, G.C. Walker, W. Siede (1995). *DNA Repair and Mutagenesis*, ASM Press, Washington, D.C.
17. I. Mellon, V.A. Bohr, C.A. Smith, P.C. Hanawalt (1986). Preferential DNA repair of an active gene in human cells. *Proc. Natl. Acad. Sci. USA*, **83**, 8878–8882.
18. I. Mellon, G. Spivak, P.C. Hanawalt (1987). Selective removal of transcription-blocking DNA damage from the transcribed strand of the mammalian *DHFR* gene. *Cell*, **51**, 241–249.
19. A.S. Balajee, V.A. Bohr (2000). Genomic heterogeneity of nucleotide excision repair. *Gene*, **30**, 15–30.
20. Q. Wei, G.M. Matanoski, E.R. Farmer, M.A. Hedayati, L. Grossman (1993). DNA repair and aging in basal cell carcinoma: a molecular epidemiology study. *Proc. Natl. Acad. Sci. USA*, **90**, 1614–1618.
21. K.H. Kraemer, D.D. Levy, C.N. Parris, E.M. Gozukara, S. Moriwaki, S. Adelberg, M.M. Seidman (1994). Xeroderma pigmentosum and related disorders: examining the linkage between defective DNA repair and cancer. *J. Inv. Dermatol.*, **103**, 96S–101S.
22. M. Hollstein, D. Sidranski, B. Vogelstein, C.C. Harris (1991). p53 mutations in human cancers. *Science*, **253**, 49–53.
23. X.W. Wang, H. Yeh, L. Schaeffer, R. Roy, V. Moncollin, J.M. Egly, Z. Wang, E.C. Freidberg, M.K. Evans, B.G. Taffe, V.A. Bohr, G. Weeda, J.H.J. Hoeijmakers,

K. Forrester, C.C. Harris (1995). p53 modulation of TFIIH-associated nucleotide excision repair activity. *Nature Genet.*, **10**, 188–195.
24. M.L. Smith, I.T. Chen, Q. Zhan, P.M. O'Connor, A.J. Fornace Jr. (1995). Involvement of the p53 tumor suppressor in repair of UV-type DNA damage. *Oncogene*, **10**, 1053–1059.
25. J.M. Ford, P.C. Hanawalt (1995). Li-Fraumeni syndrome fibroblasts homozygous for p53 mutations are deficient in global DNA repair but exhibit normal transcription-coupled repair and enhanced UV resistance. *Proc. Natl. Acad. Sci. USA*, **92**, 8876–8880.
26. B.C. McKay, M. Ljungman, A.J. Rainbow (1999). Potential roles for p53 in nucleotide excision repair. *Carcinogenesis*, **20**, 1389–1396.
27. J.P. Therrien, R. Drouin, C. Baril, E.A. Drobetsky (1999). Human cells compromised for p53 function exhibit defective global and transcription-coupled nucleotide excision repair, whereas cells compromised for Rb function are defective only in global repair. *Proc. Natl. Acad. Sci. USA*, **96**, 15038–15043.
28. S. Squires, R.T. Johnson, A.R. Collins (1982). Initial rates of DNA incision in UV-irradiated human cells: differences between normal, xeroderma pigmentosum and tumour cells. *Mutat. Res.*, **95**, 389–404.
29. A.J. Rainbow (1989). Defective repair of UV-damaged DNA in human tumor and SV40-transformed human cells but not in adenovirus-transformed human cells. *Carcinogenesis*, **10**, 1073–1077.
30. K.K. Bowman, D.M. Sicard, J.M. Ford, P.C. Hanawalt (2000). Reduced global genomic repair of ultraviolet light-induced cyclobutane pyrimidine dimers in simian virus 40-transformed human cells. *Mol. Carcinogenesis*, **29**, 17–24.
31. J. Venema, A. van Hoffen, A.T. Natarajan, A.A. van Zeeland, L.H. Mullenders (1990). The residual repair capacity of xeroderma pigmentosum complementation group C fibroblasts is highly specific for transcriptionally active DNA. *Nucleic Acids Res.*, **18**, 443–448.
32. J. Venema, A. van Hoffen, V. Karcagi, A.T. Natarajan, A.A. van Zeeland, L.H. Mullenders (1991). Xeroderma pigmentosum complementation group C cells remove pyrimidine dimers selectively from the transcribed strand of active genes. *Mol. Cell Biol.*, **11**, 4128–4134.
33. J. Venema, L.H. Mullenders, A.T. Natarajan, A.A. van Zeeland, L.V. Mayne (1990). The genetic defect in Cockayne syndrome is associated with a defect in repair of UV-induced DNA damage in transcriptionally active DNA. *Proc. Natl. Acad. Sci. USA*, **87**, 4707–4711.
34. A. van Hoffen, A.T. Natarajan, L.V. Mayne, A.A. van Zeeland, L.H. Mullenders, J. Venema (1993). Deficient repair of the transcribed strand of active genes in Cockayne's syndrome cells. *Nucleic Acids Res.*, **21**, 5890–5895.
35. L. Schaeffer, R. Roy, S. Humbert, V. Moncollin, W. Vermeulen, J.H. Hoeijmakers, P. Chambon, J.M. Egly (1993). DNA repair helicase: a component of BTF2 (TFIIH) basic transcription factor. *Science*, **260**, 58–63.
36. L. Schaeffer, V. Moncollin, R. Roy, A. Staub, M. Mezzina, A. Sarasin, G. Weeda, J.H. Hoeijmakers, J.M. Egly (1994). The ERCC2/DNA repair protein is associated with the class II BTF2/TFIIH transcription factor. *EMBO J.*, **13**, 2388–2392.
37. R. Drapkin, J.T. Reardon, A. Ansari, J.C. Huang, L. Zawel, K. Ahn, A. Sancar, D. Reinberg (1994). Dual role of TFIIH in DNA excision repair and in transcription by RNA polymerase II. *Nature*, **368**, 769–772.
38. K.A. Henning, L. Li, N. Iyer, L.D. McDaniel, M.S. Reagan, R. Legerski, R.A. Schultz, M. Stefanini, A.R. Lehmann, L.V. Mayne, E.C. Friedberg (1995). The Cockayne syndrome group A gene encodes a WD repeat protein that interacts with CSB protein and a subunit of RNA polymerase II TFIIH. *Cell*, **82**, 555–564.

39. N. Iyer, M.S. Reagan, K.J. Wu, B. Canagarajah, E.C. Friedberg (1996). Interactions involving the human RNA polymerase II transcription/nucleotide excision repair complex TFIIH, the nucleotide excision repair protein XPG, and Cockayne syndrome group B (CSB) protein. *Biochemistry*, **35**, 2157–2167.
40. C.H. Park, A. Sancar (1994). Formation of a ternary complex by human XPA, ERCC1, and ERCC4(XPF) excision repair proteins. *Proc. Natl. Acad. Sci. USA*, **91**, 5017–5021.
41. T. Matsuda, M. Saijo, I. Kuraoka, T. Kobayashi, Y. Nakatsu, A. Nagai, T. Enjoji, C. Masutani, K. Sugasawa, F. Hanaoka, A. Yasui, K. Tanaka (1995). DNA repair protein XPA binds replication protein A (RPA). *J. Biol. Chem.*, **270**, 4152–4157.
42. C. Masutani, K. Sugasawa, J. Yanagisawa, T. Sonoyama, M. Ui, T. Enomoto, K. Takio, K. Tanaka, P.J. van der Spek, D. Bootsma, J.H.J. Hoeijmakers, F. Hanaoka (1994). Purification and cloning of a nucleotide excision repair complex involving the xeroderma pigmentosum group C protein and a human homologue of yeast RAD23. *EMBO J.*, **13**, 1831–1843.
43. M.K. Evans, J.H. Robbins, M. Ganges, R.E. Tarone, R.S. Nairn, V.A. Bohr (1993). Gene-specific DNA repair in xeroderma pigmentosum complementation groups A, C, D, and F. Relation to cellular survival and clinical features. *J. Biol. Chem.*, **268**, 4839–4847.
44. R.A. Moorehead, S.A. Armstrong, A.J. Rainbow, G. Singh (1996). Nucleotide excision repair in the human ovarian carcinoma cell line (2008) and its cisplatin resistant variant (C13). *Cancer Chemotherapy and Pharmacology*, **38**, 245–253.
45. B.A. Donahue, S. Yin, J.S. Taylor, D. Reines, P.C. Hanawalt (1994). Transcript cleavage by RNA polymerase II arrested by a cyclobutane pyrimidine dimer in the DNA template. *Proc. Natl. Acad. Sci. USA*, **91**, 8502–8506.
46. M. Protic-Sabljic, K.H. Kraemer (1985). One pyrimidine dimer inactivates expression of a transfected gene in xeroderma pigmentosum cells. *Proc. Natl. Acad. Sci. USA*, **82**, 6622–6626.
47. Z.Z. Xu, V. Krougliac, J. Prevec, F.L. Graham, G.W. Both (1995). Investigations of promoter function in human and animal cells with human recombinant adenovirus expressing rotavirus antigen VP7sc. *J. Gen Virol.*, **76**, 1971–1980.
48. C.L. Addison, M.M. Hitt, D. Kunsken, F.L. Graham (1997). Comparison of the human versus murine cytomegalovirus immediate early promoters for transgene expression by adenoviral vectors. *J. Gen. Virol.*, **78**, 1653–1661.
49. K. Dorrsch-Hasler, G.M. Keil, F. Weber, M. Jasin, W. Schaffner, U.H. Koszinowski (1985). A long and complex enhancer activates transcription of the gene coding for the highly abundant immediate early mRNA in murine cytomegalovirus. *Proc. Natl. Acad. Sci. USA*, **82**, 8325–8329.
50. J. Kolls, K. Peppel, M. Silva, B. Beutler (1994). Prolonged and effective blockage of tumor necrosis activity through adenovirus-mediated gene transfer. *Proc. Natl. Acad. Sci. USA*, **91**, 215–219.
51. M.A. Morsy, E.L. Alford, A. Bett, F.L. Graham, C.T. Caskey (1993). Efficient adenoviral-mediated ornithine transcarbamylase expression in deficient mouse and human hepatocytes. *J. Clinical Investigation*, **92**, 1580–1586.
52. S.F. Barrett, J.H. Robbins, R.E. Tarone, K.H. Kraemer (1991). Evidence for defective repair of cyclobutane pyrimidine dimers with normal repair of other DNA photoproducts in a transcriptionally active gene transfected into Cockayne syndrome cells. *Mutat. Res.*, **255**, 281–291.
53. A.R. Lehmann, A. Oomen (1985). Effect of DNA damage on the expression of the chloramphenicol acetyltransferase gene after transfection into diploid human fibroblasts. *Nucleic Acids Res.*, **13**, 2087–2095.

54. H. Klocker, R. Schneider, H.J. Burtscher, B. Auer, M. Hirsch-Kauffmann, M. Schweiger (1985). Transient expression of a plasmid gene, a tool to study DNA repair in human cells: defect of DNA repair in Cockayne Syndrome; one thymine cyclobutane dimer is sufficient to block transcription. *Eur. J. Cell Biol.*, **39**, 346–351.
55. T. Stevsner, H. Frandsen, H. Autrup (1995). Repair of DNA lesions induced by ultraviolet irradiation and aromatic amines in normal and repair-deficient lymphoblastoid cell lines. *Carcinogenesis*, **16**, 2855–2858.
56. A.K. Ganesan, P.C. Hanawalt (1994). Removal of cyclobutane pyrimidine dimers from a UV-irradiated shuttle vector introduced into human cells. *Somatic Cell Mol. Genet.*, **20**, 233–242.
57. A.K. Ganesan, J. Hunt, P.C. Hanawalt (1999). Expression and nucleotide excision repair of a UV-irradiated reporter gene in unirradiated human cells. *Mutat. Res.*, **433**, 117–126.
58. J.P. Murnane, L.F. Fuller, R.B. Painter (1985). Establishment and characterisation of a permanent SV40 ori-transformed ataxia-telangiectasia cell line. *Exp. Cell Res.*, **158**, 119–126.
59. D. Canaani, T. Naiman, T. Teitz, P. Berg (1986). Immortalization of xeroderma pigmentosum cells by simian virus 40 DNA having a defective origin of replication. *Somatic Cell Mol. Genet.*, **12**, 13–20.
60. J.H. Hoeijmakers, H. Odijk, A. Westerveld (1987). Differences between rodent and human cell lines in the amount of integrated DNA after transfection. *Exp. Cell Res.*, **169**, 111–119.
61. J.M. Ford, P.C. Hanawalt (1997). Expression of wild-type p53 is required for efficient global nucleotide excision repair in UV-irradiated human fibroblasts. *J. Biol. Chem.*, **272**, 28073–28080.
62. J.M. Ford, E.L. Baron, P.C. Hanwalt (1998). Human fibroblasts expressing the human papillomavirus E6 gene are deficient in global genomic nucleotide excision repair and sensitive to ultraviolet irradiation. *Cancer Res.*, **58**, 599–603.
63. J. Renzing, D.P. Lane (1995). p53-dependent growth arrest following calcium phosphate-mediated transfection of murine fibroblasts. *Oncogene*, **10**, 1865–1868.
64. J. Siegel, M. Fritshe, S. Mai, G. Brandner, R.D. Hess (1995). Enhanced p53 activity and accumulation in response to DNA damage upon transfection. *Oncogene*, **11**, 1362–1370.
65. M.V. Blagosklonny, W.S. el-Deiry (1996). In vitro evaluation of a p53-expressing adenovirus as an anti-cancer drug. *Int. J. Cancer*, **67**, 386–392.
66. A.J. Rainbow, C. Caney, I.P. Boszko, B.C. McKay, M.A. Francis (2002). UV-enhanced reactivation of UV-damaged viruses and reporter genes in mammalian cells. Proceeding of the 30th Annual Meeting of the American Society for Photobiology, Quebec City, Canada, (abstract 39), p. 13.
67. J. Huang, N. Logsdon, F.I. Schmieg, D.T. Simmons (1998). p53-mediated transcription induces resistance of DNA to UV inactivation. *Oncogene*, **17**, 401–411.
68. M. Smith, I.T. Chen, Q. Zhan, I. Bae, C.Y. Chen, T.M. Gilmer, M.B. Kastan, P.M. O'Connor, A.J. Fornace Jr. (1994). Interaction of the p53-regulated protein Gadd45 with proliferating cell nuclear antigen. *Science*, **266**, 1376–1380.
69. M.L. Smith, H.U. Kontny, Q. Zhan, A. Sreenath, P.M. O'Connor, A.J. Fornace Jr. (1996). Antisense GADD45 expression results in decreased DNA repair and sensitizes cells to UV-irradiation or cisplatin. *Oncogene*, **13**, 2255–2263.
70. S.A. Leadon, A.B. Dunn, C.E. Ross (1996). A novel DNA repair response is induced in human cells exposed to ionizing radiation at the G1/S-phase border. *Radiat. Res.*, **146**, 123–130.

71. L.C. Chang, H.M. Sheu, Y.S. Huang, T.R. Tsai, K.W. Kuo (1999). A novel function of emodin: enhancement of the nucleotide excision repair of UV- and cisplatin-induced DNA damage in human cells. *Biochemical Pharmacology*, **58**, 49–57.
72. N. Ye, M.S. Bianchi, N.O. Bianchi, G.P. Holmquist (1999). Adaptive enhancement and kinetics of nucleotide excision repair in humans. *Mutat. Res.*, **435**, 43–61.
73. M. Germanier, M. Defais, V.A. Bohr, F. Larminat (2000). Transcription-coupled repair is inducible in hamster cells. *Nucleic Acids Res.*, **1**, 4674–4678.
74. K. Skov, B. Marples, J.B. Matthews, M. Joiner, H. Zhou (1994). A preliminary investigation into the extent of increased radioresistance or hyper-radiosensitivity in cells of hamster cell lines known to be deficient in DNA repair. *Radiat. Res.*, **138**, S126–S129.
75. M.C. Joiner, P. Lambin, E.P. Malaise, T. Robson, J.E. Arrand, K. Skov, B. Marples (1996). Hypersensitivity to very-low single radiation doses: its relationship to the adaptive response and induced radioresistance. *Mutat. Res.*, **358**, 171–183.
76. C. Caney, J.T. Bulmer, G. Singh, H. Lukka, A.J. Rainbow (1999). Pre-exposure of human squamous carcinoma cells to low-doses of gamma-rays leads to an increased resistance to subsequent low-dose cisplatin treatment. *Int. J. Radiat. Biol.*, **75**, 963–972.
77. X.C. Le, J.Z. Xing, J. Lee, S.A. Leadon, M. Weinfeld (1998). Inducible repair of thymine glycols detected by an ultrasensitive assay for DNA damage. *Science*, **280**, 1066–1069.
78. B.C. McKay, C. Winrow, A.J. Rainbow (1997). Capacity of UV irradiated human fibroblasts to support adenovirus DNA synthesis correlates with transcription coupled repair and is reduced in SV40-transformed cells and cells expressing mutant p53. *Photochem. Photobiol.*, **66**, 659–664.
79. M.L. Smith, R.S. Young (2002). P53 regulation of DNA excision repair pathways. *Mutagenesis*, **17**, 149–156.
80. B.J. Hwang, J.M. Ford, P.C. Hanawalt, G. Chu (1999). Expression of the p48 xeroderma pigmentosum gene is p53-dependent and involved in global genomic repair. *Proc. Natl. Acad. Sci. USA*, **96**, 424–428.
81. B.J. Hwang, S. Toering, U. Francke, G. Chu (1998). p48 activates a UV-damaged-DNA binding factor and is defective in xeroderma pigmentosum group E cells that lack binding activity. *Mol. Cell Biol.*, **18**, 4391–4399.
82. S.A. Amundson, K.T. Do, S. Shahab, M. Bittner, P. Meltzer, J. Trent, A.J. Fornance Jr. (2000). Identification of potential mRNA biomarkers in peripheral blood lymphocytes for human exposure to ionizing radiation. *Radiat. Res.*, **154**, 342–346.
83. S.A. Amundson, A. Patterson, K.T. Do, A.J. Fornance Jr. (2002). A nucleotide excision repair master-switch: p53 regulated coordinate induction of global genomic repair genes. *Cancer Biol. Ther.*, **1**, 145–149.
84. M. Yamaizumi, T. Sugano (1994). UV-induced nuclear accumulation of p53 is evoked through DNA damage of actively transcribed genes independent of the cell cycle. *Oncogene*, **9**, 2775–2784.
85. M. Ljungman, F. Zhang (1996). Blockage of RNA polymerase as a possible trigger for UV. light-induced apoptosis. *Oncogene*, **13**, 823–831.
86. M. Ljungman, F. Zhang, F. Chen, A.J. Rainbow, B.C. McKay (1999). Inhibition of RNA polymerase II as a trigger for the p53 response. *Oncogene*, **18**, 583–592.
87. L.J. Ko, C. Prives (1996). p53: puzzle and paradigm. *Genes Dev.*, **1**, 1054–1072.
88. B.C. McKay, M. Ljungman, A.J. Rainbow (1998). Persistent DNA damage induced by ultraviolet light inhibits p21waf1 and bax expression: implications for DNA repair, UV sensitivity and the induction of apoptosis. *Oncogene*, **17**, 545–555.

89. L.V. Mayne, A.R. Lehmann (1982). Failure of RNA synthesis to recover after UV irradiation: an early defect in cells from individuals with Cockayne's syndrome and xeroderma pigmentosum. *Cancer Res.*, **42**, 1473–1478.
90. M. Protic, S. Hirschfeld, A.P. Tsang, M. Wagner, K. Dixon, A.S. Levine (1989). Induction of a novel damage-specific DNA binding protein correlates with enhances DNA repair in primate cells. *Molecular Toxicology*, **2**, 255–270.
91. Q. Zhu, M.A. Wani, M. El-Mahdy, G. Wani, A.A. Wani (2000). Modulation of transcriptional activity of p53 by ultraviolet radiation: linkage between p53 pathway and DNA repair through damage recognition. *Mol. Carcinog.*, **28**, 215–224.
92. A.J. Rainbow, B.C. McKay, M.A. Francis (2000). Recombinant adenoviruses as expression vectors and as probes for DNA repair in human cells. *Gene Therapy and Molecular Biology*, **5**, 87–101.

Chapter 12

Transcription of p53-regulated genes under transcriptional stress: implications for nucleotide excision repair

Bruce C. McKay, Cecilia Becerril and Jennifer C. Spronck

Table of contents

Abstract

The p53 tumour suppressor gene is the most commonly altered gene in neoplastic disease. This protein is a transcription factor that regulates the expression of many genes including the nucleotide excision repair genes XPC and DDB2. The p53 protein stimulates nucleotide excision repair, presumably by regulating the level of expression of genes such as XPC and DDB2. As the level and activity of p53 protein increases greatly in response to UV light, it has been suggested that the p53-dependent induction of DDB2 and XPC contributes to the repair of UV-induced DNA damage. However, the majority of repair occurs while expression of these gene products is inhibited by transcription blocking DNA damage. We propose that p53 functions in unstressed cells prior to DNA damage to prime cells to respond to subsequent ultraviolet light challenge.

12.1 Nucleotide excision repair

DNA is susceptible to damage induced by a variety of endogenous and exogenous agents. The integrity of the genome is maintained in large part by several highly conserved DNA repair pathways. Nucleotide excision repair (NER) is the best characterized of these DNA repair pathways. NER is required to remove diverse helix distorting DNA adducts including those induced by ultraviolet (UV) light. Several distinct steps of NER have been elucidated and many recent reviews have dealt with this topic in detail (for example [1–3]). Briefly, the DNA lesion must be recognized and the repair apparatus must be recruited to the site of damage. Then, two incisions are made in the damaged strand of DNA. One is 20–22 nt to the 5′ side of the lesion and the other is 5–7 nt to the 3′ side [4]. The oligonucleotide of approximately 30 nucleotides containing the damaged bases is displaced leaving a single stranded gap. Subsequently, a nascent strand is synthesized to replace the displaced oligonucleotide. Lastly, the 3′ end of the nascent DNA is ligated to existing DNA to make a continuous DNA strand. In this way, cells are able to remove a wide range of bulky intrastrand DNA adducts.

NER can be further subdivided into two inter-related subpathways differing in the initial lesion recognition stage of NER. The repair of UV-induced DNA damage occurs more rapidly in the transcribed strands of active genes than in the remainder of the genome [5,6]. This specialization of NER is thought to involve a mechanism that directly couples RNA polymerase II transcription to NER and this has been termed transcription-coupled nucleotide excision repair (TCNER) [5,6]. Lesions induced by UV light block progression of RNA polymerase II preventing the expression of polyadenylated RNAs [7–10]. TCNER removes the impeding lesions and thus permits the rapid resumption of mRNA synthesis following UV-irradiation [11–15]. UV dimers located elsewhere in the genome are repaired by the global genomic repair (GGNER) subpathway of NER [16] and these lesions have almost no impact on the recovery of transcription following UV-irradiation [13–15,17,18]. Thus two overlapping subpathways of NER have evolved to deal with lesions with distinct biological implications.

Xeroderma pigmentosum (XP) and Cockayne syndrome (CS) are heterogeneous disorders characterized by clinical photosensitivity (reviewed in [19]). Cells isolated from CS patients fall into two complementation groups (CS-A and CS-B) whereas there are eight complementation groups of XP (XP-A through XP-G and XP-V) (reviewed in [20,21]). Cells derived from XP with the exception of XP-V and CS patients represent various combinations of NER defects. XP-A, XP-B, XP-D, XP-F and XP-G cells have defects in both subpathways of NER (summarized in Table 1) (reviewed in [19]). XP-E and XP-C fibroblasts are TCNER-proficient but GGNER-deficient whereas CS-A and CS-B cells are TCNER-deficient but GGNER-proficient (reviewed in [19]). Cells derived from patients affected with the variant form of XP (XP-V) have a defect in the process of error-free lesion bypass, not NER (reviewed in [22]). XP and CS strains have been instrumental in identifying the proteins required for both GGNER and TCNER (reviewed in [1–3]).

Each complementation group of XP and CS (except XP-E) results from mutations in the gene encoding a single protein (XPA-XPD, XPF, XPG, CSA and CSB) (reviewed in [1–3]). The only possible exception is XP-E, in which mutations in the DDB2 gene were not found in two of the patients assigned to complementation group E [23]. The XPC and DDB2 proteins are involved in the lesion recognition stages of GGNER but not TCNER whereas CSA and CSB proteins are required at an early stage of TCNER but not GGNER (reviewed in [1–3]). Cells are able to deal specifically with transcription-blocking lesions by using the processive RNA polymerase II to identify transcription-blocking DNA lesions. The remaining XP proteins are required for later stages of the repair process that are common to both repair subpathways.

12.2 The p53 tumour suppressor

The p53 tumour suppressor is the most commonly altered protein in malignancy. Inheritence of a defective allele of p53 is sufficient to confer increased

Table 1. Nucleotide excision repair and p53 induction

Strain	TCNER	GGNER	Photoproduct distribution[a]	p53
normal	+	+		−
XP-A	−	−	TS, NTS	+
XP-B	−	−	TS, NTS	+
XP-C	+	−	NTS	−
XP-D	−	−	TS, NTS	+
XP-F	−	−	TS, NTS	+
XP-G	−	−	TS, NTS	+
CS-A	−	+	TS	+
CS-B	−	+	TS	+

[a] The distribution of UV-induced photoproducts remaining following repair. “TS” indicates the transcribed strand of active genes whereas “NTS” indicates the non-transcribed strand of active genes and non-transcribed regions of the genome.

[b] ++ indicates that p53 accumulates in the indicated cells following exposure to lower doses of UV light than normal diploid fibroblasts [12,17,32].

susceptibility to a wide range of cancers (reviewed in [24]). The tumour suppressive function of p53 involves the regulation of many cellular processes including nucleotide excision repair [25], multiple cell cycle checkpoints (notably G_1 arrest) [26] and apoptosis [27,28]. The p53 protein is a transcription factor that positively regulates the expression of a large number of genes that contain p53 response elements that consist of inverted repeats of a 5′-PuPuPuC(A/T)-3′ sequence [29]. Many p53-responsive genes have been identified that contribute to the anti-neoplastic activity of p53 [30]. It has been proposed that many of these target genes are important in various aspects of the cellular response to DNA damage.

In actively growing cells, the half-life of p53 is short because p53 is ubiquitinated and degraded in the 26S proteasome (reviewed in [31]). In response to cellular stresses including DNA damage induced by UV light, the proteasome-mediated turnover of p53 is blocked and p53 accumulates in cells (reviewed in [31]). As indicated in table 1, p53 protein levels accumulate in TCNER-deficient fibroblasts following exposure to lower doses of UV light compared to TCNER-proficient strains [12,17,32–35]. Thus, lesions remaining in the transcribed strand of active genes result in the sustained induction of p53 [12,31,32]. Similarly, persistent lesions in active genes lead to p53 induction following exposure to the chemotherapeutic drug cisplatin [36,37]. Furthermore, inhibitors of transcription trigger p53 induction without inducing detectable DNA damage [38]. Taken together, it appears that p53 responds to blocked transcription rather than the presence of UV-induced DNA lesions *per se*. This poses an interesting problem for the cell because the p53 transcriptional response is induced at a time when transcription is attenuated. This conundrum and its relevance to the regulation of nucleotide excision repair will be discussed in more detail below.

12.3 p53 and nucleotide excision repair

It has been recognized for some time that p53-dependent G_1 arrest would be effective in concert with NER to allow time for repair prior to the onset of DNA synthesis [39]. More recently, it has been reported that p53 affects the repair of UV-induced DNA damage through a more direct mechanism [25,40–53]. There is compelling evidence that p53 regulates GGNER [25,40,43,44,49–53] whereas the involvement of p53 in TCNER has been controversial [45–47,49].

UV-irradiation causes an inhibition of mRNA synthesis that is transient in TCNER-proficient cells but is sustained in TCNER-deficient strains [11,12,15,17]. Therefore the *r*ecovery of m*R*NA *s*ynthesis (RRS) is a biologically relevant endpoint that is dependent on TCNER. We have assessed the role of p53 in the RRS following UV-irradiation in several pairs of isogenic cell lines. Normal fibroblasts, XP-C fibroblasts and RKO colon carcinoma cells in which p53 protein was targeted for proteasome-mediated degradation by expressing the human papillomavirus E6 (HPV-E6) protein were delayed in their capacity to recover mRNA synthesis compared to isogenic controls [13–15,48,54]. Similarly, five strains of cells carrying 3 different mutations in the p53 gene failed to recover mRNA synthesis with normal kinetics [13,15,48,54] (and unpublished data). Importantly, conditional expression of p53 restored the ability of Li-Fraumeni syndrome (LFS) fibroblasts and mutant p53-expressing HT29 colon carcinoma cells to recover mRNA synthesis efficiently [15,54]. Clearly, p53 either directly or indirectly stimulates the RRS in a manner consistent with the involvement of p53 in TCNER.

At least two mechanisms could be responsible for the effect of p53 on NER and RRS. First, p53 binds to the DNA repair proteins XPB, XPD and CSB [42,55]. Although these interactions may play a role in the cellular response to DNA damage, there is presently no evidence *in vivo* or *in vitro* that p53 directly participates in either GGNER or TCNER. Second, p53 is a transcription factor that regulates the level of expression of the XP-C and DDB2 DNA repair genes in human cells [56,57]. Both genes contain p53 response elements that are bound by wildtype p53 *in vitro* [57,58]. Thus proteins involved in the lesion recognition stage of GGNER are p53-regulated. Interestingly, proteins involved in other stages of NER are also p53-responsive. For example, the ribonucleotide reductase p53R2 is regulated by p53 [59–62]. It is thought that this protein is required to permit DNA repair synthesis, a step common to both TCNER and GGNER [59–62]. As p53 is thought to be relatively inactive in the absence of cellular stresses, it has been proposed that in response to DNA damage p53 upregulates NER factors in a manner analogous to the bacterial SOS response.

12.4 A p53 paradox

As already discussed, there is now a large body of evidence indicating that p53 stimulates NER, primarily the GGNER subpathway of NER [25,40–53]. In addition, p53 stimulates the ability of cells to recover transcription following UV-irradiation [13–15,48,54]. The effect of p53 on GGNER and RRS appears to depend at least in part on the p53-regulated protein DDB2 [23,56,63] (our unpublished results). The DDB2 protein forms a heterodimeric complex with DDB1 that has specific DNA damage binding activity [64,65]. This activity is absent in most cell lines derived from individuals in XP complementation group E (XP-E cells) [64,65]. Both the NER phenotype and the RRS phenotype of XP-E and LFS cells are virtually indistinguishable [56,63,66] (and our unpublished data). The simplest interpretation of the available data is that p53 stimulates NER and the RRS following UV-irradiation by upregulating DDB2 protein levels and thus DNA damage binding activity.

However, several separate laboratories have reported detecting defects in NER in XP-E (DDB2-deficient) and LFS (p53- and thus DDB2-deficient) cells within 2 hours following UV-irradiation [25,42,44,63,66,67]. Similarly, we find that RRS is complete within 6 hours following exposure to 10 J/m^2 and this recovery is dependent on both p53 and DDB2 [13–15,48,54,66] (and unpublished results). Paradoxically, DDB2 mRNA and DDB2 DNA damage binding activity do not accumulate until 16 to 36 hours following UV-irradiation [64] (and our unpublished results). UV light induces a rapid redistribution of DDB2 into tight association with UV-damaged chromatin [68–71]. Therefore, the time required for the induction of

DDB2 is too slow to account for its effect on NER, its effect on RRS or its rapid relocalization following UV-irradiation. This implies that at early times following UV-irradiation, cells must rely on basal not UV-induced levels of DDB2. Importantly, the level of DDB2 mRNA is p53-regulated prior to DNA damage [56] (and our unpublished results). In addition, we find that p53 regulates the level of expression of many other gene products in the absence of exogenous cellular stresses.

12.5 UV dimers impede the p53 transcriptional response

The discordance with the timing of DDB2 activation has led us to hypothesize that the primary role of p53 in NER and RRS is the regulation of DNA repair genes prior to DNA damage. To test this, HT29 colon carcinoma cells expressing a temperature sensitive allele of p53 were used to determine whether functional p53 was required prior to UV-irradiation to stimulate RRS [54]. At the restrictive temperature, this allele of p53 is in mutant conformation in the cytoplasm and thus cannot act as a transcriptional activator. At the permissive temperature, the p53 is in wildtype conformation in the nucleus and activates p53-dependent gene expression within 2 hours [54,72–75]. Importantly, returning cells to the restrictive temperature eliminated p53 activity. Therefore, this model system permitted rapid and reversible regulation of p53 activity [54].

A series of experiments was performed to determine whether functional p53 was required prior to or following UV-irradiation to stimulate RRS [54]. Consistent with our hypothesis, we found that expression of functional p53 exclusively following UV-irradiation was not sufficient to stimulate RRS (Table 2). However, RRS was stimulated in a p53-dependent manner when cells were maintained at the permissive temperature [54]. The requirement for several hours of functional p53 expression to stimulate RRS suggested that the expression of one or more p53-regulated genes had to be upregulated prior to DNA damage [54]. In fact, we found that under these conditions UV-irradiation blocked the induction of p53 responsive genes including WAF1 and DDB2 when expression of functional p53 was limited to the post-UV incubation period [54] (and unpublished results) suggesting that the p53 transcriptional response is not efficient when transcription is

Table 2. Timing of p53 expression

Treatment[a]	Before UV[b]	After UV[c]	RRS[d]
R-UV-R	Mutant p53	Mutant p53	–
R-UV-P	Mutant p53	Wildtype p53	–
P-UV-R	Wildtype p53	Mutant p53	+
P-UV-P	Wildtype p53	Wildtype p53	+

Data summarized from [54].

[a] Combination of restrictive (R) and permissive (P) temperatures relative to the UV exposure.

[b] The conformation of p53 during the 24 hours immediately before UV-irradiation.

[c] The conformation of p53 following UV-irradiation.

[d] '+' indicates that cells recovered mRNA synthesis within 24 hours following exposure to 20 J/m^2 whereas '–' indicates that they did not.

Table 3. Effect of UV dose on $p21^{WAF1}$ induction

		Low dose		High dose	
Strain	**TCNER**[a]	**p53**[b]	**p21**[b]	**p53**[b]	**p21**[b]
normal	+	−	−	+	+
XP-A	−	+	+	+	−
XP-B	−	+	+	+	ND
XP-C	+	−	−	+	−
XP-D	−	+	+	+	ND
XP-F	−	+	+	+	ND
XP-G	−	+	+	+	ND
CS-A	−	+	+	+	−
CS-B	−	+	+	+	−

[a] + indicates that the fibroblast strain is proficient in transcription-coupled nucleotide excision-repair (reviewed in [19]).
[b] + denotes that either p53 or p21WAF1 (p21) accumulated in the fibroblast strain listed following exposure to either 5 (low dose) or 30 J/m^2 (high dose) of UV light, as reported [32].

impaired by UV-induced DNA damage. Using primary human fibroblasts derived from XP and CS patients, we found that the induction of the p53-regulated genes WAF1 and Bax required TCNER under conditions in which transcription is inhibited in all cells, indicating that expression of these genes was blocked by unrepaired UV-induced dimers (Table 3) [32].

UV-induced photoproducts pose a complete block to the progression of RNA polymerase II [8–10] and the likelihood that a gene sustains damage is proportional to gene size [7]. Hence the larger the gene the greater the probability that expression is blocked by UV-induced lesions. This concept forms the fundamental basis of the ultraviolet light transcription unit mapping technique developed in the 1970s [7]. Based on estimates of lesion frequencies determined from a number of laboratories for both (6-4)phoroproducts and cyclobutane pyrimidine dimers, one should expect approximately 8.6×10^3 lesions/kb/J/m^2 [25,76–80]. Exposure of cells to a modest dose of 10 J/m^2 UV light would be expected to induce 2.1 lesions per allele of DDB2 (24278 bases) so assuming a Poisson distribution only 12% of DDB2 alleles would be lesion free immediately following irradiation (Table 4). This means that most cells cannot transcriptionally increase the expression of

Table 4. UV-induced lesions and p53 regulated genes

Gene	**Gene size**[a]	**Lesions/gene**[b]	**Probability**[c]
WAF1	8600	0.7	0.50
DDB2	24300	2.1	0.12
PTEN	102900	8.9	1.4×10^{-4}

[a] Gene size in bases obtained from Locuslink website (www.ncbi.nlm.nih.gov/LocusLink).
[b] Number of lesions ((6-4)phoroproducts+pyrimidine dimers) per gene using the value 8.6×10^3 lesions/kb/J/m^2 [25,76–80].
[c] Probability that any single copy of the gene is lesion free determined using the Poisson distribution. The Poisson distribution solved for the zero class is: $P = e^{-\text{lesions per gene}}$.

DDB2 prior to TCNER of the DDB2 gene. In fact, even with 50% repair of the transcribed strand within 6 hours, there would still be an average of one lesion per gene remaining. The kinetics of repair and the size of the DDB2 gene likely contribute strongly to the delayed kinetics of DDB2 induction. Using HT29 cells expressing the temperature sensitive allele of p53, we assessed the expression of several p53-regulated genes including DDB2 in the presence and absence of UV-induced DNA damage. We found that gene size negatively correlated with the induction of gene expression at the permissive temperature, suggesting that gene size is a constraint to the p53 response following UV-irradiation.

12.6 Summary

The activity of the p53 tumour suppressor increases in response to UV-irradiation and p53 stimulates the repair of UV-induced DNA damage. However, the induction of genes involved in GGNER and RRS are induced too late to account for the p53-dependent activity of these gene products. This implies that the basal activity of p53 (i.e. prior to exogenous DNA damage) is required to regulate GGNER and RRS. This raises a couple of questions. Why would human cells utilize a DNA damage-inducible transcription factor to regulate the basal levels of expression of DNA repair genes? Why would cells increase the expression of p53 regulated DNA repair genes after repair is complete? One can speculate that cells may benefit from increased transcription of genes in anticipation of further exposure to genotoxic agents even if this increase in expression is too late to contribute to the repair of the initial damage. In this way, p53 could be envisaged to participate in an adaptive response to chronic daily exposure to UV light. It is likely that this situation is most relevant from an evolutionary perspective.

References

1. R.D. Wood (1997). Nucleotide excision repair in mammalian cells. *J. Biol. Chem.*, **272**, 23465–23468.
2. J.E. Cleaver, E. Crowley (2002). UV damage, DNA repair and skin carcinogenesis. *Front Biosci.*, **7**, d1024–d1043.
3. J.H. Hoeijmakers (2001). DNA repair mechanisms. *Maturitas*, **38**, 17–22; discussion 22–3.
4. J.C. Huang, et al. (1992). Human nucleotide excision nuclease removes thymine dimers from DNA by incising the 22nd phosphodiester bond 5′ and the 6th phosphodiester bond 3′ to the photodimer. *Proc. Natl. Acad. Sci. USA*, **89**, 3664–3668.
5. V.A. Bohr, et al. (1985). DNA repair in an active gene: removal of pyrimidine dimers from the DHFR gene of CHO cells is much more efficient than in the genome overall. *Cell*, **40**, 359–369.
6. I. Mellon, et al. (1986). Preferential DNA repair of an active gene in human cells. *Proc. Natl. Acad. Sci. USA*, **83**, 8878–8882.
7. W. Sauerbier, K. Hercules (1978). Gene and transcription mapping by radiation effects. *Annu. Rev. Genet.*, **12**, 329–363.
8. S. Tornaletti, P. Hanawalt (1999). Effect of DNA lesions on transcription elongation. *Biochimie*, **81**, 139–146.

9. S. Tornaletti, et al. (1997). Nucleotide sequence context effect of a cyclobutane pyrimidine dimer upon RNA polymerase II transcription. *J. Biol. Chem.*, **272**, 31719–31724.
10. B.A. Donahue, et al. (1994). Transcript cleavage by RNA polymerase II arrested by a cyclobutane pyrimidine dimer in the DNA template. *Proc. Natl. Acad. Sci. USA*, **91**, 8502–8506.
11. L. Mayne, A. Lehmann (1982). Failure of RNA synthesis to recover after UV-irradiation: an early defect in cells from individuals with Cockayne's Syndrome and xeroderma pigmentosum. *Cancer Res.*, **42**, 1473–1478.
12. M. Ljungman, F. Zhang (1996). Blockage of RNA polymerase as a possible trigger for uv light-induced apoptosis. *Oncogene*, **13**, 823–831.
13. B.C. McKay, M. Ljungman (1999). Role for p53 in the recovery of transcription and protection against apoptosis induced by ultraviolet light. *Neoplasia*, **1**, 276–284.
14. B.C. McKay, C. Becerril, M. Ljungman (2001). P53 plays a protective role against UV- and cisplatin-induced apoptosis in transcription-coupled repair proficient fibroblasts. *Oncogene*, **20**, 6805–6808.
15. B.C. McKay, et al. (2001). UV light-induced degradation of RNA polymerase II is dependent on the Cockayne's syndrome A and B proteins but not p53 or MLH1. *Mutat. Res.*, **485**, 93–105.
16. E. Friedberg, G. Walker, W. Siede (1995). DNA repair and mutagenesis., Washington, D.C.: ASM Press.
17. M. Yamaizumi, T. Sugano (1994). Uv-induced nuclear accumulation of p53 is evoked through DNA damage of actively transcribed genes independent of the cell cycle. *Oncogene*, **9**, 2775–2784.
18. M. Ljungman (1996). Effect of differential gene expression on the chromatin structure of the DHFR gene domain *in vivo. Biochim. Biophys. Acta*, **1307**, 171–177.
19. K.H. Kraemer, et al. (1994). Xeroderma pigmentosum and related disorders: Examining the linkage between defective DNA repair and cancer. *J. Invest. Dermatol.*, **103**, Suppl. S96-S101.
20. G. Chu, L. Mayne (1996). Xeroderma pigmentosum, Cockayne syndrome and trichothiodystrophy: Do the genes explain the diseases? *Trends Genet*, **12**, 187–192.
21. H. van Steeg, K.H. Kraemer (1999). Xeroderma pigmentosum and the role of UV-induced DNA damage in skin cancer. *Mol. Med. Today*, **5**, 86–94.
22. B.S. Tung, et al. (1996). Comparison of the rate of excision of major UV photoproducts in the strands of the human HPRT gene of normal and xeroderma pigmentosum variant cells. *Mutat. Res.-DNA Repair*, **362**, 65–74.
23. J. Tang, G. Chu (2002). Xeroderma pigmentosum complementation group E and UV-damaged DNA-binding protein. *DNA Repair*, **1**, 601–616.
24. S. Jin, A. Levine (2001). The p53 functional circuit. *J. Cell Sci.*, **114**, 4139–4140.
25. J.M. Ford, P.C. Hanawalt (1997). Expression of wild-type p53 is required for efficient global genomic nucleotide excision repair in UV-irradiated human fibroblasts. *J. Biol. Chem*, **272**, 28073–28080.
26. M. Kastan, et al. (1991). Participation of p53 protein in the cellular response to DNA damage. *Cancer Res.*, **51**, 6304–6311.
27. E. Yonish-Rouach, et al. (1991). Wild-type p53 induces apoptosis of myeloid leukaemic cells that is inhibited by interleukin-6. *Nature*, **352**, 345–347.
28. S. Lowe, et al. (1993). p53 is required for radiation-induced apoptosis in mouse thymocytes. *Nature*, **362**, 847–849.
29. W. El-Deiry, et al. (1992). Definition of consensus binding site for p53. *Nature Genetics*, **1**, 45–49.

30. R. Zhao, et al. (2000). Analysis of p53-regulated gene expression patterns using oligonucleotide arrays. *Genes Dev.* **14**, 981–993.
31. M. Ljungman (2000). Dial-911 for p53: Mechanisms of p53 Activation by Cellular Stress. *Neoplasia*, **2**, 208–225.
32. B.C. McKay, M. Ljungman, A.J. Rainbow (1998). Persistent DNA damage induced by ultraviolet light inhibits p21(waf1) and bax expression: implications for DNA repair, UV sensitivity and the induction of apoptosis. *Oncogene*, **17**, 545–555.
33. S. Queille, et al. (2001). Effects of XPD Mutations on Ultraviolet-Induced Apoptosis in Relation to Skin Cancer-Proneness in Repair-Deficient Syndromes. *J. Invest. Dermatol.*, **117**, 1162–1170.
34. N. Dumaz, et al. (1998). Recovery of the normal p53 response after UV treatment in DNA repair-deficient fibroblasts by retroviral-mediated correction with the XPD gene. *Carcinogenesis*, **19**, 1701–1704.
35. N. Dumaz, et al. (1997). Prolonged p53 protein accumulation in trichothiodystrophy fibroblasts dependent on unrepaired pyrimidine dimers on the transcribed strands of cellular genes. *Mol. Carcinogen.*, **20**, 340–347.
36. L. Andera, B. Wasylyk (1997). Transcription abnormalities potentiate apoptosis of normal human fibroblasts. *Mol. Med.*, **3**, 852–863.
37. M. Ljungman (1999). Recovery of RNA synthesis from the DHFR gene following UV-irradiation precedes the removal of photolesions from the transcribed strand. *Carcinogenesis*, **20**, 395–399.
38. M. Ljungman, et al. (1999). Inhibition of RNA polymerase II as a trigger for the p53 response. *Oncogene*, **18**, 583–592.
39. D. Lane (1992). *p53,* guardian of the genome. *Nature*, **358**, 15–16.
40. M.L. Smith, et al. (2000). p53-mediated DNA repair responses to UV radiation: studies of mouse cells lacking p53, p21, and/or gadd45 genes. *Mol. Cell Biol.*, **20**, 3705–3714.
41. M.L. Smith, et al. (1995). Involvement of the p53 tumor suppressor in repair of u.v.-type DNA damage. *Oncogene*, **10**, 1053–1059.
42. X.W. Wang, et al. (1995). p53 modulation of TFIIH-associated nucleotide excision repair activity. *Nat. Genet.*, **10**, 188–195.
43. J.M. Ford, E.L. Baron, P.C. Hanawalt (1998). Human fibroblasts expressing the human papillomavirus E6 gene are deficient in global genomic nucleotide excision repair and sensitive to ultraviolet irradiation. *Cancer Res.*, **58**, 599–603.
44. J.M. Ford, P.C. Hanawalt (1995). Li-Fraumeni syndrome fibroblasts homozygous for p53 mutations are deficient in global DNA repair but exhibit normal transcription-coupled repair and enhanced UV resistance. *Proc. Natl. Acad. Sci. USA*, **92**, 8876–8880.
45. B.C. McKay, M.A. Francis, A.J. Rainbow (1997). Wildtype p53 is required for heat shock and ultraviolet light enhanced repair of a UV-damaged reporter gene. *Carcinogenesis*, **18**, 245–249.
46. B.C. McKay, C. Winrow, A.J. Rainbow (1997). Capacity of UV-irradiated human fibroblasts to support adenovirus DNA synthesis correlates with transcription-coupled repair and is reduced in SV40-transformed cells and cells expressing mutant p53. *Photochem. Photobiol.*, **66**, 659–664.
47. B.C. McKay (1997). The relationship between the repair of ultraviolet light induced DNA damage in human cells and the p53 tumour suppressor, in *Department of Biology*. (p. 206). McMaster University: Hamilton, Canada.
48. B.C. McKay, M. Ljungman, A.J. Rainbow (1999). Potential roles for p53 in nucleotide excision repair. *Carcinogenesis*, **20**, 1389–1396.
49. J.-P. Therrien, et al. (1999). Human cells compromised for p53 function exhibit defective global- and transcription coupled-repair whereas cells compromised for

pRb function are defective only in global repair. *Proc. Natl. Acad. Sci. USA*, **96**, 15038–15043.
50. M.A. Wani, et al. (1999). Influence of p53 tumor suppressor protein on bias of DNA repair and apoptotic response in human cells. *Carcinogenesis*, **20**, 765–772.
51. Q. Zhu, et al. (2000). Decreased DNA repair efficiency by loss or disruption of p53 function preferentially affects removal of cyclobutane pyrimidine dimers from non-transcribed strand and slow repair sites in transcribed strand. *J. Biol. Chem.*, **275**, 11482–11497.
52. M.A. El-Mahdy, et al. (2000). p53-degradation by HPV-16 E6 preferentially affects the removal of cyclobutane pyrimidine dimers from non-transcribed strand and sensitizes mammary epithelial cells to UV-irradiation. *Mutat. Res.*, **459**, 135–145.
53. M. El-Mahdy, et al. (2000). p53-degradation by HPV-16 E6 preferentially affects the removal of cyclobutane pyrimidine dimers from non-transcribed strand and sensitizes mammary epithelial cells to UV-irradiation. *Mutat. Res.*, **459**, 135–145.
54. B.C. McKay, et al. (2000). The tumor suppressor p53 can both stimulate and inhibit ultraviolet light-induced apoptosis. *Mol. Biol. Cell*, **11**, 2543–2551.
55. X. Wang, et al. (1994). Hepatitus B virus X protein inhibits p53 sequence-specific DNA binding, transcriptional activity, and association with transcription factor ERCC3. *Proc. Natl. Acad. Sci. USA*, **91**, 2230–2234.
56. B.J. Hwang, et al. (1999). Expression of the p48 xeroderma pigmentosum gene is p53-dependent and is involved in global genomic repair. *Proc. Natl. Acad. Sci. USA*, **96**, 424–428.
57. S. Adimoolam, J.M. Ford (2002). p53 and DNA damage-inducible expression of the xeroderma pigmentosum group C gene. *Proc. Natl. Acad. Sci. USA*, **99**, 12985–12990.
58. T. Tan, G. Chu (2002). p53 Binds and activates the xeroderma pigmentosum DDB2 gene in humans but not mice. *Mol. Cell Biol.*, **22**, 3247–3254.
59. O. Guittet, et al. (2001). Mammalian p53R2 protein forms an active ribonucleotide reductase in vitro with the R1 protein, which is expressed both in resting cells in response to DNA damage and in proliferating cells. *J. Biol. Chem.*, **276**, 40647–40651.
60. T. Yamaguchi, et al. (2001). p53R2-dependent pathway for DNA synthesis in a p53-regulated cell cycle checkpoint. *Cancer Res.*, **61**, 8256–8262.
61. H. Tanaka, et al. (2000). A ribonucleotide reductase gene involved in a p53-dependent cell-cycle checkpoint for DNA damage. *Nature*, **404**, 42–49.
62. K. Nakano, et al. (2000). A ribonucleotide reductase gene is a transcriptional target of p53 and p73. *Oncogene*, **19**, 4283–4289.
63. J.Y. Tang, et al. (2000). Xeroderma pigmentosum p48 gene enhances global genomic repair and suppresses UV-induced mutagenesis. *Mol. Cell.*, **5**, 737–744.
64. A.F. Nichols, et al. (2000). Human damage-specific DNA-binding protein p48. Characterization of XPE mutations and regulation following UV irradiation. *J. Biol. Chem.*, **275**, 21422–21428.
65. B.J. Hwang, et al. (1998). p48 Activates a UV-damaged-DNA binding factor and is defective in xeroderma pigmentosum group E cells that lack binding activity. *Mol. Cell Biol.*, **18**, 4391–4399.
66. R.D.C. Barley, et al. (1998). Aberrant p21(WAF1)-dependent growth arrest as the possible mechanism of abnormal resistance to ultraviolet light cytotoxicity in Li-Fraumeni syndrome fibroblast strains heterozygous for TP53 mutations. *Oncogene*, **17**, 533–543.
67. R. Mirzayans, et al. (1996). Faulty DNA polymerase delta/epsilon-mediated excision repair in response to gamma radiation or ultraviolet light in p53-deficient fibroblast strains from affected members of a cancer-prone family with Li-Fraumeni syndrome. *Carcinogenesis*, **17**, 691–698.

68. M. Wakasugi, et al. (2002). DDB accumulates at DNA damage sites immediately after UV irradiation and directly stimulates nucleotide excision repair. *J. Biol. Chem.*, **277**, 1637–1640.
69. W. Liu, et al. (2000). Nuclear transport of human DDB protein induced by ultraviolet light. *J. Biol. Chem.*, **275**, 21429–21434.
70. V.R. Otrin, et al. (1997). Translocation of a UV-damaged DNA binding protein into a tight association with chromatin after treatment of mammalian cells with UV light. *J. Cell Sci*, **110**, 1159–1168.
71. E. Martinez, et al. (2001). Human STAGA complex is a chromatin-acetylating transcription coactivator that interacts with pre-mRNA splicing and DNA damage-binding factors in vivo. *Mol. Cell Biol.*, **21**, 6782–6795.
72. J. Gannon, D. Lane (1991). Protein synthesis required to anchor a mutant p53 protein which is temperature-sensitive for nuclear transport. *Nature*, **349**, 802–806.
73. J. Martinez, et al. (1991). Cellular localization and cell cycle regulation by a temperature-sensitive p53 protein. *Genes Dev.*, **5**, 151–159.
74. D. Ginsberg, et al. (1991). Induction of growth arrest by a temperature-sensitive p53 mutant is correlated with increased nuclear localization and decreased stability of the protein. *Mol. Cell. Biol.*, **11**, 582–585.
75. D. Michalovitz, O. Halevy, M. Oren (1990). Conditional inhibition of transformation and of cell proliferation by a temperature-sensitive mutant of p53. *Cell*, **62**, 671–680.
76. J. Venema, et al. (1990). The genetic defect in Cockayne syndrome is associated with a defect in repair of UV-induced DNA damage in transcriptionally active DNA. *Proc. Natl. Acad. Sci. USA*, **87**, 4707–4711.
77. J. Venema, et al. (1990). The residual repair capacity of xeroderma pigmentosum complementation group C fibroblasts is highly specific for transcriptionally active DNA. *Nuc. Acids. Res.*, **18**, 443–448.
78. A. van Hoffen, et al. (1995). Transcription-coupled repair removes both cyclobutane pyrimidine dimers and 6-4 photoproducts with equal efficiency and in a sequential way from transcribed DNA in xeroderma pigmentosum group C fibroblasts. *EMBO J.*, **14**, 360–367.
79. M.K. Evans, et al. (1993). Gene-specific DNA repair in xeroderma-pigmentosum complementation group-A, group-C, group-D, and group-F - relation to cellular survival and clinical features. *J. Biol. Chem.*, **268**, 4839–4847.
80. D.M. Tolbert, G.J. Kantor (1996). Definition of a DNA repair domain in the genomic region containing the human p53 gene. *Cancer Res*, **56**, 3324–3330.

Chapter 13

What a difference a wavelength makes: the role of p53 in nucleotide excision repair of UV-induced DNA damage

Julie Desnoyers, Dindial Ramotar, Géraldine Mathonnet, Caroline Léger and Elliot A. Drobetsky

Table of contents

Abstract

Nucleotide excision repair (NER) inhibits skin cancer development by eliminating highly-genotoxic cyclobutane pyrimidine dimers (CPDs) induced in DNA by natural sunlight. NER is comprised of global and transcription-coupled subpathways which remove CPD from the genome overall, and uniquely from the transcribed strand of active genes, respectively. We recently employed the ligation-mediated PCR technique to elucidate, at nucleotide resolution, the role of the p53 tumor suppressor in NER. Specifically we demonstrated in an isogenic human system at two chromosomal loci (*hprt* and *c-jun*) that functional p53 is required for GNER and TCNER of CPD following irradiation with polychromatic UVB (290–320-nm). However for cells exposed to 254-nm UV, p53 was found to be essential for GNER but completely dispensable for TCNER. The disclosure of this novel UV-wavelength dependent phenomenon reconciles previous conflicting studies on the role of p53 in NER, and, of fundamental interest regarding experimental photobiology, provides a prominent example where the "nonsolar" model mutagen 254-nm UV does not faithfully reproduce the effects of environmentally-relevant UVB.

13.1 Nucleotide excision repair

13.1.1 General considerations

The nucleotide excision repair (NER) pathway forestalls the accumulation of genetic mutations by eliminating highly-genotoxic DNA adducts that characteristically distort the sugar-phosphate backbone and block transcription. Such adducts include dipyrimidine-type photoproducts, i.e., cyclobutane pyrimidine dimers (CPDs) and 6-4 photoproducts (6-4PPs), which arise through covalent linkage of adjacent pyrimidine bases subsequent to direct absorption of UV photons by DNA. It is now well established that exposure to solar UVB (290–320-nm) constitutes the single most important determinant in skin cancer development, primarily *via* the induction of dipyrimidine photoproducts and subsequent mutations in critical growth-control genes. Moreover, studies on the DNA sequence specificity of UV-induced damage and mutation in human tumours and cultured cells have indicated that the powerful promutagenic/carcinogenic potential of UVB can be mostly attributed specifically to CPDs [1–4] which, relative to 6-4PPs, form in substantially higher yield and are repaired much more slowly [5,6].

NER is comprised of two subpathways: global NER (GNER) removes dipyrimidine photoproducts or other types of helix-distorting lesions from virtually anywhere within the genome, whereas transcription-coupled NER (TCNER) proceeds more rapidly than GNER and removes such lesions exclusively along the transcribed strand (TS) of active genes (See ref. [7] for extensive review). GNER and TCNER are characterized by entirely distint and mechanistically-separable modes of initiation (i.e., of lesion recognition), although all subsequent steps (comprising the "core" NER pathway) are essentially identical (Figure 1). The initiating signal for

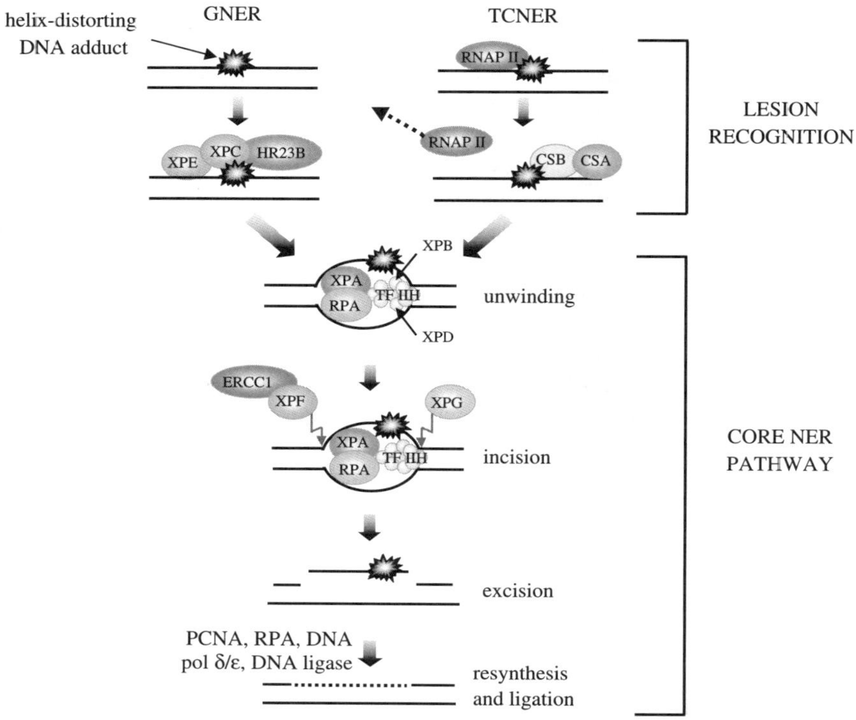

Figure 1. Schematic representation of the human nucleotide excision repair pathway.

GNER is provided by the helical distortion introduced into DNA by e.g. CPDs, which results in binding of the GNER-specific proteins XP-E and XP-C/hHR23B to the damaged site [8]. In contrast, TCNER is triggered when the progression of RNA polymerase II becomes blocked at adducted positions along the TS of active genes, whereupon the TCNER-specific proteins CS-B and -A converge ostensibly to facilitate removal or displacement of the polymerase [9]. Following these unique lesion recognition events in the case of either GNER or TCNER, the core NER pathway is recruited and accomplishes error-free restoration of the DNA through sequential steps of (i) strand unwinding mediated by the XPB and XPD helicases; (ii) incision on either side of the lesion *via* the endonuclease activities of XP-G and XP-F/ERCC1, (iii) excision of the lesion as part of a single-stranded oligonucleotide approximately 30 bp in length; and finally (iv) DNA resynthesis (NER gap-filling) and ligation, using normal DNA replication factors and the intact complementary strand as template.

NER is highly functionally-conserved in evolution and, with very few known exceptions [10], appears to constitute the sole means by which higher eukaryotes are able to repair a host of carcinogenic DNA adducts including CPD. The importance of NER in cancer prevention is highlighted by individuals afflicted with *Xeroderma pigmentosum* (XP), who are deficient in either GNER only (i.e., mutated in the

GNER-specific genes *XP-E* or *-C*) or in both GNER and TCNER (i.e., mutated in any of the core NER pathway genes *XP-A, -B, -D, -F,* or *-G*). The inability of XP patients to efficiently remove CPDs leads to extreme UV photosensitivity and hypermutability, in addition to a vastly increased incidence of sunlight-induced melanoma and nonmelanoma skin cancers [11]. On the other hand, inactivating mutations in either of the known TCNER-specific genes (i.e., *CS-B* or *-A*) causes Cockayne's syndrome (CS) which is completely distinguishable from XP in that all CS patients are deficient in TCNER but completely proficient in GNER and, furthermore, display no predisposition to cutaneous tumors [12]. Moreover CS is characterized by other striking phenotypic characteristics not observed for XP including severe developmental abnormalities and mental retardation [13]. These latter differences may be explained by the demonstration that CS but not XP patients exhibit yet another profound defect namely in transcription-coupled removal of oxidized DNA bases (e.g., thymine glycol and 8-hydroxyguanine) presumably *via* the base excision repair pathway [14].

13.1.2 Measurement of NER rates

Since the discovery of NER in the early 1960's, diverse experimental approaches have been developed for evaluating the efficiency of this pathway in the genome of UV-exposed living cells (See ref. [15] for an exhaustive overview.) These approaches include: (i) unscheduled DNA synthesis which directly measures NER gapfilling by monitoring incorporation of labelled nucleotides into repair patches; (ii) immunoassay-based methods where repair is correlated with the disappearance of binding sites for lesion-specific antibodies, e.g., directed against either CPD or 6-4PP; and, (iii) alkaline sucrose-gradient sedimentation analysis, which quantifies the removal of T4 bacteriophage endonuclease V (TEV)-sensitive sites from bulk DNA. This latter method reflects CPD repair exclusively, because TEV efficiently incises DNA immediately adjacent to this particular DNA photoproduct with extremely high specificity. Each of the above assays constitutes a sensitive and reliable indicator of global DNA repair capacity, and is still employed to the present day. However, a highly significant advance was achieved during the 1980's with the emergence of a powerful method to directly measure CPD removal at the level of single copy chromosomal genes [16], and subsequently at the level of the DNA strand of such genes [17]. This technique employs strand-specific DNA probes in conjunction with Southern blotting to quantify the removal of TEV-sensitive sites from either the TS or the nontranscribed strand (NTS) of active genes (usually *dihydrofolate reductase*; *dhfr*), thereby reflecting the rate of CPD removal *via* TCNER or GNER, respectively. In addition, the bacterial UvrABC protein complex, which exhibits much broader specificity for lesion-recognition and -incision compared with TEV, has been used to investigate gene- and strand-specific repair of UV-induced 6-4PP [18] as well as other helix-distorting adducts generated by exposure to e.g., benzo(a)pyrene diolepoxide [19], N-acetoxy-2-acetylaminofluorene [20], or 4-nitroquinolone-1-oxide [21].

13.1.3 Measurement of NER rates in living cells at nucleotide resolution: the ligation-mediated polymerase chain reaction (LMPCR)

LMPCR is a genomic sequencing method which has been adapted to measure strand-specific repair at the DNA sequence level along endogenous genes. Any DNA adduct that can be revealed either chemically or enzymatically as a ligatable strand break is amenable to LMPCR analysis. The CPD represents an ideal model lesion for such studies, as this photoproduct can be specifically incised by TEV in genomic DNA from UV-irradiated cells, followed by digestion of the DNA with *E. coli* photolyase to create ligatable 5' phosphate groups at the termini of incised strands. The basic LMPCR protocol employed in our laboratory to investigate CPD repair rates in human cells along the *c-jun*, *p53*, or *hprt* genes has been described in detail [22], and is schematized in Figure 2. Briefly, replicate cultures are irradiated with either 20 J/m^2 254-nm UV or 450 J/m^2 UVB (290–320-nm), and incubated for various times (0–24 hr) to allow repair. At these UV doses, during the time period that repair is monitored, semiconservative synthesis is suppressed such that there is no significant accumulation of newly-replicated DNA which may otherwise bias repair rate determinations. Total genomic DNA is then purified, and digested firstly by TEV followed by *E. coli* photolyase. Following denaturation of the digested

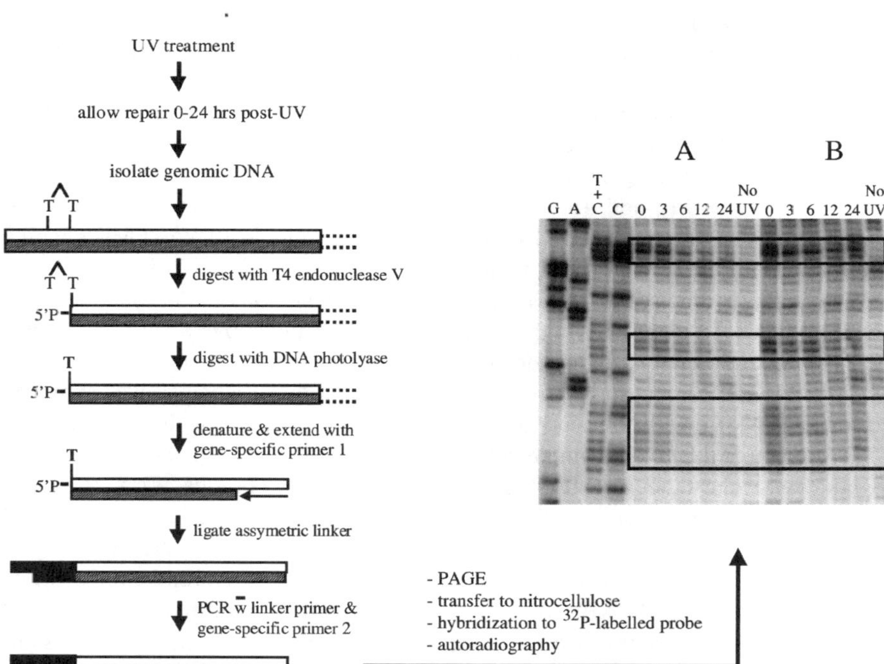

Figure 2. Outline of the ligation-mediated PCR technique. The boxed portions on the representative autoradiogram on the right depict pyrimidine runs which were quantified by phosphorimager analysis.

DNA, a gene-specific oligonucleotide is annealed downstream of the region to be analyzed, and a set of genomic cleavage products, i.e., each terminating precisely at a site where TEV had incised the DNA immediately adjacent to CPDs, is generated *via* primer extension with cloned *Pfu* polymerase. An asymmetric double-stranded oligonucleotide linker is ligated to the phosphate groups at the fragment termini, providing a common sequence at the 5' end of all fragments. A primer complementary to this linker, in conjunction with another gene-specific primer, is then used in a PCR reaction to amplify the gene-specific cleavage products of interest. These products are subjected to electrophoresis on 8% polyacrylamide gels alongside a Maxam and Gilbert sequencing ladder, transferred to nylon membranes, hybridized to a ^{32}P-labelled gene-specific probe, and visualized by autoradiography. All bands located within sequences of two or more pyrimidines, and yielding a measurable signal above background (as determined by comparison with unirradiated DNA digested with TEV/photolyase), are quantified using a Storm 840 phosphorimager (Amersham). Each band, when matched against the Maxam-Gilbert sequencing ladder, represents a precise nucleotide position where a ligateable strand break had been introduced by CPD cleavage with TEV/photolyase. Furthermore the diminishment in band intensity over time, relative to a corresponding band in the zero time lane (100% of CPD remaining), reflects the rate of disappearance of TEV-sensitive sites, and therefore the rate of CPD repair, at nucleotide resolution. We emphasize that a highly critical aspect of the LMPCR technique, with respect to proper quantification of the data, relates to variations in DNA sample loading. These are carefully controlled for by assessing *via* phosphorimager analysis the relative intensities of non-specific signals between lanes. We have experimentally determined that these assessments are accurate and reproducible as long as the initial deviation between individual samples does not exceed a factor of two.

13.2 The p53 tumour suppressor

13.2.1 General considerations

Following exposure to UV light or other diverse DNA damaging agents, the p53 tumor suppressor protein transiently accumulates *via* inhibition of its ubiquitin-mediated proteolysis, and becomes further activated by way of structural modifications such as phosphorylation and acetylation [23]. This enables p53 to maintain genomic stability and inhibit neoplastic transformation through transactivation of genes, and through protein-protein interactions, that regulate cell cycle arrest, apoptosis, and DNA repair. The activity of p53 is somehow abrogated in the vast majority of human tumours, usually *via* mutations either in *p53* itself or in other genes implicated in the p53 regulatory pathway [24], underscoring the pivotal role of this tumour suppressor in cancer prevention. In addition, for preneoplastic cells carrying a wild-type *p53* gene, functional inactivation at the protein level can be achieved through intracellular expression of various viral oncoproteins including human papilloma virus (HPV-) E6. This gene product binds and efficiently stimulates the proteasomal degradation of p53 [25], thereby strongly inhibiting the latter's ability to accumulate and transactivate downstream effectors during periods of genotoxic stress [26]. As such, HPV-E6 expression in cultured cells has been extensively employed as a model for characterizing the phenotypic consequences of p53 inactivation, although it is becoming increasingly clear that this viral gene product modulates the activity of many cellular proteins aside from p53 [27]. Indeed, the extent to which exogenous HPV-E6 expression represents a true p53-null situation following mutagen exposure has been questioned [28].

The interaction between natural sunlight and p53 leading to skin cancer development has become a paradigm for multistage carcinogenesis. A series of experiments over the last decade established that initial exposures to solar UV can initiate nonmelanoma tumours through induction of CPD, leading to mutations in the *p53* gene of keratinocytes residing in the basal layer of the epidermis (tumour initiation) [29,30]. Subsequent sunlight exposures then favour clonal expansion (tumour promotion) specifically of these p53-mutated skin cells which, relative to surrounding wild-type counterparts, are much less efficiently eliminated by apoptosis (manifested as sunburn and peeling) following the acquisition of solar UV-induced DNA damage [31,32]. Notably, it appears that p53 may not protect against skin cancer through one of its otherwise hallmark functions, i.e., the initiation of growth arrest at G1/S following exposure to DNA damaging agents, since the p53 pathway has been shown to be nonessential for this latter process in UV-exposed cultured human cells [33,34].

13.2.2 p53 and NER

Another apparently significant contribution of p53 inactivation to genomic instability and multistage carcinogenesis was revealed by the discovery that this tumour suppressor plays a critical role in NER. Indeed it was shown in a variety of systems, using the well-characterized Southern blot-based assay described earlier, that functional p53 is required for GNER of CPD at the *dhfr* and *p53* loci following exposure to the model mutagen 254-nm UV [35–38]. However, at the same time, these studies consistently showed that p53 was completely dispensable for CPD removal along the TS of active genes *via* TCNER. Consistent with these observations were follow-up demonstrations that upregulation of the GNER-specific proteins XP-E and -C are p53-dependent in 254-nm UV-irradiated human cells [39,40]. However levels of other NER proteins were not investigated in the latter studies, nor is it clear that higher-than-basal expression of XP-E or -C, while existant, would be strictly required for GNER. An alternative mechanistic basis for p53-regulated GNER may relate to the well-established requirement for chromatin remodelling during DNA repair [41]. Specifically, it was recently demonstrated that efficient DNA photoproduct removal in UV-exposed human cells depends upon the ability of p53 to interact with the p300 histone acetyltransferase, resulting in the hyperacetylation and consequent relaxation of damaged heterochromatin [42]. In addition, it had been previously shown in murine embryonic fibroblasts that upregulation of the p53 downstream effector gadd45 may be necessary to promote physical access of the repair complex to sites of UV-induced DNA damage [38]. Such p53-dependent mechanisms of chromatin remodelling would be strictly required for GNER, but completely nonessential for TCNER where a "prexisiting" open DNA conformation is presumably bestowed by the basal transcription machinery.

13.2.3 Wavelength-dependent regulation of TCNER in p53-deficient human cells

We initially exploited the LMPCR technique to investigate the effect of p53 inactivation on strand-specific NER at nucleotide resolution in the same set of human fibroblast strains which were previously employed to convincingly demonstrate, at the level of the gene, an essential role for this tumour suppressor in GNER but not in TCNER [35–37]. We observed that relative to wild-type counterparts, p53-null and HPV-E6-expressing fibroblasts were clearly deficient in the removal of CPD along both the TS and NTS of the *c-jun* and *p53* target genes following irradiated with polychromatic UVB (290–320-nm), i.e., demonstrating in our hands that functional p53 is required for TCNER as well as GNER [43]. Although somewhat unexpected given the previous results of Ford and Hanawalt, our finding was nonetheless consistent with preceding investigations suggesting that p53 might regulate both NER subpathways. For example, p53 had been shown (i) to interact with the XP-B and XP-D helicases which mediate strand separation during GNER and TCNER, as well as with the CS-B protein that participates in lesion recognition during TCNER only [44] and, (ii) to transactivate a gene encoding the ribonucleotide reductase subunit p53R2 which is essential for the DNA resynthesis step common to GNER and TCNER [45]. In addition, p53-null human fibroblasts are deficient in the recovery of mRNA synthesis after UV treatment, indicating the requirement for a p53-dependent process to efficiently clear transcription-blocking CPD from the TS of active genes [46].

We were interested in understanding the basis for the apparent divergence between our results and those of Ford and Hanawalt regarding p53-dependence for transcription-coupled repair. One possibility centered on the use of distinct DNA repair assays, i.e., LMPCR vs. the Southern blot-based approach. We reasoned that this was unlikely, because both techniques have been rigorously validated as a means to measure strand-specific repair and, furthermore, each essentially quantifies the same event, i.e., the lesion recognition/incision step of NER as manifested by the disappearance of TEV-sensitive sites. Another formal explanation for the discrepancy in experimental outcomes concerned the use of different UV sources, either 254-nm UV or polychromatic UVB in the case of studies showing no role, or an important role, respectively, for p53 in TCNER. This again may have seemed unlikely at first, since CPDs underlie the vast majority of mutations induced by either 254-nm UV or UVB, and furthermore these wavelengths have been shown in many cases to elicit similar patterns of intracellular signaling and gene activation [47].

However, it is also well established that 254-nm UV and UVB manifest some important differences in their respective cellular effects (see below), which prompted us to rigorously investigate whether p53 might regulate NER in a wavelength-dependent manner. For this purpose, we employed the p53 +/+ human lymphoblastoid strain TK6 and two isogenic p53-deficient counterparts: NH32, carrying a homozygous knockout of *p53*; and TK6–5E, which constitutively expresses the HPV-E6 oncoprotein. Following treatment with either polychromatic UVB or 254-nm UV, using LMPCR, we compared the rate of CPD removal in each strain at nucleotide resolution along the TS and NTS of the autosomal *c-jun* protooncogene and the X-linked *hprt* locus. We were able to clearly show, using this well-defined isogenic human system, that loss of functional p53 significantly reduces the efficiency of GNER and TCNER in human cells exposed to polychromatic UVB; however following treatment with monochromatic 254-nm UV, p53 was shown to be required for efficient GNER but absolutely nonessential for TCNER [48]. This finding apparently reconciles the important controversy discussed above regarding the role of p53 in TCNER.

The mechanistic basis for wavelength-dependent regulation of TCNER in p53-deficient cells remains unknown. We speculate that 254-nm UV, but not UVB, may compensate for loss of p53 by stimulating the induction/activation of one or more proteins (normally also regulated by p53) that are required for TCNER. Such a process could be part of the so-called "mammalian UV-response", which comprises a series of signalling cascades initiated independently of nuclear DNA damage *via* UV-induced autophosphorylation of plasma membrane-associated growth-factor receptors [49]. In support of this general model, it has been demonstrated that UV-induced signal transduction originating at the plasma membrane can indeed be subject to wavelength-dependent regulation. For example, 254-nm UV was shown to be considerably more effective than UVB in the activation of various mitogen-activated protein kinases (including ERK and JNK) in human and murine cells [50–52]. An explanation for such phenomena may reflect the fact that exposure to UVB, to a much greater extent than to 254-nm UV, alters the cellular redox state through production of reactive oxygen species [53]. This, in turn, may be expected to engender some significant differences between 254-nm UV and UVB with respect to activation or inhibition of major stress-responsive signalling pathways which are known to be modulated by levels of oxidative stress [54]. In any case, the challenging task of characterizing the mechanistic foundation for wavelength-dependent regulation of TCNER may be expected to provide some of the "missing links" in our understanding of the relationship between DNA repair and the transcriptional machinery.

The revelation here that efficient removal of UV-induced DNA photoproducts can depend on incident wavelength has important implications for experimental photobiology. Monochromatic 254-nm UV is attenuated at least 10^7-fold by the earth's atmosphere, and therefore does not represent a biologically-significant component of the solar wavelength spectrum. On the other hand UVB is only partially blocked by stratospheric ozone, and clearly accounts for most of the pathological burden imposed by sunlight [55]. Nonetheless, irradiation with 254-nm UV has traditionally been widely employed to model the genotoxic and carcinogenic effects of solar UV. This situation has been encouraged by the convenience and ready availability of fluorescent lamps emitting 254-nm UV, coupled with the popular (but perhaps too-often erroneous) assumption that 254-nm UV and UVB produce equivalent genotoxic responses. Our results therefore highlight the need to employ environmentally-relevant UV sources, rather than the "nonsolar" model mutagen 254-nm UV, specifically for investigating the basis of sunlight-induced mutation and cancer. However the use of 254-nm UV may be more appropriate when probing mechanisms of nucleotide excision repair *per se,* i.e., as applies to the removal of helix-distorting lesions other than those induced by UV light. Indeed, it has been found that TCNER of bulky adducts induced by benzo(a)pyrene diolepoxide, in accord with the situation for 254-nm UV but not for UVB, does not depend on the presence of functional p53 [56]. On the other hand, in light of the possibility that TCNER efficiency depends on cellular redox potential, UVB may still constitute the better model for elimination of bulky adducts in an oxidizing environment, e.g., for lesions induced by the liver-specific carcinogen aflatoxin B1 during periods of chronic hepatic inflammation.

Acknowledgements

E.A.D. is supported by operating grants from the National Cancer Institute of Canada (with funds from the Canadian Cancer Society) and the Canadian Institutes of Health Research, and by a scholarship from the Fonds de la Recherche en Santé du Québec. J.D. holds a graduate student fellowship from the Fonds de la Recherche en Santé du Québec.

References

1. S. Tornaletti, G.P. Pfeifer (1994). Slow repair of pyrimidine dimers at p53 mutation hotspots in skin cancer. *Science*, **263**, 1436–1438.
2. R. Drouin, J.P. Therrien (1997). UVB-induced cyclobutane pyrimidine dimer frequency correlates with skin cancer mutational hotspots in p53. *Photochem. Photobiol.*, **66**, 719–726.
3. Y.H. You, P.E. Szabo, G.P. Pfeifer (2000). Cyclobutane pyrimidine dimers form preferentially at the major p53 mutational hotspot in UVB-induced mouse skin tumors. *Carcinogenesis*, **21**, 2113–2117.
4. Y.H. You, D.H. Lee, J.H. Yoon, S. Nakajima, A. Yasui, G.P. Pfeifer (2001). Cyclobutane pyrimidine dimers are responsible for the vast majority of mutations induced by UVB irradiation in mammalian cells. *J. Biol. Chem.* **276**, 44688–44694.
5. D.L. Mitchell, C.A. Haipek, J.M. Clarkson (1985). (6-4)Photoproducts are removed from the DNA of UV-irradiated mammalian cells more efficiently than cyclobutane pyrimidine dimmers. *Mutat. Res.*, **143**, 109–112.
6. D.L. Mitchell, R.S. Nairn (1989). The biology of the (6-4) photoproduct. *Photochem. Photobiol.*, **49**, 805–819.
7. W.L. de Laat, N.G. Jaspers, J.H. Hoeijmakers (1999). Molecular mechanism of nucleotide excision repair, *Genes Dev.*, **13**, 768–785.

8. S. Kugasawa, T. Okamoto, Y. Shimizu, C. Masutani, S. Iwai, F.A. Hanaoka (2001). Multistep damage recognition mechanism for global genomic nucleotide excision repair. *Genes Dev.*, **15**, 507–521.
9. J.Q. Svejstrup (2002). Mechanisms of transcription-coupled DNA repair. *Nat. Rev. Mol. Cell Biol.*, **3**, 21–29.
10. J.S. Cook, J.D. Regan (1969). Photoreactivation and photoreactivating enzyme activity in an order of mammals (Marsupialia). *Nature*, **223**, 1066–1067.
11. K.H. Kraemer, M.M. Lee, J. Scotto (1984). DNA repair protects against cutaneous and internal neoplasia: evidence from xeroderma pigmentosum. *Carcinogenesis*, **5**, 511–514.
12. J. Venema, L.H. Mullenders, A.T. Natarajan, A.A. van Zeeland, L.V. Mayne (1990). The genetic defect in Cockayne syndrome is associated with a defect in repair of UV-induced DNA damage in transcriptionally active DNA. *Proc. Natl. Acad. Sci.*, **87**, 4707–4711.
13. M.A. Nance, S.A. Berry (1992). Cockayne syndrome: review of 140 cases. *Am. J. Med. Genet.*, **42**, 68–84.
14. F. Le Page, E.E. Kwoh, A. Avrutskaya, A. Gentil, S.A. Leadon, A. Sarasin, P.K. Cooper (2001). Transcription-coupled repair of 8-oxoguanine: requirement for XPG, TFIIH, and CSB and implications for Cockayne syndrome. *Cell*, **101**, 59–171.
15. E.C. Friedberg, G.C. Walker, W. Siede (1995). *DNA Repair and Mutagenesis*. ASM Press, Washington D.C.
16. V.A. Bohr, C.A. Smith, D.S. Okumoto, P.C. Hanawalt (1985). DNA repair in an active gene: removal of pyrimidine dimers from the DHFR gene of CHO cells is much more efficient than in the genome overall. *Cell*, **40**, 359–369.
17. I. Mellon, G. Spivak, P.C. Hanawalt (1987). Selective removal of transcription-blocking DNA damage from the transcribed strand of the mammalian DHFR gene. *Cell*, **51**, 241–249.
18. M.P. Vreeswijk, A. van Hoffen, B.E. Westland, H. Vrieling, A.A. van Zeeland, L.H. Mullenders (1994). Analysis of repair of cyclobutane pyrimidine dimers and pyrimidine 6-4 pyrimidone photoproducts in transcriptionally active and inactive genes in Chinese hamster cells. *J. Biol. Chem.*, **269**, 31858–31863.
19. M.S. Tang, J.B. Zheng, M.F. Denissenko, G.P. Pfeifer, Y. Zheng (1999). Use of UvrABC nuclease to quantify benzo[a]pyrene diolepoxide-DNA adduct formation at methylated versus unmethylated CpG sites in the p53 gene. *Carcinogenesis*, **20**, 1085–1089.
20. M.S. Tang, V.A. Bohr, X.S. Zhang, J. Pierce, P.C. Hanawalt (1989). Quantification of aminofluorene adduct formation and repair in defined DNA sequences in mammalian cells using the UvrABC nuclease. *J. Biol. Chem.*, **264**, 14455–14462.
21. E.G. Snyderwine, V.A. Bohr (1992). Gene- and strand-specific damage and repair in Chinese hamster ovary cells treated with 4-nitroquinoline 1-oxide. *Cancer Res.*, **52**, 4183–4189.
22. R. Drouin, J.P. Therrien, M. Angers, S. Ouellette (2001). In: T. Moss (Ed.). *Methods in Molecular Biology: DNA-Protein Interactions, Principles, and Protocols* (pp. 175–219) Humana Press, Totawa, N.J.
23. N.D. Lakin, S.P. Jackson (1999). Regulation of p53 in response to DNA damage. *Oncogene*, **18**, 7644–7655.
24. X.W. Wang, C.C. Harris (1997). p53 tumor-suppressor gene: clues to molecular carcinogenesis. *J. Cell Physiol.* **173**, 247–255.
25. M. Scheffner, B.A. Werness, J.M. Huibregtse, A.J. Levine, P.M. Howley (1990). The E6 oncoprotein encoded by human papillomavirus types 16 and 18 promotes the degradation of p53. *Cell*, **63**, 1129–1136.

26. T.D. Kessis, R.J. Slebos, W.G Nelson, M.B. Kastan, B.S. Plunkett, S.M. Han, A.T. Lorincz, L. Hedrick, K.R. Cho (1993). Human papillomavirus 16 E6 expression disrupts the p53-mediated cellular response to DNA damage. *Proc. Natl. Acad. Sci.*, **90**, 3988–3992.
27. M. Thomas, D. Pim, L. Banks (1999). The role of the E6-p53 interaction in the molecular pathogenesis of HPV. *Oncogene*, **18**, 7690–7700.
28. C. Leger, E.A. Drobetsky (2002). Modulation of the DNA damage response in UV-exposed human lymphoblastoid cells through genetic-versus functional-inactivation of the p53 tumor suppressor. *Carcinogenesis*, **23**, 1631–1640.
29. D.E. Brash, J.A. Rudolph, J.A. Simon, A. Lin, G.J. McKenna, H.P. Baden, A.J. Halperin, J. Ponten (1991). A role for sunlight in skin cancer: UV-induced p53 mutations in squamous cell carcinoma. *Proc. Natl. Acad. Sci.*, **88**, 10124–10128.
30. A. Ziegler, D.J. Leffell, S. Kunala, H.W. Sharma, M. Gailani, J.A. Simon, A.J. Halperin, H.P. Baden, P.E. Shapiro, A. E. Bale, D.E. Brash (1993). Mutation hotspots due to sunlight in the p53 gene of nonmelanoma skin cancers. *Proc. Natl. Acad. Sci.*, **90**, 4216–4220.
31. A. Ziegler, A.S. Jonason, D.J. Leffell, J.A. Simon, H.W. Sharma, J. Kimmelman, L. Remington, T. Jacks, D.E. Brash (1994). Sunburn and p53 in the onset of skin cancer. *Nature*, **372**, 773–776.
32. A.S. Jonason, S. Kunala, G.J. Price, R.J. Restifo, H.M. Spinelli, J.A. Persing, D.J. Leffell, R.E. Tarone, D.E. Brash (1996). Frequent clones of p53-mutated keratinocytes in normal human skin. *Proc. Natl. Acad. Sci.*, **93**, 14025–14029.
33. M. Loignon, R. Fetni, A.J. Gordon, E.A. Drobetsky (1997). A p53-independent pathway for induction of p21waf1cip1 and concomitant G1 arrest in UV-irradiated human skin fibroblasts. *Cancer Res.*, **57**, 3390–3394.
34. M. Loignon, E.A. Drobetsky (2002). The initiation of UV-induced G(1) arrest in human cells is independent of the p53/p21/pRb pathway but can be attenuated through expression of the HPV E7 oncoprotein. *Carcinogenesis*, **23**, 35–45.
35. J.M. Ford, P.C. Hanawalt (1995). Li-Fraumeni syndrome fibroblasts homozygous for p53 mutations are deficient in global DNA repair but exhibit normal transcription-coupled repair and enhanced UV resistance. *Proc. Natl. Acad. Sci.*, **92**, 8876–8880.
36. J.M. Ford, P.C. Hanawalt (1997). Expression of wild-type p53 is required for efficient global genomic nucleotide excision repair in UV-irradiated human fibroblasts. *J. Biol. Chem.*, **272**, 28073–28080.
37. J.M. Ford, E.L. Baron, P.C. Hanawalt (1998). Human fibroblasts expressing the human papillomavirus E6 gene are deficient in global genomic nucleotide excision repair and sensitive to ultraviolet irradiation. *Cancer Res.*, **58**, 599–603.
38. M.L. Smith, J.M. Ford, M.C. Hollander, R.A. Bortnick, S.A. Amundson, Y.R. Seo, C.X. Deng, P.C. Hanawalt, A.J. Fornace Jr. (2000). p53-Mediated DNA Repair Responses to UV Radiation: Studies of Mouse Cells Lacking p53, p21, and/or gadd45 Genes. *Mol. Cell Biol.*, **20**, 3705–3714.
39. B.J. Hwang, J.M. Ford, P.C. Hanawalt, G. Chu (1999). Expression of the p48 xeroderma pigmentosum gene is p53-dependent and is involved in global genomic repair. *Proc. Natl. Acad. Sci.*, **96**, 424–428.
40. S. Adimoolam, J.M. Ford (2002). p53 and DNA damage-inducible expression of the xeroderma pigmentosum group C gene. *Proc. Natl. Acad. Sci.*, **99**, 12985–12990.
41. M.J. Smerdon, A. Conconi (1999). Modulation of DNA damage and DNA repair in chromatin. *Prog. Nucleic Acid Res. Mol. Biol.*, **62**, 227–255.
42. C.P. Rubbi, J. Milner (2003). p53 is a chromatin accessibility factor for nucleotide excision repair of DNA damage. *EMBO J.*, **22**, 975–986.

43. J.P. Therrien, R. Drouin, C. Baril, E.A. Drobetsky (1999). Human cells compromised for p53 function exhibit defective global and transcription-coupled nucleotide excision repair, whereas cells compromised for pRb function are defective only in global repair. *Proc. Natl. Acad. Sci.*, **96**, 15038–15043.
44. X.W. Wang, H. Yeh, L. Schaeffer, R. Roy, V. Moncollin, J.M. Egly, Z. Wang, E.C. Freidberg, M.K. Evans, B.G. Taffe, V.A. Bohr, G. Weeda, J.H.J. Hoeijmakers, K. Forrester, C.C. Harris (1995). p53 modulation of TFIIH-associated nucleotide excision repair activity. *Nature Genet.,* **10** 188–195.
45. H. Tanaka, H. Arakawa, T. Yamaguchi, K. Shiraishi, S. Fukuda, K. Matsui, Y. Takei, Y. Nakamura (2000). A ribonucleotide reductase gene involved in a p53-dependent cell-cycle checkpoint for DNA damage. *Nature*, **404**, 42–49.
46. B.C McKay, M.A. Francis, A.J. Rainbow (1997). Wildtype p53 is required for heat shock and ultraviolet light enhanced repair of a UV-damaged reporter gene. *Carcinogenesis*, **18**, 245–249.
47. R.M. Tyrrell (1996). UV activation of mammalian stress proteins. Exs. **77**, 255–271.
48. G. Mathonnet, C. Leger, J. Desnoyers, R. Drouin, J.P. Therrien, E.A. Drobetsky (2003). UV wavelength-dependent regulation of transcription-coupled nucleotide excision repair in p53-deficient human cells. *Proc. Natl. Acad. Sci. USA*, **100**, 7219–7224.
49. K. Bender, C. Blattner, A. Knebel, M. Iordanov, P. Herrlich, H.J. Rahmsdorf (1997). UV-induced signal transduction. *J. Photochem. Photobiol. B – Biology*, **37**, 1–17.
50. V. Adler, S.Y. Fuchs, J. Kim, A. Kraft, M.P. King, J. Pelling, J., Z. Ronai (1995). Jun-NH2-terminal kinase activation mediated by UV-induced DNA lesions in melanoma and fibroblast cells. *Cell Growth Differ.*, **6**, 1437–1446.
51. E.E. Medrano, S. Im, F. Yang, Z.A. Abdel-Malek (1995). Ultraviolet B light induces G1 arrest in human melanocytes by prolonged inhibition of retinoblastoma protein phosphorylation associated with long-term expression of the p21Waf-1/SDI-1/Cip-1 protein. *Cancer Res.*, **55**, 4047–4052.
52. K.R. Dhanwada, M. Dickens, R. Neades, R. Davis, J.C. Pelling (1995). Differential effects of UV-B and UV-C components of solar radiation on MAP kinase signal transduction pathways in epidermal keratinocytes. *Oncogene*, **11**, 1947–1953.
53. R.M. Tyrrell (1996). Activation of mammalian gene expression by the UV component of sunlight- from models to reality, Bioessays. **18**, 139–148.
54. V. Adler, Z. Yin, K.D. Tew, Z. Ronai (1999). Role of redox potential and reactive oxygen species in stress signaling. *Oncogene*, **18**, 6104–6111.
55. Y. Matsumura, H.N. Ananthaswamy (2002). Molecular mechanisms of photocarcinogenesis. *Front Biosci.*, **7**, 765–783.
56. M.A Wani, Q. Zhu, M. El-Mahdy, S. Venkatachalam, A.A. Wani (2000). Enhanced sensitivity to anti-benzo(a)pyrene-diol-epoxide DNA damage correlates with decreased global genomic repair attributeable to abrogated p53 function in human cells. *Cancer Res.*, **60**, 2273–2280.

Chapter 14

p53 and p33^{ING1}: Role in nucleotide excision repair of UV-damaged DNA

K-John Cheung Jr. and Gang Li

Table of contents

Abstract

The tumor suppressor p53 is considered to be a crucial molecule in the regulation of cellular responses to various DNA-damaging events. One of the well-established functions of p53 is the enhancement on repair of UV-damaged DNA. Recent studies indicate that p53 cooperates with, among other genes, the tumor suppressor ING1 to exert its biological functions. Among the many functions such as growth arrest, senescence, apoptosis, anchorage-dependent growth, and chemosensitivity, p33^{ING1} (one of *ING1* isoforms) has been found to be involved in nucleotide excision repair after UV irradiation. In this review, we will examine what is known about p53 in relations to repair of UV-damaged DNA and how p33^{ING1} became a recent addition to the p53-signaling pathway.

14.1 Nucleotide excision repair

Exposure to solar ultraviolet radiation (UVA and UVB) that penetrates the earth's atmosphere causes DNA damage in the skin [1]. If the lesions are not properly removed, mutations and ultimately carcinogenic transformation may occur. Consequently, an array of DNA repair systems has evolved in the cell to maintain the integrity and stability of the genetic material. One of the most important and versatile repair mechanisms is the nucleotide excision repair (NER), which eliminates a wide variety of DNA damage, including the common UV-induced lesions, cyclobutane pyrimidine dimers (CPD) and (6-4) photoproducts [2]. The sequence of the NER process consists of multi-steps: lesion recognition, strand incision, damaged nucleotide displacement, gap filing by DNA polymerization, and ligation [3]. Current evidence suggests that there are two NER subpathways: global genome repair (GGR) and transcription-coupled repair (TCR). Specifically, GGR operates genome-wide and is capable of removing DNA lesions from all locations in the genome at any moment in the cell cycle. TCR, on the other hand, acts on the transcribed strand of active genes, where it rapidly removes elongation-blocking lesions [3]. Three rare autosomal recessive inherited human disorders are associated with dysfunctional NER activity: xeroderma pigmentosum, cockayne syndrome, and trichothiodystrophy, all of which are characterized by a high sensitivity to sunlight [3,4].

14.2 *p53* Tumor suppressor gene

Genetic analysis of NER mutants and biochemical studies have provided evidence for the involvement of over 20 genes in the repair process. One of the key players in the NER network is the tumor suppressor p53. The *p53* gene, first described in 1979, maps to 17p13.1 and encodes a 393-amino acid, 53 kDa nuclear phosphoprotein [5–8]. It has been shown to be frequently mutated in sporadic cancer [9]. Specifically, more than 50% of human malignancies of epithelial, mesenchymal, haematopoietic, lymphoid, and nervous origin analyzed to date were shown to contain an altered *p53* gene. Germline mutations in p53 result in Li-Fraumeni syndrome, a hereditary cancer susceptibility syndrome predisposing individuals various cancers [9]. Since its isolation, p53 has been shown to exhibit a wide range of molecular functions, such as cell-cycle arrest, apoptosis, differentiation, angiogenesis, genetic stability, and DNA repair [10]. To further investigate the role of p53 in carcinogenesis, two transgenic models were put forth. Lavigueur et al. [11] generated transgenic mice by introducing mutant p53 gene fragments from tumour cell lines into mouse embryos. Donehower et al. [12] created p53-deficient mice by introducing a null mutation of the p53 gene in murine embryonic stem cells. Mice lacking the endogenous p53 gene or carrying a mutant p53 transgene have a near normal embryonic development but are prone to the spontaneous development of a variety of neoplasms [11,12]. After chronic UVB exposure, both mutant p53 transgenic and p53-deficient mice developed more tumors than wild-type (wt) control mice [13,14]. This outcome was later attributed to decreased DNA repair efficiency and reduced apoptosis in these transgenic mice [15–17].

14.3 *p53* Involvement in NER

Loss of p53 function, as occurs in more than 50% of human cancers, leads to decreased NER of specific lesions. Several independent groups observed that cells lacking functional p53 exhibited defective repair of UV damage and were more sensitive to UV irradiation than their wild-type p53 counterparts [18–27]. For instance, disruption of wt p53 function with either the human papillomavirus E6 oncoprotein or a mutant p53 transgene results in reduced repair of UV-induced DNA damage [18,23]. Furthermore, DNA repair efficiency is reduced in fibroblasts from Li-Fraumeni patients with germline p53 mutations [19]. p53 has also been shown to enhance γ-radiation-induced DNA repair [28]. Using p53 transgenic mouse models, we demonstrated that repair of UV-induced DNA damage is decreased in mouse keratinocytes isolated from transgenic mice which carry mutant alleles of p53 and p53 knockout mice, both *in vivo* and *in vitro* [15,16].

One of the shortcomings of most studies showing a role for p53 in NER is the reliance on tissue culture systems, which may not accurately reflect the *in vivo* conditions. We employed the *in situ* method to answer the question as to whether p53 regulated NER in the epidermis by monitoring the removal of UV-induced photoproducts. Consistent with previous findings, reduced repair efficiency of both CPDs and 6-4 photoproducts was observed in the epidermis of p53-null mice [17].

Several types of experimental techniques have been used to assay for the role of p53 in NER [29]. One approach utilized T4 endonuclease mapping of genomic DNA in UV-irradiated cells which showed that p53 enhanced the removal of UV-induced CPDs from the non-transcribed strand of the endogenous dihydrofolate reductase (DHFR) gene, while the transcribed strand was unaffected [19,30,31]. It was later discovered that p53 may also have a role in TCR [32,33; also see next paragraph]. Another approach used host cell reactivation (HCR) of UV-irradiated reporter plasmids in isogenic matched cells carrying wt or mutant p53 [18].

The assay relies on the ability of the cell to repair the transiently transfected plasmid DNA and thereby restore reporter gene expression [34]. Cells carrying wt p53 showed 3-fold higher levels of reactivation than mutant p53 cells [18]. Yet another approach utilized UV-damaged shuttle vectors, which were introduced into wt p53 or p53-deficient cells and then retrieved by transformation of competent *Escherichia coli*. Colonies carrying mutations within the shuttle plasmid are identified by a blue-white screen, so that mutation frequencies for each cell line may be calculated. Normal p53 resulted in a decreased mutation frequency which commensurated with increased DNA repair. In addition, the correct sequence information was restored in the presence of functional p53 [35]. These independent assays solidly established p53 as an enhancer of NER of UV-damaged DNA.

Information on whether p53 is involved in GGR or TCR can also be gathered from the fact that different techniques measure different aspects or subpathways of NER. For instance, some researchers view that the HCR assay measures TCR efficiencies (of the expressed reporter gene). The observed enhancement of DNA repair by p53 using this type of assay may therefore be attributed to TCR. This is substantiated by evidence presented by Therrien et al. [32] and Zhu et al. [33]. These authors found that, using the technique called ligation-mediated PCR (LMPCR), p53 is capable of modulating the repair of transcribed strands, as well as the repair of non-transcribed strands. LMPCR is highly sensitive, due to the fact that it is PCR-based, and measures repair at nucleotide resolution, i.e., at a number of different sites simultaneously, compared with only one overall assessment at the gene level in previous studies. A comprehensive study of HCR conducted by Ganesan et al. [34] further highlights the degree of contribution between TCR and GGR to HCR. They showed that, using TCR- or GGR-defective cell lines, both TCR and GGR contribute to HCR. They reported that the strength of the promoter element was one factor that affected TCR and, at least in some cell types, GGR may be occurring at a sufficient rate to effectively mask TCR. Hence, the HCR and shuttle vector experiments measure DNA repair attributed to both TCR and GGR pathways, consistent with the data that p53 largely mediates GGR [31], but also affects TCR, though to a lesser degree [32,33]. Finally, it should be cautioned that factors such as the genetic targets being assessed (DHFR vs. p53), the different UV wavelengths used for producing DNA damage, and cell types, may give rise to discrepancies in terms of the extent of TCR and GGR involvement [32].

The exact mechanism by which p53 promotes DNA repair remains to be elucidated. A likely possibility is the activation of, among other genes, the downstream target, GADD45. The gene was originally cloned on the basis of rapid induction in cells by UV light [36]. Later it was shown that GADD45 enhances NER [37]. Most evidence suggests that the GADD45 response is p53-dependent [21,38]. Smith et al. [39]. demonstrated that GADD45 stimulates nucleotide excision repair and inhibits entry of cells into S phase. Furthermore, antisense DNA directed against GADD45 mRNA reduced the repair of UV-damaged DNA [40]. Data from Hall and colleagues have suggested that the DNA repair response is mediated by the binding of GADD45 to PCNA, a protein involved in cell cycle regulation and DNA repair [41]. Other lines of evidence suggest that p53 may directly participate in NER through a protein-protein interaction with components such as XPB and

XPD, subunits of TFIIH involved in NER, or by binding to sites of DNA damage [30,42]. Bakalkin and colleagues showed that p53 has the ability to bind to single-stranded DNA ends and catalyze DNA renaturation and strand transfer [43]. *In vivo* evidence demonstrated that when Xpc homozygous mutant ($Xpc^{-/-}$) mice were crossed with $p53^{-/-}$ mutants, the rate of development of UV-induced skin cancer correlated with genotype as follows: $Xpc^{-/-}/p53^{-/-} > Xpc^{-/-}/p53^{+/-} > Xpc^{-/-}/p53^{+/+}$. Hence, inactivation of both p53 alleles augments the development of skin cancer in mice [44]. More recently, studies suggest that there exists a physical association between p53 and BRCA1 and BRCA2, both of which are important for cellular response to DNA damage by interacting with RAD50 and RAD51 [45–47].

14.4 p33^{ING1}: A tumor suppressor

Recently, the p33^{ING1} has been shown to cooperate with p53 in a variety of tumor suppressive functions [48]. The gene *ING1* (*in*hibitor of *g*rowth 1) was initially identified using subtractive hybridization between cDNAs from a non-transformed mammary epithelial cell line and breast cancer cell lines and subsequent selection of short cDNA fragments capable of promoting neoplastic transformation upon injection into nude mice [49].

The human *ING1* gene contains three exons, named 1a, 1b, and 2, and two introns [50]. It produces four mRNA variants from three different promoter regions (50–52) (Figure 1). The p47^{ING1} transcript consists of exons 1b and 2, while the p33^{ING1} transcript consists of exons 1a and 2. The third spliced form, p24^{ING1}, is composed of a truncated p47^{ING1} messenger with the first ATG codon in exon 2. A fourth *ING1* variant, p27^{ING1}, was recently detected [50,51]. Its transcript contains the part of the coding sequence in exon 1b and exon 2. All four *ING1* protein isoforms share an identical C-terminus with a conserved PHD finger motif (a C4HC3-type zinc finger spanning 50–80 aa residues), which is implicated in transcriptional regulation (50–53). Two possible nuclear localization signal motifs (which generally consist of positively charged aa, such as lysine and arginine) in the isoforms are also present [48].

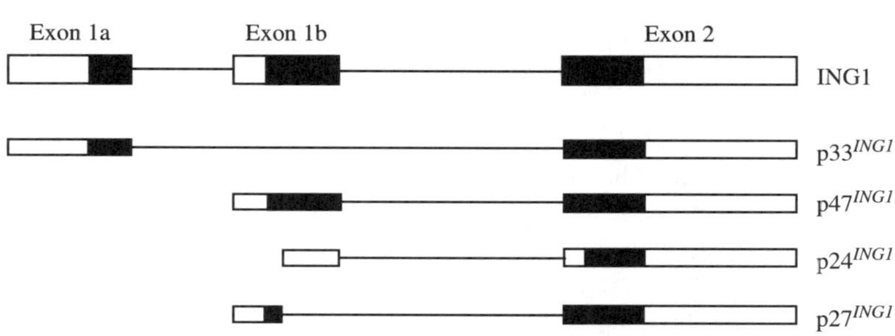

Figure 1. Genomic structure of human *ING1* and its alternatively spliced mRNA variants, p47^{ING1}, p33^{ING1}, p27^{ING1}, and p24^{ING1}, ■ Coding sequence in exon. □ Noncoding sequence in exon.

Fluorescent *in situ* hybridization (FISH) localized *ING1* to 13q33–34, while radiation hybrid mapping showed that *ING1* is linked to the cytogenetic marker SHGC-5819, which resides at 13q34 [54,55]. Immunofluorescence using the ING1 antibody indicated that ING1 protein products are primarily localized in the nucleus [54].

In *ING1* mutational and expression analyses, higher expression and rearrangement of *ING1* were reported in neuroblastoma cell lines [49]. Decreased or loss of expression of *ING1* was seen in lymphoid tumour cell lines [56], breast cancer primaries and cell lines [57,58], gastric cancers [59], and esophageal squamous cell cancer [60]. Conversely, we observed increased expression of *ING1* in human melanoma cell lines [61]. According to two other reports, the increased expression was observed mainly in the cytoplasm, suggesting that there is a failure of translocation of the protein from the cytoplasm to the nucleus [62,63]. Several missense and silent mutations in *ING1* were also detected in head and neck squamous cell carcinomas [50], melanoma cell lines [61], and esophageal squamous cell cancer [60]. Findings from these studies strongly support the notion that *ING1* is a tumour suppressor gene and plays a significant role in the process of carcinogenesis.

In terms of tumor suppressive functions, the p24^{ING1} isoform has been shown to inhibit cell growth in the G1 phase by transactivating the cyclin-dependent kinase inhibitor p21^{waf1} in the presence of *p53* [49,64]. p24^{ING1} can also sensitize cells to stress agents, such as etoposide and γ-irradiation, in wild-type (wt) but not *p53*-deficient cell lines [64]. Furthermore, *ING1* appears to play a role in senescence, as senescent cells express higher levels of ING1 and antisense *ING1* can increase cell's replicative life [65]. Overexpression of p24^{ING1} and p33^{ING1} enhances serum starvation-induced cell death and UV-induced apoptosis, respectively [66–68]. Two research groups have independently shown that apoptosis can be synergistically enhanced by adenovirus-mediated transfer of both p33^{ING1} and p53 in glioma and esophageal carcinoma cells [69,70]. Accumulating evidence suggests that p33^{ING1} is involved in chromatin remodeling functions by functionally and physically associating with histone acetyltransferases (HATs) and histone deacetylases (HDACs) [71–73].

14.5 Role of p33^{ING1} in stress response to UV

Scott et al. [67] recently demonstrated that *ING1* possesses two distinct nucleolar targeting sequences (NTS) within the nuclear localization signal region, which promotes the translocation of its encoded products to the nucleolus after UVR. The isoform p33^{ING1} was also found to contain a common octapeptide motif called the PCNA-interacting-protein (PIP) domain at the amino terminus, through which it binds competitively to the interdomain connector loop of PCNA upon UV irradiation [68,74,75]. These authors showed that human fibroblasts overexpressing p33^{ING1} have higher percentage of apoptosis compared to vector controls. Furthermore, Vieyra et al. [73] showed that up-regulation of p33^{ING1} sensitizes young but not senescent human fibroblasts to UV irradiation- and hydrogen peroxide-mediated apoptosis, suggesting that there is an age-dependent effect. In order to investigate the molecular pathways of p33^{ING1} enhancement in UV-induced apoptosis in biologically relevant cells, we transfected p33^{ING1} and antisense p33^{ING1} expression vectors into melanoma cells and irradiated with various doses of UVB [76]. Overexpression of p33^{ING1} was found to enhance the apoptosis rate of melanoma cells after UVB irradiation and p53 appeared to have a synergistic enhancement effect. A change in mitochondrial membrane potential and an increase in Bax expression in UV-irradiated p33^{ING1}-overexpressing cells were also observed [76], suggesting that p33^{ING1} enhances UV-induced apoptosis via the mitochondrial cell death pathway in melanoma cells.

To investigate if the expression of p33^{ING1} is induced by UV irradiation, we exposed normal human keratinocytes and a keratinocyte cell line HaCaT to UVB. Western blot analysis showed that p33^{ING1} protein is induced by UVB in a dose-dependent manner [77]. Furthermore, we demonstrated that the induction of p33^{ING1} by UVB is at the transcriptional level [77]. It appears that p33^{ING1} induction by UVB is cell-type specific as this induction was also observed in human melanoma cells but not in mouse dermal fibroblasts [77,78].

The numerous findings that the tumor suppressor p33^{ING1} shares similar biological functions with p53 [78] and that p33^{ING1} is induced by UVB led us to hypothesize that p33^{ING1} may also, like p53, participate in NER of UV-damaged DNA. To test this hypothesis, we co-transfected the human melanoma MMRU cells with vector, p33^{ING}, or antisense p33^{ING1} with a UV-damaged pCMVcat plasmid which carries a chloramphenicol acetyltransferase (CAT) reporter gene. The activity of the reporter gene was used as an indicator of the extent of repair. Our results demonstrated that cells overexpressing the p33^{ING1} gene had 2–4-fold increase in the rate of repair of UV-damaged plasmid DNA compared with the vector and antisense controls. The results were further confirmed by the radioimmunoassay (RIA), which measures the efficiency of GGR by monitoring the amount of remaining CPD photoproducts in the genomic DNA over a period of time. Moreover, p33^{ING1} enhancement on DNA repair was dramatically suppressed in cells transfected with a dominant-negative *p53* construct but restored in wt *p53*-transfected cells, suggesting a synergistic relationship between p33^{ING1} and p53. As mentioned, p53 upregulates GADD45 and binds to and modulates XPB and XPD [30], two components of the TFIIH transcription unit which possesses helicase, ATPase and kinase activity [42]. To further investigate if p33^{ING1} affects the expression of the XP genes and GADD45, we analyzed the levels of XPA, XPB, and GADD45 proteins in cells transfected with p33^{ING1} or antisense p33^{ING1} by Western blotting. Our results show that p33^{ING1} does not regulate the expression of XPA, XPB, and GADD45. However, we found that p33^{ING1} physically binds to GADD45, but not XPA or XPB, by immunoprecipitation, suggesting that p33^{ING1} may be a crucial component in the GADD45-mediated NER pathway. The fact that GADD45 is upregulated by p53 and that p33^{ING1} requires the participation of functional p53 in DNA repair further supports the close association of p33^{ING1} and GADD45. Increasing evidence has indicated that GADD45 is essential in UV-damaged DNA repair and genome stability [37,40,79]. Recently, an interesting report shows that GADD45 can recognize UV-altered chromatin state and modulate DNA accessibility to repair proteins such as DNase I and T4 endonuclease V [80].

It would be of interest to exploit the mechanistic role of p33^{ING1} in this GADD45-mediated repair process.

14.6 Concluding remark

Over the past few years, numerous studies using a variety of assays have demonstrated that p53 plays a crucial role in the repair of UV-damaged DNA. However, the exact molecular pathway as to how p53 enhances NER is still not completely understood. One of the questions frequently raised concerning the relationship between p53 and NER is to what degree p53 is directly involved in DNA repair, versus p53 transcriptionally regulated gene products that contribute to the process of NER [40]. For instance, p53 has been shown to participate in the recognition of primary DNA damage [81,82], and associate with TFIIH, a multiprotein complex required in NER [30,83], while genes implicated in repair, such as XPC, XPE, and GADD45, are p53-regulated [80,84,85]. Our findings that p33^{ING1} cooperates with p53 to enhance NER and that p33^{ING1} physically binds to GADD45 further add to the complexity of the NER pathway and should stimulate further interest in the mechanism of p53-mediated DNA repair and the search for new genes involved in this network.

References

1. H.R. Griffiths, P. Mistry, K.E. Herbert, J. Lunec (1998). Molecular and cellular effects of ultraviolet light-induced genotoxicity. *Crit. Rev. Clin. Lab. Sci.*, **35**, 189–237.
2. T. Maeda, P.P. Chua, M.T. Chong, A.B. Sim, O. Nikaido, V.A. Tron (2001). Nucleotide excision repair genes are upregulated by low-dose artificial ultraviolet B: evidence of a photoprotective SOS response? *J. Invest. Dermatol.*, **117**, 1490–1497.
3. S. Emmert, N. Kobayashi, S.G. Khan, K.H. Kraemer (2000). The xeroderma pigmentosum group C gene leads to selective repair of cyclobutane pyrimidine dimers rather than 6-4 photoproducts. *Proc. Natl. Acad. Sci. USA*, **97**, 2151–2156.
4. M. van Oosten, H. Rebel, E.C. Friedberg, H. van Steeg, G.T. van der Horst, H.J. van Kranen, A. Westerman, A.A. van Zeeland, L.H. Mullenders, F.R. de Gruijl (2000). Differential role of transcription-coupled repair in UVB-induced G2 arrest and apoptosis in mouse epidermis. *Proc. Natl. Acad. Sci. USA*, **97**, 11268–11273.
5. M. Kress, E. May, R. Cassingena, P. May (1979). Simian virus 40-transformed cells express new species of proteins precipitable by anti-simian virus 40 tumor serum. *J. Virol.*, **31**, 472–483.
6. D.P. Lane, L.V. Crawford (1979). T antigen is bound to a host protein in SV40-transformed cells. *Nature*, **278**, 261–263.
7. D.I. Linzer, A.J. Levine (1979). Characterization of a 54K dalton cellular SV40 tumor antigen present in SV40-transformed cells and uninfected embryonal carcinoma cells. *Cell*, **17**, 43–52.
8. C.A. Finlay, P.W. Hinds, A.J. Levine (1989). The *p53* proto-oncogene can act as a suppressor of transformation. *Cell*, **57**, 1083–1093.
9. R.R. Wallace-Brodeur, S.W. Lowe (1999). Clinical implications of *p53* mutations. *Cell. Mol. Life Sci.*, **55**, 64–75.

10. A.J. Levine (1997). p53, the cellular gatekeeper for growth and division. *Cell*, **88**, 323–331.
11. A. Lavigueur, V. Maltby, D. Mock, J. Rossant, T. Pawson, A. Bernstein (1989). High incidence of lung, bone, and lymphoid tumors in transgenic mice overexpressing mutant alleles of the p53 oncogene. *Mol. Cell. Biol.*, **9**, 3982–3991.
12. L.A. Donehower, M. Harvey, B.L. Slagle, M.J. McArther, C.A. Montgomery Jr., J.S. Butel, A. Bradley (1992). Mice deficient for p53 are developmentally normal but susceptible to spontaneous tumours. *Nature*, **356**, 215–221.
13. G. Li, V.C. Ho, K. Berean, V.A. Tron (1995). Ultraviolet radiation induction of squamous cell carcinomas in p53 transgenic mice. *Cancer Res.*, **55**, 2070–2074.
14. G. Li, V. Tron, V. Ho (1998). Induction of squamous cell carcinoma in p53-deficient mice after ultraviolet irradiation. *J. Invest. Dermatol.*, **110**, 72–75.
15. G. Li, D.L. Mitchell, V.C. Ho, J.C. Reed, V.A. Tron(1996). Decreased DNA repair but normal apoptosis in UV-exposed keratinocytes of p53 transgenic mice. *Am. J. Pathol.*, **148**, 1113–1123.
16. G. Li, V.C. Ho, D.L. Mitchell, M.J. Trotter, V.A. Tron (1997). Differentiation-dependent p53 regulation of nucleotide excision repair in keratinocytes. *Am. J. Pathol.*, **150**, 1457–1464.
17. V.A. Tron, M.J. Trotter, L. Tang, M. Krajewska, J.C. Reed, V.C. Ho, G. Li (1998). p53-regulated apoptosis is differentiation-dependent in UVB irradiated mouse keratinocytes. *Am. J. Pathol.*, **153**, 579–585.
18. M.L. Smith, I.T. Chen, Q. Zhan, P.M. O'Connor, A.J. Fornace Jr. (1995). Involvement of the p53 tumor suppressor in repair of u.v.-type DNA. *Oncogene*, **10**, 1053–1059.
19. J.M. Ford, P.C. Hanawalt (1995). Li-Fraumeni syndrome fibroblasts homozygous for p53 mutations are deficient in global DNA repair but exhibit normal transcription-coupled repair and enhanced UV resistance. *Proc. Natl. Acad. Sci. USA*, **92**, 8876–8880.
20. P.A. Havre, J. Yuan, L. Hedrick, K.R. Cho, P.M. Glazer (1995). p53 inactivation by HPV16 E6 results in increased mutagenesis in human cells. *Cancer Res.*, **55**, 4420–4424.
21. Q. Zhan, S. Fan, I. Bae, M.L. Smith, K. Yu, I. Alamo, P.M. O'Conner, A.J. Fornace Jr. (1996). Abrogation of p53 function affects gadd gene responses to DNA base-damaging agents and starvation. *DNA & Cell Biol.*, **15**, 805–815.
22. G. Li, V.C. Ho (1998). p53-dependent DNA repair and apoptosis respond differently to high and low dose ultraviolet radiation. *Br. J. Dermatol.*, **139**, 3–10.
23. J.M. Ford, E.L. Baron, P.C. Hanawalt (1998). Human fibroblasts expressing the human papillomavirus E6 gene are deficient in global genomic nucleotide excision repair and sensitive to ultraviolet irradiation. *Cancer Res.*, **58**, 599–603.
24. C.A. Cistulli, W.K. Kaufmann (1998). p53-dependent signaling sustains DNA replication and enhances clonogenic survival in 254 nm ultraviolet-irradiated human fibroblasts. *Cancer Res.*, **58**, 1993–2002.
25. D. Chang, F. Chen, F. Zhang, B.C. McKay, M. Ljungman (1999). Dose-dependent effects of DNA-damaging agents on p53-mediated cell cycle arrest. *Cell Growth Differ.*, **10**, 155–162.
26. M.A. El-Mahdy, F.M. Hamada, M.A. Wani, Q. Zhu, A.A. Wani (2000). p53-degradation by HPV-16 E6 preferentially affects the removal of cyclobutane pyrimidine dimers from non-transcribed strand and sensitizes mammary epithelial cells to UV-irradiation. *Mutat. Res.*, **459**, 135–145.
27. B.C. McKay, F. Chen, C.R. Perumalswami, F. Zhang, M. Ljungman (2000). The tumor suppressor p53 can both stimulate and inhibit ultraviolet light-induced apoptosis. *Mol. Biol. Cell*, **11**, 2543–2551.

28. S.M. Mallya, M.O. Sikpi (1998). Evidence of the involvement of p53 in gamma-radiation-induced DNA repair in human lymphoblasts. *Int. J. Rad. Biol.*, **74**, 231–238.
29. M.L. Smith, Y.R. Seo (2002). p53 regulation of DNA excision repair pathways. *Mutagenesis*, **17**, 149–156.
30. X.W. Wang, H. Yeh, L. Schaeffer, R. Roy, V. Moncollin, J.M. Egly, Z. Wang, E.C. Freidberg, M.K. Evans, B.G. Taffe (1995). p53 modulation of TFIIH-associated nucleotide excision repair activity. *Nature Genet.*, **10**, 188–195.
31. J.M. Ford, P.C. Hanawalt (1997). Expression of wild-type p53 is required for efficient global genomic nucleotide excision repair in UV-irradiated human fibroblasts. *J. Biol. Chem.*, **272**, 28073–28080.
32. J.P. Therrien, R. Drouin, C. Baril, E.A. Drobetsky (1999). Human cells compromised for p53 function exhibit defective global and transcription-coupled nucleotide excision repair, whereas cells compromised for pRb function are defective only in global repair. *Proc. Natl. Acad. Sci. USA*, **96**, 15038–15043.
33. Q. Zhu, M.A. Wani, M. El-Mahdy, A.A. Wani (2000). Decreased DNA repair efficiency by loss or disruption of p53 function preferentially affects removal of cyclobutane pyrimidine dimers from non-transcribed strand and slow repair sites in transcribed strand. *J. Biol. Chem.*, **275**, 11492–11497.
34. A.K. Ganesan, J. Hunt, P.C. Hanawalt (1999). Expression and nucleotide excision repair of a UV-irradiated reporter gene in unirradiated human cells. *Mutat. Res.*, **433**, 117–126.
35. J. Yuan, T.M. Yeasky, P.A. Havre, P.M. Glazer (1995). Induction of p53 in mouse cells decreases mutagenesis by UV radiation. *Carcinogenesis*, **16**, 2295–2300.
36. A.J. Fornace Jr., I.J. Alamo, M.C. Hollander (1998). DNA damage-inducible transcripts in mammalian cells. *Proc. Natl. Acad. Sci USA*, **85**, 8800–8804.
37. M.L. Smith, J.M. Ford, M.C. Hollander, R.A. Bortnick, S.A. Amundson, Y.R. Seo, C.X. Deng, P.C. Hanawalt, A.J. Fornace Jr. (2000). p53-mediated DNA repair responses to UV radiation: studies of mouse cells lacking p53, p21, and/or gadd45 genes. *Mol. Cell Biol.*, **20**, 3705–3714.
38. M.B. Kastan, Q. Zhan, W.S. El-Deiry, F. Carrier, T. Jacks, W.V. Walsh, B.S. Plunkett, B. Vogelstein, A.J. Fornace Jr. (1992). A mammalian cell cycle checkpoint pathway utilizing p53 and GADD45 is defective in ataxia-telangiectasia. *Cell*, **71**, 587–597.
39. M.L. Smith, I. Chen, Q. Zhan, I. Bae, C.Y. Chen, T.M. Gilmer, M.B. Kastan, P.M. O'Conner, A.J. Fornace Jr. (1994). Interaction of the p53-regulated protein GADD45 with proliferating cell nuclear antigen. *Science*, **266**, 1376–1380.
40. M.L. Smith, H.U. Kontny, Q. Zhan, A. Sreenath, P.M. O'Connor, A.J. Fornace Jr. (1996). Antisense GADD45 expression results in decreased DNA repair and sensitizes cells to u.v.-irradiation or cisplatin. *Oncogene*, **13**, 2255–2263.
41. P.A. Hall, J.M. Kearsey, P.J. Coates, D.G. Norman, E. Warbrick, L.S. Cox (1995). Characterisation of the interaction between PCNA and Gadd45. *Oncogene*, **10**, 2427–2433.
42. X.W. Wang, K. Forrester, H. Yeh, M.A. Feitelson, J.R. Gu, C.C. Harris (1994). Hepatitis B virus X protein inhibits p53 sequence-specific DNA binding, transcriptional activity, and association with transcription factor ERCC3. *Proc. Natl. Acad. Sci. USA*, **91**, 2230–2234.
43. G. Bakalkin, T. Yakovleva, G. Selivanova, K.P. Magnusson, L. Szekely, E. Kiseleva, G. Klein, L. Terenius, K.G. Wiman (1994). p53 binds single-stranded DNA ends and catalyzes DNA renaturation and strand transfer. *Proc. Natl. Acad. Sci. USA*, **91**, 413–417.
44. E.C. Friedberg (2001). How nucleotide excision repair protects against cancer. *Nat. Rev. Cancer*, **1**, 22–33.

45. Q. Zhong, C. Chen, S. Li, Y. Chen, C. Want, J. Xiao, P. Chen, Z. Sharp, W.H. Lee (1999). Association of BRCA1 with the hRad50-hMre11-p95 complex and the DNA damage response. *Science*, **285**, 747–750.
46. Y.L. Chai, J. Cui, N. Shao, E. Shyam, P. Reddy, V.N. Rao. (1999). The second BRCT domain of BRCA1 proteins interacts with p53 and stimulates transcription from the p21waf1/cip1 promoter. *Oncogene*, **18**, 263–268.
47. L.Y. Marmorstein, T. Ouchi, S.A. Aaronson (1998). The BRCA2 gene product functionally interacts with p53 and RAD51. *Proc. Natl. Acad. Sci. USA*, **95**, 13869–13874.
48. K.J. Cheung Jr., G. Li (2001). The tumor suppressor ING1: structure and function. Exp. *Cell Res.*, **268**, 1–6.
49. I. Garkavtsev, A. Kazarov, A. Gudkov, K. Riabowol (1996). Suppression of the novel growth inhibitor $p33^{ING1}$ promotes neoplastic transformation. *Nat. Genet.*, **14**, 415–420.
50. M. Gunduz, M. Ouchida, K. Fukushima, H. Hanafusa, T. Etani, S. Nishioka, K. Nishizaki, K. Shimizu (2000). Genomic structure of the human *ING1* gene and tumor-specific mutations detected in head and neck squamous cell carcinomas. *Cancer Res.*, **60**, 3143–3146.
51. D. Jager, E. Stockert, M.J. Scanlan, A.O. Gure, E. Jager, A. Knuth, L.J. Old, Y.T. Chen (1999). Cancer-testis antigens and *ING1* tumor suppressor gene product are breast cancer antigens: characterization of tissue-specific *ING1* transcripts and a homologue gene. *Cancer Res.*, **59**, 6197–6204.
52. A. Saito, T. Furukawa, S. Fukushige, S. Koyama, M. Hoshi, Y. Hayashi, A. Horii (2000). p24/*ING1*-ALT1 and p47/*ING1*-ALT2, distinct alternative transcripts of p33*ING1*. *J. Hum. Genet.*, **45**, 177–181.
53. I. Garkavtsev, A. Kazarov, A. Gudkov, K. Riabowol (1999). Correction. *Nat. Genet.*, **23**, 373.
54. I. Garkavtsev, D. Demetrick, K. Riabowol (1997). Cellular localization and chromosome mapping of a novel candidate tumor suppressor gene (*ING1*). *Cytogenet. Cell Genet.*, **76**, 176–178.
55. M. Zeremski, S.K. Horrigan, I.A. Grigorian, C.A. Westbrook, A.V. Gudkov (1997). Localization of the candidate tumor suppressor gene *ING1* to human chromosome 13q34. *Somat. Cell Mol. Genet.*, **23**, 233–236.
56. M. Ohmori, M. Nagai, T. Tasaka, H.P. Koeffler, K. Riabowol, J. Takahara (1999). Decreased expression of $p33^{ING1}$ mRNA in lymphoid malignancies. *Am. J. Hematol.*, **62**, 118–119.
57. T. Toyama, H. Iwase, P. Watson, H. Muzik, E. Saettler, A. Magliocco, L. DiFrancesco, P. Forsyth, I. Garkavtsev, S. Kpbayashi, K. Riabowol (1999). Suppression of *ING1* expression in sporadic breast cancer. *Oncogene*, **18**, 5187–5193.
58. E. Tokunaga, Y. Maehara, E. Oki, K. Kitamura, Y. Kakeji, S. Ohno, K. Sugimachi (2000). Diminished expression of *ING1* mRNA and the correlation with p53 expression in breast cancers. *Cancer Lett.*, **152**, 15–22.
59. E. Oki, Y. Maehara, E. Tokunaga, Y. Kakeji, K. Sugimachi (1999). Reduced expression of $p33^{ING1}$ and the relationship with p53 expression in human gastric cancer. *Cancer Lett.*, **147**, 157–162.
60. L. Chen, N. Matsubara, T. Yoshino, T. Nagasaka, N. Hoshizima, Y. Shirakawa, Y. Naomoto, H. Isozaki, K. Riabowol, N. Tanaka (2001). Genetic alterations of candidate tumor suppressor ING1 in human esophageal squamous cell cancer. *Cancer Res.*, **61**, 4345–4349.
61. E.I. Campos, K.-J. Cheung Jr., A. Murray, S. Li, G. Li (2002). The novel tumour suppressor gene ING1 is overexpressed in human melanoma cell lines. *Br. J. Dermatol.*, **146**, 574–580.

62. G.S. Nouman, J.J. Anderson, M.E. Mathers, N. Leonard, S. Crosier, J. Lunec, B. Angus (2002). Nuclear to cytoplasmic compartment shift of the p33^{ING1b} tumour suppressor protein is associated with malignancy in melanocytic lesions. *Histopathology*, **40**, 360–366.
63. G.S. Nouman, B. Angus, J. Lunec, S. Crosier, A. Lodge, J.J. Anderson (2002). Comparative assessment expression of the inhibitor of growth 1 gene (ING1) in normal and neoplastic tissues. *Hybrid Hybridomics*, **21**, 1–10.
64. I. Garkavtsev, I.A. Grigorian, V.S. Ossovskaya, M.V. Chernov, P.M. Chumakov, A.V. Gudkov (1998). The candidate tumour suppressor p33^{ING1} cooperates with p53 in cell growth control. *Nature*, **391**, 295–298.
65. I. Garkavtsev, K. Riabowol (1997). Extension of the replicative life span of human diploid fibroblasts by inhibition of the p33^{ING1} candidate tumor suppressor. *Mol. Cell. Biol.*, **17**, 2014–2019.
66. C.C. Helbing, C. Veillette, K. Riabowol, R.N. Johnston, I. Garkavtsev (1997). A novel candidate tumor suppressor, *ING1*, is involved in the regulation of apoptosis. *Cancer Res.*, **57**, 1255–1258.
67. M. Scott, F.M. Boisvert, D. Vieyra, R.N. Johnston, D.P. Bazett-Jones, K. Riabowol (2001). UV induces nucleolar translocation of *ING1* through two distinct nucleolar targeting sequences. *Nucleic Acids Res.*, **29**, 2052–2058.
68. M. Scott, P. Bonnefin, D. Vieyra, F.M. Boisvert, D. Young, D.P. Bazett-Jones, K. Riabowol (2001). UV-induced binding of *ING1* to PCNA regulates the induction of apoptosis. *J. Cell Sci.*, **114**, 3455–3462.
69. N. Shinoura, Y. Muramatsu, M. Nishimura, Y. Yoshida, A. Saito, T. Yokoyama, T. Furukawa, A. Horii, M. Hashimoto, A. Asai, T. Kirino, H. Hamada (1999). Adenovirus-mediated transfer of p33^{ING1} with p53 drastically augments apoptosis in gliomas. *Cancer Res.*, **59**, 5521–5528.
70. H. Shimada, T.L. Liu, T. Ochiai, T. Shimizu, Y. Haupt, H. Hamada, T. Abe, M. Oka, M. Takiguchi, T. Hiwasa (2002). Facilitation of adenoviral wild-type p53-induced apoptotic cell death by overexpression of p33^{ING1} in T.Tn human esophageal carcinoma cells. *Oncogene*, **21**, 1208–1216.
71. D. Skowyra, M. Zeremski, N. Neznanov, M. Li, Y. Choi, M. Uesugi, C. Hauser, W. Gu, A. Gudkov, J. Qin (2001). Differential association of products of alternativ transcripts of the candidate tumor suppressor *ING1* with the mSin3/HDAC1 transcriptional corepressor complex. *J. Biol. Chem.*, **276**, 8734–8739.
72. A. Kuzmichev, Y. Zhang, H. Erdjument-Bromage, P. Tempst, D. Reinberg (2002). Role of the Sin3-histone deacetylase complex in growth regulation by the candidate tumor suppressor p33^{ING1}. *Mol. Cell. Biol* , **22**, 835–848.
73. D. Vieyra, T. Toyama, Y. Hara, D. Boland, R. Johnston, K. Riabowol (2002). ING1 isoforms differentially affect apoptosis in a cell age-dependent manner. *Cancer Res.*, **62**, 4445–4452.
74. E. Warbrick (1998). PCNA binding through a conserved motif. *Bioessays*, **20**, 195–199.
75. T. Tsurimoto (1999). PCNA binding proteins. *Front Biosci.*, **4**, D849–858.
76. K.-J. Cheung, G. Li (2002). p33^{ING1} enhances UVB-induced apoptosis in melanoma cells. *Exp. Cell Res.*, In press.
77. K.-J. Cheung, J.A. Bush, W. Jia, G. Li (2000). Expression of the novel tumour suppressor p33^{ING1} is independent of p53. *Br. J. Cancer*, **83**, 1468–1472.
78. K.-J. Cheung, D. Mitchell, P. Lin, G. Li (2001). The novel tumor suppressor p33^{ING1} mediates repair of UV-damaged DNA. *Cancer Res.*, **61**, 4974–4977.
79. M.C. Hollander, M.S. Sheikh, D.V. Bulavin, K. Lundgren , L. Augeri-Henmueller, R. Shehee , T.A. Molinaro, K.E. Kim, E. Tolosa , J.D. Ashwell, M.P. Rosenberg,

Q. Zhan, P.M. Fernandez-Salguero, W.F. Morgan, C.X. Deng, A.J. Fornace Jr. (1999). Genomic instability in Gadd45a-deficient mice. *Nat. Genet.*, **23**, 176–184.
80. F. Carrier, P.T. Georgel, P. Pourquier, M. Blake, H.U. Kontny, M.J. Antinore, M. Gariboldi, T.G. Myers, J.N. Weinstein, Y. Pommier, A.J. Fornace Jr. (1999). Gadd45, a p53-responsive stress protein, modifies DNA accessibility on damaged chromatin. *Mol. Cell. Biol.*, **19**, 1673–1685.
81. S. Lee, B. Elenbaas, A. Levine, J. Griffith (1995). p53 and its 14 kDa C-terminal domain recognize primary DNA damage in the form of insertion/deletion mismatches. *Cell*, **81**, 1013–1020.
82. M. Reed, B. Woelker, P. Wang, Y. Wang, M.E. Anderson, P. Tegtmeyer (1995). The C-terminal domain of p53 recognizes DNA damaged by ionizing radiation. *Proc. Natl. Acad. Sci. USA*, **92**, 9455–9459.
83. T. Leveillard, L. Andera, N. Bissonnette, L. Schaeffer , L. Bracco, J.M. Egly, B. Wasylyk (1996). Functional interactions between p53 and the TFIIH complex are affected by tumour-associated mutations. *EMBO J.*, **15**, 1615–1624.
84. S.A. Amundson, K.T. Do, S. Shahab, M. Bittner, P. Meltzer, J. Trent, A.J. Fornace Jr. (2000). Identification of potential mRNA biomarkers in peripheral blood lymphocytes for human exposure to ionizing radiation. *Radiat. Res.*, **154**, 342–346.
85. B.J. Hwang, J.M. Ford, P.C. Hanawalt, G. Chu (1999). Expression of the p48 xeroderma pigmentosum gene is p53-dependent and is involved in global genomic repair. *Proc. Natl. Acad. Sci. USA*, **96**, 424–428.

Chapter 15

Nuclear and non-nuclear signals leading to UV-induced apoptosis

Momchil D. Vodenicharov and Girish M. Shah

Table of contents

Abstract

Cells that have been irreversibly damaged by solar ultraviolet (UV) irradiation are eliminated by apoptosis, which is best exemplified by appearance of sunburn cells, i.e., apoptotic keratinocytes in the skin overexposed to sunlight. UV causes a variety of changes in the cell, and some of these nuclear and sub-nuclear events that serve as initiating signals for apoptosis are reviewed here. UV-induced DNA damage, in conjunction with activation of tumor suppressor protein p53, has long been considered as the predominant signal that leads to apoptosis. Recent studies have shown that apoptosis can also be initiated in the UV damaged cells by non-nuclear events, such as Fas or TNF receptor aggregation at the cell surface or activation of MAP kinase family members, such as JNK and p38 MAP kinase in the cytoplasm. Finally, reactive oxygen species formed in the cytoplasm of UV exposed cells can initiate apoptosis either directly or indirectly through DNA damage or MAP kinase signalling cascade. The presence of multiple independent apoptosis-initiating pathways ensures a timely elimination of UV damaged cell before it can turn cancerous.

15.1 Introduction

Solar ultraviolet radiation (UV), specifically UVB radiation (280–315 nm), is recognized as a major etiological factor in the development of skin cancer [1–3]. Despite continuous solar UV irradiation of vast surface of our skin, majority of us enjoy cancer-free existence, which is a testimony to our capacity to either repair and restore the UV damaged cells or to eliminate by apoptosis those cells that have been damaged beyond repair. A resistance to apoptotic death is one of the early events in development of skin cancer [3], and therefore, understanding the molecular mechanisms by which UV irradiated cells undergo apoptosis will permit us to prevent or cure skin cancers.

Mammalian cells respond to UV irradiation by initiating a vast array of events inside the nucleus as well as at the cell membrane and in the cytoplasm (Figure 1). In the nucleus, UV radiation causes a variety of direct and indirect DNA damages, with bulk of the damage appearing as base modifications in the form of cyclobutane pyrimidine dimers (CPD) and 6,4-photoproducts. The damaged DNA is repaired by different DNA repair pathways, principal among which is the nucleotide excision repair (NER) pathway that eliminates CPD and 6,4-PP damages [1,4]. Cells respond to the DNA damage by increased expression or activation of many early responsive genes, such as tumor-suppressor p53, Gadd45, c-fos, c-jun and NF-*k*B [5–7]. Interestingly, some of these genes, such as NF-*k*B can also be activated by signals that arise outside the nucleus [8], indicating a degree of overlap between nuclear and non-nuclear signalling events. The activities of these genes participate in cell survival, cell cycle arrest or cell death, and tumor suppressor gene p53 plays a central role in these responses [3]. At the plasma membrane, UV irradiated cells activate Fas or TNF receptors, which recruit and activate upstream caspase 8 leading to apoptotic death [9–11]. In the cytoplasm, different members of the

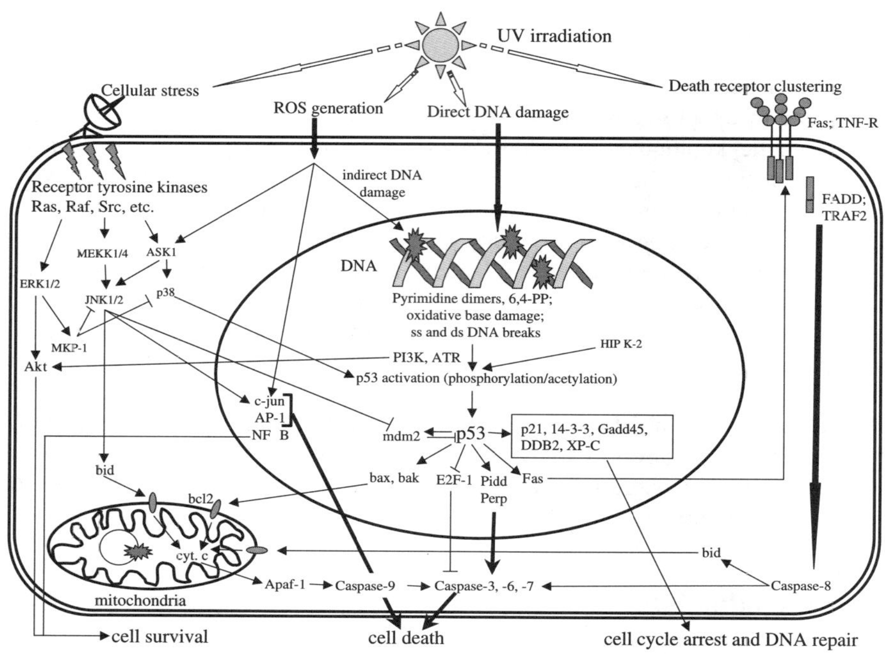

Figure 1. Cellular responses to UV. Mammalian cells respond to UV irradiation with a diverse array of reactions originating in the nucleus, at the cell membrane and in the cytoplasm. A net-sum of these pro-survival and pro-apoptotic reactions determines the eventual fate of the cell, namely, DNA repair and survival, mutation and cancer or cell death by apoptosis.

mitogen-activated protein kinase (MAP kinase) family [12] are activated, which participate in either pro-survival (e.g., ERK) [13] or pro-apoptotic (e.g., JNK and p38-MAP kinase) functions [8,12,14,15] (Figure 1). Finally, reactive oxygen species (ROS), generated in the cytoplasm [1,4,16] can induce different pro- or anti-apoptotic events, by mediating its effect through oxidative DNA damage, activation of kinase signalling cascade or activation of different genes (Figure 1). A net-sum of all of these events determines the eventual fate of UV irradiated cell, i.e., repair and recovery, mutation and cancer or cell death by apoptosis.

Cells have evolved intricate defence mechanisms to counteract the harmful effects of UV irradiation. The DNA repair and recovery from lower levels of UV damage is the first line of defense against UV by prokaryotes as well as eukaryotes. However, if not properly repaired, persisting DNA damage can result in mutations in oncogenes and tumor suppressor genes, leading to permanent alterations in signalling pathways that control cell proliferation and differentiation, ultimately resulting in cancer [17]. While lower organisms can survive with a mutated DNA, higher organisms use apoptosis as a second line of defense, also referred to as "cellular proofreading" [6,18] to rapidly eliminate severely UV damaged cells before they develop altered genotype and phenotype. Indeed, formation of apoptotic sunburn cells in epidermis of many mammalian species is one of the hallmark events in the skin exposed to solar UVB radiation [19]. Sunburn cells are apoptotic keratinocytes, which show typical morphological features, such as shrunken eosinophilic cytoplasm with a condensed pyknotic nucleus and contain DNA double strand breaks typical of apoptotic DNA fragmentation [20]. Sunburn cells are produced weakly after UVA (315–400 nm) irradiation [21], whereas UVC (100–280 nm) is ineffective due to absorption in the keratin. In the UVB exposed skin, sunburn cells occur as early as 8 h after UV exposure, peak between 1–2 days and disappear by 3–4 days (reviewed in [22]). Since apoptosis is a universal phenomenon deployed by all metazoan cells, it is not surprising that UV-induced apoptosis has been observed not only in keratinocytes but also in fibroblasts of skin and cells from other organs. In the past few years, rapid advances have occurred in our knowledge about UV-induced apoptosis, and we invite the readers to refer to earlier excellent reviews on UV-induced apoptosis by Kulms and Schwarz [22,23] and Kaina [24]. Here, we will review the recent work on nuclear and non-nuclear molecular signals and pathways that lead to UV-induced apoptosis.

15.2 DNA Damage as a nuclear signal for UV-induced apoptosis

UV irradiation causes different types of DNA damage, and the evidence that these DNA lesions are initiating signals for apoptosis comes from two complementary sets of observations: (i) Defects in DNA repair, specifically NER that repairs major UV-induced DNA damages, modulates susceptibility to apoptosis; and (ii) augmenting DNA repair capacity decreases apoptosis.

15.2.1 DNA repair defect and UV-induced apoptosis

Upon absorption, UV light energy induces various types of lesions in DNA; more than 99% of UV-induced damage consists of chemical base modifications, with cyclobutane pyrimidine dimers (CPD) [25] and 6,4-pyrimidine pyrimidone photoproducts (6-4PP) [26] as the most frequent types. CPDs are considered to be more carcinogenic than 6-4PPs, since they occur three times more often and are repaired about 5–10 times slower than 6-4PPs [27–29]. Apart from these two major types of DNA damage, UV light induces a much wider spectrum of other minor damages, such as oxidative base damage, SSB, DNA-protein cross-links, which may also contribute to UV carcinogenesis [1,4,30].

UV-induced DNA damages are repaired by different DNA repair pathways, of which NER is specifically important because it repairs CPD and 6,4-PP, the principal UV-induced DNA lesions. That unrepaired DNA damage is a signal for induction of apoptosis is confirmed by inverse correlation between DNA repair capacity and susceptibility to apoptosis, as seen in the cells derived from patients with NER deficiency diseases, such as Xeroderma pigmentosum (XP) or Cockayne syndrome (CS) or trichothiodystrophy (TTD). Table 1 provides an overview of susceptibility to UV-induced apoptosis in NER deficient models, using few

Table 1. Susceptibility to UV-induced apoptosis in NER-deficient cells

Gene	Model	UV type	Susceptibility to apoptosis (assay)	Ref*
XPA	Mouse (−/−) skin	UVB	increased (caspase -3)	[31]
	Mouse (−/−) skin	UVB	increased (sunburn cells)	[32]
	Mouse (−/−) MEF	UVC	increased	[33]
	human skin keratinocytes	UVB	increased (DNA fragmentation)	[34]
	primary human skin fibroblasts	UVC	increased	[35]
XPB	primary human skin fibroblasts	no UV	apoptosis when p53 is injected	[36]
XPC	Mouse (−/−) skin	UVB	no change (caspase -3)	[31]
	Mouse (−/−) skin	UVB	no change (sunburn cells)	[32]
	Mouse (−/−) MEF	UVC	increased	[37]
	human skin keratinocytes	UVB	increased (DNA fragmentation) or no change	[34]
	primary human skin fibroblasts	UVC	increased	[35]
XPD	human skin keratinocyte	UVB	increased (DNA breaks)	[34]
	primary human skin fibroblasts	UVC	increased	[35]
	human lymphoblastoid cells	UVC	increased in one mutant (Asn312-homozygous)	[38]
	human lymphoblastoid cells	UVC	no change (another mutant at 712 residue)	[38]
XPE/DDB2	human primary fibroblasts (low p53)	UVC	decreased (survival): p53 addition restored apoptosis susceptibility	[40]
	Mouse (−/−) MEF	UVC	decreased	[39]
XPF	Mouse (mutant) skin	UVC	increased (viability)	[42]
XPG	Mouse (mutant) skin	UVB	increased (sunburn cells)	[41]
	Mouse (mutant) MEF	UVC	increased	[41]
CSA	Mouse (−/−) MEF	UVC	increased	[43]
CSB	Mouse (−/−) skin	UVB	increased (caspase -3)	[31]
	Mouse (−/−) skin	UVB	increased (sunburn cells)	[32]
	Mouse (−/−) MEF	UVC	increased	[33]
	Mouse (−/−) ES	UVC	increased	[33]
TTD	Mouse (mutant) MEF	UVC	increased (survival)	[44]

*A small selection of references in each category of NER genes has been shown here to provide an overview of NER defects and apoptosis.

examples in each category [31–44]. It is noteworthy that these studies evaluated apoptosis in different models ranging from intact animal skin to different cells, such as fibroblasts, keratinocytes and lymphoblastoid cells derived from patients or mice with defects in a given NER gene. Moreover, apoptosis was induced by UVB or UVC, and apoptotic end-points examined in these studies ranged from appearance of sunburn cells in skin to activation of caspase 3 and apoptotic DNA fragmentation. Despite these differences, there was a remarkable consensus for the eventual effect of NER deficiency on apoptosis. Overall, it can be seen that defect in all the NER genes (except XPE) resulted in an increased susceptibility to UV-induced apoptosis (Table 1). For XPC cells, which are deficient in global genomic repair of NER (GGR), but not in transcription-coupled NER (TCR), some studies showed an increased susceptibility to apoptosis [34,35,37], whereas others showed no change in susceptibility to apoptosis [31,32]. The major exception to the rule was XPE/DDB2, which plays a key role in the first step of recognition of the DNA lesion in case of GGR. The fibroblasts derived from XPE patients or from DDB2 mice were resistant to UVC-induced apoptosis [39,40].

15.2.2 Characteristics of DNA lesions that serve as pro-apoptotic signals

From above studies with NER deficient models, three generalizations can be made about the characteristics of DNA lesions that induce apoptosis. (i) DNA lesions in the transcribed strand of the genome, which are repaired by TCR are the key lesions which induce apoptosis. This was elegantly shown by Brash et al. [32] and van Oosten et al. [31] by measuring sunburn cells in UVB exposed XPA, XPC and CSB mice. XPA are deficient in both GGR and TCR, whereas XPC or CSB are deficient only in GGR or TCR, respectively. They showed that while XPA and CSB mice were 7–10 fold more susceptible than controls to develop sunburn cells, apoptosis in XPC mice was unaltered. This has been confirmed in other studies involving XPA and CSB mice (see Table 1). These studies also support an earlier argument that accumulation of unrepaired CPD in transcribed strand can block the transcription and that stalled RNA polymerase II may constitute a specific signal for the induction of apoptosis [45–47]. Although it is important to note that for XPC model, some studies have also reported an increased susceptibility to UVB or UVC-induced apoptosis (Table 1). Hence, factors associated with or downstream from the DNA lesion, such as p53 activation, may play a key role in modulating apoptotic signal ensuing from DNA damage. (ii) To become pro-apoptotic, DNA lesion needs to be recognized by DNA repair proteins. The decreased susceptibility to apoptosis in XPE/DDB2 [39,40], which perform the first step of recognition of lesion during GGR, indicates that DNA lesion needs to be recognized and bound by the lesion recognizing proteins to initiate apoptosis. In case of TCR, this may be accomplished by stalled RNA pol II or the TFIIH. (iii) The third generalization concerns the point in DNA repair process upto which DNA lesion remains pro-apoptotic. It appears that DNA lesion stops being pro-apoptotic only when it is completely removed from DNA, since XPF and XPG continue to show increased susceptibility to UV-induced apoptosis (Table 1). These two genes code for nucleases that excise the DNA on either side of the DNA damage to finally remove the lesion from DNA. Thus, unrepaired DNA lesions due to defective DNA repair can induce apoptotic death.

15.2.3 Exogenous DNA repair enzymes decrease UV-induced apoptosis

If defects in DNA repair can increase susceptibility to UV-induced apoptosis, as a corollary, augmentation of DNA repair capacity should decrease apoptosis. Indeed, topical application of exogenous repair enzyme T4 endoV, delivered through liposomes has been shown to reduce the number of sunburn cells and incidence of skin cancer in both mouse and human models [48–50]. Similarly, enhancement of endogenous DNA repair by encapsulated photolyase resulted in significant reduction of CPD and suppression of UVB-induced apoptosis [51]. These results were confirmed by the generation of a transgenic mice expressing marsupial CPD photolyase [52]. These animals displayed a dramatic reduction of acute UV effects like sunburn and hyperplasia and dermal fibroblasts showed superior survival when treated with photoreactivating light.

15.2.4 Signalling pathway between DNA lesion and apoptotic machinery

DNA lesion can simultaneously initiate pro-survival as well as pro-apoptotic functions, and this may serve an important purpose of preventing killing of the cells that can be easily repaired. We will discuss later the major pathway of p53 induction through which DNA lesion conveys both pro-survival (cell cycle arrest, DNA repair) and pro-apoptotic signals. In this context, cancer cells can arise if p53 is mutated and its pro-apoptotic function is inactivated; and this is often seen in UV-induced skin cancers. It is important to bear in mind that many of the NER factors also participate in other cellular metabolisms, such as transcription. Therefore, it can not be excluded that other roles of NER factors may also contribute towards apoptotic response to UV. Moreover, cells deficient in the genes implicated in non-nuclear responses to UV have normal DNA repair, but still show higher susceptibility to UV-induced apoptosis (see below). Thus, unrepaired DNA lesion is only one of the signals that lead to UV-induced apoptosis.

15.3 Non-nuclear signals for UV-induced apoptosis

Although nuclear DNA damage clearly plays a crucial role in sunburn cell formation, several pieces of evidence argue for the existence of other apoptotic pathways which are independently and concomitantly induced by UV light. It has been noted that 4% of keratinocytes in sun-exposed skin carry p53 mutations and thus should be protected form undergoing apoptosis, if only p53-mediated pathway controlled apoptosis. However, far less than 4% of cells develop into pre-cancer or cancer stages [53], suggesting that other pathways are involved in apoptosis of damaged keratinocytes [54]. It became evident that such a parallel pathway might be initiated in a DNA damage-independent fashion because enhancement of DNA repair by photoreactivation was unable to completely suppress apoptosis [51]. Furthermore, cells like HeLa and HaCaT, which are compromised for p53 function, and p53 deficient primary mouse fibroblasts still undergo apoptosis in response to UV irradiation [51,55,56]. Here, we will discuss three major extra-nuclear events that are known to mediate UV-induced apoptosis.

15.3.1 Role of membrane receptors Fas and TNF

The tumor necrosis factor (TNF) superfamily of death receptors, residing in the plasma membrane, initiate apoptosis upon binding to their cognate ligands [57,58]. The signal transduction cascade is initiated by receptor trimerization and recruitment of cytoplasmic adaptor proteins which bind to the intracellular region of the receptor, called death domain (Figure 1). The adaptor proteins recruit and activate caspase 8, which cleaves both caspase 3 and the pro-apoptotic protein Bid to initiate the execution phase of apoptosis. Truncated Bid is translocated to mitochondria where it interacts with anti-apoptotic Bcl-2 resulting in pore formation and cytochrome c release [58].

UV exposure has been shown to up-regulate two members of the TNF receptor family, i.e., TNF-R1 and Fas, as well as their respective ligands, TNFα and FasL [9,59–61]. Mice deficient in either Fas, FasL or TNF-R1 displayed remarkably decreased levels of keratinocyte apoptosis in response to UV [9,62]. In a model of skin carcinogenesis induced by chronic UVB exposure of SKH-1 mice, loss of FasL expression and gain of p53 mutation along with decreased apoptosis were seen as key events that preceded initiation of skin cancers [63]. Chronic exposure to UV caused 70% of FasL-deficient mice and only 5 percent wild-type mice to accumulate p53 mutations in the epidermis [9]. Moreover, FasL-deficient mice exposed to UV lack systemic UV-induced suppression of immune responses that involve host antigen presentation, suggesting an interrelationship between UV-induced DNA damage and FasL in this phenomenon [64]. Thus, FasL-mediated apoptosis is important for skin homeostasis and dysregulation of Fas-FasL interactions may be central to the development of skin cancer. Defects in Fas signalling, mostly associated with somatic mutations in the death domain, has been reported in melanoma and squamous cell carcinoma [65,66].

Inactivation of TNFα in vivo by gene knockout or in vitro by anti-TNFα antibody has been shown to partially or completely block UV-induced apoptosis [11,62,67]. TNFα is important in the early stages of tumor promotion and is required for de novo carcinogenesis, because TNFα (−/−) mice are resistant to development of malignant skin tumors [68]. Since TNF has a function in the induction of UV-induced immuno-suppression [69,70], resistance of mice lacking TNF expression to carcinogenesis might be due to maintenance of normal immune responses [71,72]. In the same vein, it has been demonstrated that TNFα is incapable of inducing apoptosis by itself [62,67], suggesting that another UV-related signal is implicated in diminished apoptotic response of these animals. The reduced cancer incidence seen in TNFα (−/−) mice may be due to combined effects of decreased apoptosis and preservation of normal immune response. These data suggest that reduction of pro-inflammatory cytokine (TNFα) production may be useful in cancer treatment and prevention.

15.3.1.1 Death receptor activation in absence of ligands

Although both the Fas and TNF receptors as well as their ligands go up in response to UV, it has been shown that receptor activation and apoptosis can also occur in the absence of ligand [11,55,73]. In a recent study examining apoptosis in UVB-exposed human epidermis, apoptotic DNA fragmentation and clustering of Fas was observed in the same subpopulation of epidermal cells, although there was no associated increase in FasL in these cells [10].

15.3.1.2 Downstream apoptotic events after death receptor activation

The expression levels of different adaptor and effector molecules involved in transmission of death signal may be as critical for the execution of the apoptotic program as the receptors themselves. It was shown that UVB irradiation increased in a dose-dependent manner the expression of FADD adaptor protein, which augmented UV-induced keratinocyte apoptosis [74]. The expression levels of signal adaptor protein TRAF2 also increased after UV irradiation but this was coincident with resistance of melanoma cells to UV-induced apoptosis [75]. Comparison of changes in the expression of TRAF2 during early- and late-stage melanoma cells revealed that low expression of TRAF2 in early-stage melanoma coincided with poor resistance to UV-induced, TNF-mediated apoptosis. Conversely, forced expression of the dominant negative form of TRAF2 in late-stage melanoma cells reduced NF-kappaB activity and decreased Fas expression, resulting in a lower degree of UV-induced, Fas-mediated cell death [76]. This suggests that UV-induced melanoma cell death depends on the nature of the apoptotic cascade (TNF or Fas) and on the availability and competition between adaptor molecules whose expression may fluctuate during melanoma progression. Similarly, overexpression of wild-type Daxx, Fas receptor binding protein, also inhibited UV-induced apoptosis, suggesting that Daxx competes for Fas receptor binding sites with other pro-apoptotic factors such as FADD [77].

The death receptors-mediated apoptotic pathway may be in a dynamic crosstalk with the p53-dependent pathway. Expression of Fas is enhanced by wild type p53 in response to UV radiation [61,78,79] and p53 mutants that do not induce apoptosis can still bind to p53-responsive element in the Fas gene and modulate death receptor-induced apoptosis [78,80]. Moreover, Fas expression is diminished or absent in basal cell and squamous cell carcinomas which usually contain multiple mutations in p53 [81,82].

15.3.2 Role of MAP kinases: JNK and p38 MAPK

Mitogen activated proteins kinases (MAPK) are important signal transducing enzymes, unique to eukaryotes, that are involved in many aspects of cell regulation from cell survival and proliferation to cell death. MAPK connect cell surface receptors to regulatory targets within the cells. MAPK are activated in response to external stimuli as well as stress, including UV radiation [12]. The principal cause for activation of MAP kinases, such as ERK is UV-induced activation of receptor tyrosine kinases at the cell surface (Figure 1). However, non-receptor mediated activation of some members of the MAPK family has also been observed, e.g., activation of p38 MAPK by UV-induced oxidants (see below). MAPK function in three tiers: MAPK, MAPKK and MAPKKK. These modules are activated by small GTP binding proteins or STE20 kinases and are inactivated by MAPK phosphatases. MAPK family members can be divided in four distinctly regulated groups: 1) Extracellular signal related kinase (ERK) 1 or 2; 2) c-Jun-NH2 terminal kinase (JNK) 1, 2 or 3; 3) p38 kinase (or p38 MAPK) p38α, β, γ or δ; and 4) ERK5. By and large, activation of ERK1 or 2 are linked to cell survival, whereas JNK and p38 are linked to apoptosis [12,83]. However, this is an oversimplification and actual role of these kinases depends on the cell type, and signalling event [12,83]. Here, we will discuss only the pro-apoptotic functions of JNK and p38 MAPK.

15.3.2.1 JNK

UV-induced activation of c-Jun-NH2 terminal kinase (JNK) has been implicated in both survival and cell death functions in response to UV. Chen et al. [14] showed that persistent activation of JNK was associated with apoptosis. Use of gene-deletion model for JNK provided a clear proof for the role of JNK in UV-induced apoptosis. Davis's group first showed that JNK (−/−) primary MEF are resistant to UVC-induced apoptosis [84]. Recently, they examined the mechanism by which activated JNK transmits the signal to apoptotic machinery in the mitochondria [85]. They showed that UV activated JNK phosphorylates Bim/Bmf, the pro-apoptotic members of the bcl-2 family, which causes their release from the complex with dynein and myosin V in the cytoskeleton. The pro-apoptotic Bim/Bmf then induce apoptotic events in the cell.

15.3.2.2 P38 MAPK

Several recent reports suggest that activation of p38 MAPK, more than JNK activation, plays a key role in UV-induced apoptosis, and that this is achieved through induction of p53. Fornace's team showed that, unlike ionizing radiation that activate p53 through ATM pathway, UV exposed cells phosphorylate and activate p53 through p38 MAPK, which leads to apoptosis [15]. Recently, this team showed that application of p38 MAPK-inhibitor, and not JNK-inhibitor, on the skin of mice decreases UVB-induced apoptosis in the skin [86]. It has been suggested that UVB-induced activation of p38 MAPK activates p53, which activates DNA damage-responsive Gadd45 gene. The latter loops back to up-regulate MAPK activation, resulting in amplified p53 response and cell death [7,86].

15.3.3 Reactive oxygen species in UV-induced apoptosis

UV irradiation is a potent inducer of ROS, such as superoxide radical, hydrogen peroxide and hydroxyl radical, which are implicated in various inflammatory disorders, cutaneous ageing and malignant transformation [87,88]. That UV-induced ROS play a role in apoptosis has been mainly derived from inhibition of UV-induced apoptosis by anti-oxidants. For example, formation of sunburn cells in skin of UVB exposed mice has been shown to be suppressed by anti-oxidant plant flavonoid silymarin [89]. Kulms et al. [90] showed that radical scavenger pyrrolidene-dithiocarbamate suppressed UVB-induced apoptosis in human epithelial carcinoma cells. They also showed that when DNA repair and ROS are simultaneously inhibited, there is a complete suppression of UV-induced apoptosis. On the other hand, anti-oxidants have been shown to stimulate UV-induced apoptosis, suggesting that ROS can also activate pro-survival pathways. For example, green tea polyphenols have been shown to stimulate apoptotic sunburn cell formation in the skin of UVB-irradiated mice [91]. Peus et al. [13] showed that UVB-induced oxidant can activate pro-survival ERK pathway and pro-apoptotic p38 MAP kinase pathway at different time and in dose-dependent manner. They showed that selective inhibition of ERK can increase UV-induced apoptosis.

The exact mechanism by which ROS induce apoptosis in UV irradiated cells is not fully characterized. ROS have been directly implicated in release of cytochrome c from mitochondria [90,92], possibly due to lipid peroxidation and changes in membrane integrity [23]. It is also possible that UV-induced ROS can indirectly cause apoptosis via induction of p38 MAPK. For example, Decraene et al. [93] recently showed that anti-oxidant mimetic seleno-manganese complex can significantly reduce UVB-induced activation of p38 MAPK and p53 accumulation in primary human keratinocytes. Thus, role of UV-induced ROS in apoptosis may be a result of net balance of its pro-apoptotic and pro-survival influences on the cell.

15.4 The Central Role of p53 in UV-induced apoptosis

An important advance in our understanding of UV-induced apoptosis, i.e., formation of sunburn cells, and its impact on prevention of skin carcinogenesis came from studies demonstrating that mice lacking the tumor suppressor gene p53 were impaired in formation of sunburn cells [6,20]. These studies lead to a new concept that severely damaged keratinocytes that fail to repair UV-induced damages will be eliminated via p53-mediated apoptosis, a process known as "cellular proofreading" [18]. Consistently, mutational inactivation of p53 has been identified at an early stages of development of skin cancer in actinic keratoses, as well as in late cancer stages in melanomas and squamous cell carcinomas [18,94,95]. The persistence of UV-induced p53 mutations in NER-deficient individuals is associated with decreased apoptosis, increased keratinocytes proliferation and increased risk to develop skin cancer [96]. Thus, p53 specifically targets heavily damaged cells for destruction, however cells with this p53 function inactivated by UV-induced mutation, acquire a growth advantage over the neighbouring cells and may expand to give rise to cancer [53]. In UV irradiated cells, signal for activation of p53 can come from damaged DNA or MAP kinase signalling pathway. Thus, nuclear and some of the non-nuclear signals converge on p53 to mediate UV-induced apoptosis.

15.4.1 Activation of p53 by phosphorylation

The p53 stabilization and activation involves specific post-translational modification of particular amino acid residues. Phosphorylation of several serine and threonine residues in N- and C-terminal portions of p53 in response to DNA damage are known to contribute to its stabilization and accumulation due to decreased binding of Mdm2 oncoprotein, a ubiquitin E3 ligase responsible for p53 degradation [97,98]. In particular, UV-induced p53 activation is accompanied with phosphorylation of Ser6, Ser9, Ser15, Ser20, Ser37 and Ser46 and threonine 18 [99–101]. Phosphorylation at Ser46 seems to be functionally relevant to apoptosis-inducing properties of p53 because mutation at this site attenuated sensitivity towards UV-induced apoptosis [15,102].

15.4.2 Activation of p53 in response to DNA damage

Different cellular kinases, which respond to UV-induced DNA damage have been proposed to mediate p53 phosphorylation. Obvious candidates are ATR kinase, a member of ATM/ATR sensor family of PI3-Kinases that is specifically induced by UV irradiation, and its downstream kinase Chk1 [103]. Both ATR and Chk1 have been implicated in p53 phosphorylation after UV and γ-irradiation at Ser15, a site specifically modified following DNA damage [104–106]. In addition, ATR was found to preferentially bind to and be activated by UV damaged DNA [107], thereby reinforcing the notion that it could be directly involved in sensing UV-induced photodamage, p53 activation and possibly induction of apoptosis.

15.4.3 Activation of p53 in response to p38 MAPK

The p38-MAPK has also been proposed to contribute to phosphorylation and activation of p53. Inhibition of p38 activation after UV irradiation decreased p53 phosphorylation at Ser33, Ser37 and Ser15, and also markedly reduced UV-induced apoptosis in a p53-dependent manner [15]. UV-induced phosphorylation of p53 at ser389 was shown to be mediated by p38 kinase, because a specific inhibitor of p38 blocked p53 phosphorylation at this site, reduced its DNA binding activity and decreased p53-dependent transcription [108]. In another study, p38 MAPK was shown to physically associate with p53; and the inhibition of p38-mediated phosphorylation of the NH2-terminal transactivation domain of p53 at Ser33 interfered with apoptosis induction, suggesting a role for p38 activation in p53-dependent apoptotic response [109]. Obviously proteins phosphatase, such as Wip1 can suppress UV-induced apoptosis by attenuating p53 phosphorylation by p38 MAPK [110].

JNKs are other members of MAPK family that are directly involved in the phosphorylation of p53 at Ser20, a residue that has been shown to be required for the activation of p53 following UV radiation. A defective UVB-induced apoptosis has been observed in cells that are Jnk1 (−/−), Jnk2 (−/−), or p53 (−/−) or p53 (−/−)

containing inactive Ser20 mutant [111]. These findings suggest that JNKs are the major direct signalling mediators of UVB-induced p53 phosphorylation at Ser20.

Thus, several independent cascades cooperate in p53 phosphorylation-dependent activation in response to UV radiation. The mechanism by which activated p53 participates in cell cycle arrest, trans-activation of genes and in apoptosis is beyond the scope of this review. The mechanism by which p53 induces apoptosis in UV irradiated cells may not be different from the two distinct mechanisms by which p53 is known to induce apoptosis, namely via transcriptional activation of pro-apoptotic protein Bax [112] or via direct interaction with mitochondria [113,114] and release of cytochrome c [114], leading to assembly of apoptosome, activation of caspase 9, followed by activation of the executioner caspases and cell death [58].

15.5 Perspectives

We have reviewed the key nuclear and non-nuclear signalling events that initiate apoptosis in UV irradiated cell. It does not diminish the importance of many other molecules, enzymes or metabolic pathways that play a role in the long process of apoptotic elimination of the cell. Many factors have been specifically shown to intervene in p53-mediated apoptosis in UV irradiated cells, such as Galectin -7 [115], ING1 [116,117], Gadd45 [7] and E2F1 [118]. Some of these factors, such as Gadd45 and E2F1 may act upstream or parallel to p53 during apoptosis, and may exercise control over the action of p53. Similarly, UV-induced DNA damage as a signal for apoptosis has been discussed here as a key initiating event for apoptosis, but several genes that are immediately activated in response to DNA damage are known to play a crucial role in apoptosis. For example, gene-deletion for c-fos [56,119,120] or c-jun [120] makes cells more sensitive to UV-induced apoptosis, whereas Gadd45 (−/−) cells are resistant to apoptosis [7].

Finally, a recent study of UV-induced apoptosis in Drosophila embryo at different developmental stages shows that cells can rapidly and dramatically change pathways that mediate apoptosis under different circumstances [121]. In midstage embryos, UVC, and not UVB, induced apoptosis through DNA damage-initiated pathway, which involves mei-41, a Drosophila ATM homolog, Drosophila p53 and apoptotic Reaper protein. Whereas in early stage embryos, both UVB and UVC induced DNA damage-independent apoptosis through unknown cytoplasmic targets, but mediated through hac-1, a drosophila Apaf-1; and E2F1 may play a role in transcriptional activation of hac-1 [121]. Therefore, we should anticipate that future studies may reveal other distinct pathways that can induce apoptosis in UV irradiated cell.

Acknowledgment

GMS is the recipient of FRSQ Senior scientist award, and work related to this subject in his laboratory is supported by a research grant from the National Cancer Institute of Canada with funds from the Canadian Cancer Society.

References

1. H.S. Black, F.R. deGruijl, P.D. Forbes, J.E. Cleaver, H.N. Ananthaswamy, E.C. deFabo, S.E. Ullrich, R.M. Tyrrell (1997). Photocarcinogenesis: an overview. *J. Photochem. Photobiol. B*, **40**, 29–47.
2. F. Urbach (1997). Ultraviolet radiation and skin cancer of humans. *J. Photochem. Photobiol. B*, **40**, 3–7.
3. D.E. Brash (1997). Sunlight and the onset of skin cancer. *Tr. Genetics*, **13**, 410–414.
4. J. Cadet, M. Berger, T. Douki, B. Morin, S. Raoul, J.L. Ravanat, S. Spinelli (1997). Effects of UV and visible radiation on DNA-final base damage. *Biol. Chem.*, **378**, 1275–1286.
5. P. Herrlich, C. Blattner, A. Knebel, K. Bender, H.J. Rahmsdorf (1997). Nuclear and non-nuclear targets of genotoxic agents in the induction of gene expression. Shared principles in yeast, rodents, man and plants. *Biol. Chem.*, **378**, 1217–1229.
6. D.E. Brash, A. Ziegler, A.S. Jonason, J.A. Simon, S. Kunala, D.J. Leffell (1996). Sunlight and sunburn in human skin cancer: p53, apoptosis, and tumor promotion. *J. Investig. Dermatol. Symp. Proc.*, **1**, 136–142.
7. J. Hildesheim, D.V. Bulavin, M.R. Anver, W.G. Alvord, M.C. Hollander, L. Vardanian, A.J. Fornace Jr. (2002). Gadd45a protects against UV irradiation-induced skin tumors, and promotes apoptosis and stress signaling via MAPK and p53. *Cancer Res.*, **62**, 7305–7315.
8. Y. Devary, C. Rosette, J.A. DiDonato, M. Karin (1993). NF-kappa B activation by ultraviolet light not dependent on a nuclear signal. *Science*, **261**, 1442–1445.
9. L.L. Hill, A. Ouhtit, S.M. Loughlin, M.L. Kripke, H.N. Ananthaswamy, L.B. Owen-Schaub (1999). Fas ligand: a sensor for DNA damage critical in skin cancer etiology. *Science*, **285**, 898–900.
10. B. Bang, R. Gniadecki, J.K. Larsen, O. Baadsgaard, L. Skov (2003). In vivo UVB irradiation induces clustering of Fas (CD95) on human epidermal cells. *Exp. Dermatol.*, **12**, 791–798.
11. M.S. Sheikh, M.J. Antinore, Y. Huang, A.J. Fornace Jr. (1998). Ultraviolet-irradiation-induced apoptosis is mediated via ligand independent activation of tumor necrosis factor receptor 1. *Oncogene*, **17**, 2555–2563.
12. L. Chang, M. Karin (2001). Mammalian MAP kinase signalling cascades. *Nature*, **410**, 37–40.
13. D. Peus, R.A. Vasa, A. Beyerle, A. Meves, C. Krautmacher, M.R. Pittelkow (1999). UVB activates ERK1/2 and p38 signaling pathways via reactive oxygen species in cultured keratinocytes. *J. Invest. Dermatol.*, **112**, 751–756.
14. Y.R. Chen, X. Wang, D. Templeton, R.J. Davis, T.H. Tan (1996). The role of c-Jun N-terminal kinase (JNK) in apoptosis induced by ultraviolet C and gamma radiation. Duration of JNK activation may determine cell death and proliferation. *J. Biol. Chem.*, **271**, 31929–31936.
15. D.V. Bulavin, S. Saito, M.C. Hollander, K. Sakaguchi, C.W. Anderson, E. Appella, A.J. Fornace Jr. (1999). Phosphorylation of human p53 by p38 kinase coordinates N-terminal phosphorylation and apoptosis in response to UV radiation. *EMBO J.*, **18**, 6845–6854.
16. T. Douki, D. Perdiz, P. Grof, Z. Kuluncsics, E. Moustacchi, J. Cadet, E. Sage (1999). Oxidation of guanine in cellular DNA by solar UV radiation: biological role. *Photochem. Photobiol.*, **70**, 184–190.
17. D. Hanahan, R.A. Weinberg (2000). The hallmarks of cancer. *Cell*, **100**, 57–70.

18. D.E. Brash (1996). Cellular proofreading. *Nat. Med.*, **2**, 525–526.
19. K. Danno, T. Horio (1987). Sunburn cell: factors involved in its formation. *Photochem. Photobiol.*, **45**, 683–690.
20. A. Ziegler, A.S. Jonason, D.J. Leffell, J.A. Simon, H.W. Sharma, J. Kimmelman, L. Remington, T. Jacks, D. Brash (1994). Sunburn and p53 in the onset of skin cancer. *Nature*, **372**, 773–776.
21. R. Lavker, K. Kaidbey (1997). The spectral dependence for UVA-induced cumulative damage in human skin. *J. Invest. Dermatol.*, **108**, 17–21.
22. D. Kulms, T. Schwarz (2000). Molecular mechanisms of UV-induced apoptosis. *Photodermatol. Photoimmunol. Photomed.*, **16**, 195–201.
23. D. Kulms, T. Schwarz (2002). Independent contribution of three different pathways to ultraviolet-B-induced apoptosis. *Biochem. Pharmacol.*, **64**, 837–841.
24. B. Kaina (2003). DNA damage-triggered apoptosis: critical role of DNA repair, double-strand breaks, cell proliferation and signaling. *Biochem. Pharmacol.*, **66**, 1547–1554.
25. R.B. Setlow (1966). Cyclobutane-type pyrimidine dimers in polynucleotides. *Science*, **153**, 379–386.
26. D.L. Mitchell, R.S. Nairn (1989). The biology of the (6–4) photoproduct. *Photochem. Photobiol.*, **49**, 805–819.
27. B.S. Rosenstein, D.L. Mitchell (1987). Action spectra for the induction of pyrimidine (6–4)pyrimidone photoproducts and cyclobutane pyrimidine dimers in normal human skin fibroblasts. *Photochem. Photobiol.*, **45**, 775–780.
28. E. Eveno, F. Bourre, X. Quilliet, O. Chevallier-Lagente, L. Roza, A.P. Eker, W.J. Kleijer, O. Nikaido, M. Stefanini, J.H. Hoeijmakers, D. Boostma, J. Cleaver, A. Sarasin, M. Mezzina (1995). Different removal of ultraviolet photoproducts in genetically related xeroderma pigmentosum and trichothiodystrophy diseases. *Cancer Res.*, **55**, 4325–4332.
29. Y.H. You, D.H. Lee, J.H. Yoon, S. Nakajima, A. Yasui, G.P. Pfeifer (2001). Cyclobutane pyrimidine dimers are responsible for the vast majority of mutations induced by UVB irradiation in mammalian cells. *J. Biol. Chem.*, **276**, 44688–44694.
30. K. Ito, S. Kawanishi (1997). Site-specific DNA damage induced by UVA radiation in the presence of endogenous photosensitizer. *Biol. Chem.*, **378**, 1307–1312.
31. M. van Oosten, H. Rebel, E.C. Friedberg, H. van Steeg, G.T. van der Horst, H.J. van Kranen, A. Westerman, A.A. van Zeeland, L.H. Mullenders, F.R. de Gruijl (2000). Differential role of transcription-coupled repair in UVB-induced G2 arrest and apoptosis in mouse epidermis. *Proc. Natl. Acad. Sci. USA*, **97**, 11268–11273.
32. D.E. Brash, N.M. Wikonkal, E. Remenyik, G.T. van der Horst, E.C. Friedberg, D.L. Cheo, H. van Steeg, A. Westerman, H.J. van Kranen (2001). The DNA damage signal for Mdm2 regulation, Trp53 induction, and sunburn cell formation in vivo originates from actively transcribed genes. *J. Invest. Dermatol.*, **117**, 1234–1240.
33. H. de Waard, J. de Wit, T.G. Gorgels, G. van den Aardweg, J.O. Andressoo, M. Vermeij, H. van Steeg, J.H. Hoeijmakers, G.T. van der Horst (2003). Cell type-specific hypersensitivity to oxidative damage in CSB and XPA mice. *DNA Repair (Amst)*, **2**, 13–25.
34. C. Petit-Frere, E. Capulas, J.E. Lowe, L. Koulu, R.J. Marttila, N.G. Jaspers, P.H. Clingen, M.H. Green, C.F. Arlett (2000). Ultraviolet-B-induced apoptosis and cytokine release in xeroderma pigmentosum keratinocytes. *J. Invest. Dermatol.*, **115**, 687–693.
35. S. Queille, C. Drougard, A. Sarasin, L. Daya-Grosjean (2001). Effects of XPD mutations on ultraviolet-induced apoptosis in relation to skin cancer-proneness in repair-deficient syndromes. *J. Invest. Dermatol.*, **117**, 1162–1170.
36. X.W. Wang, W. Vermeulen, J.D. Coursen, M. Gibson, S.E. Lupold, K. Forrester, G. Xu, L. Elmore, H. Yeh, J.H. Hoeijmakers, C.C. Harris (1996). The XPB and XPD DNA

helicases are components of the p53-mediated apoptosis pathway. *Genes Dev.*, **10**, 1219–1232.
37. J.M. Ng, H. Vrieling, K. Sugasawa, M.P. Ooms, J.A. Grootegoed, J.T. Vreeburg, P. Visser, R.B. Beems, T.G. Gorgels, F. Hanaoka, J.H. Hoeijmakers, G.T. van der Horst (2002). Developmental defects and male sterility in mice lacking the ubiquitin-like DNA repair gene mHR23B. *Mol. Cell Biol.*, **22**, 1233–1245.
38. H. Seker, D. Butkiewicz, E.D. Bowman, M. Rusin, M. Hedayati, L. Grossman, C.C. Harris (2001). Functional significance of XPD polymorphic variants: attenuated apoptosis in human lymphoblastoid cells with the XPD 312 Asp/Asp genotype. *Cancer Res.*, **61**, 7430–7434.
39. T. Itoh, D. Cado, R. Kamide, S. Linn (2004). DDB2 gene disruption leads to skin tumors and resistance to apoptosis after exposure to ultraviolet light but not a chemical carcinogen. *Proc. Natl. Acad. Sci. USA*, **101**, 2052–2057.
40. T. Itoh, C. O'Shea, S. Linn (2003). Impaired regulation of tumor suppressor p53 caused by mutations in the xeroderma pigmentosum DDB2 gene: mutual regulatory interactions between p48(DDB2) and p53. *Mol. Cell Biol.*, **23**, 7540–7553.
41. M. Tian, D.A. Jones, M. Smith, R. Shinkura, F.W. Alt (2004). Deficiency in the nuclease activity of xeroderma pigmentosum G in mice leads to hypersensitivity to UV irradiation. *Mol. Cell Biol.*, **24**, 2237–2242.
42. M. Tian, R. Shinkura, N. Shinkura, F.W. Alt (2004). Growth retardation, early death, and DNA repair defects in mice deficient for the nucleotide excision repair enzyme XPF. *Mol. Cell Biol.*, **24**, 1200–1205.
43. G.T. van der Horst, L. Meira, T.G. Gorgels, J. de Wit, S. Velasco-Miguel, J.A. Richardson, Y. Kamp, M.P. Vreeswijk, B. Smit, D. Bootsma, J.H. Hoeijmakers, E.C. Friedberg (2002). UVB radiation-induced cancer predisposition in Cockayne syndrome group A (Csa) mutant mice. *DNA Repair (Amst)*, **1**, 143–157.
44. J. de Boer, J.O. Andressoo, J. de Wit, J. Huijmans, R.B. Beems, H. van Steeg, G. Weeda, G.T. van der Horst, W. van Leeuwen, A.P. Themmen, M. Meradji, J.H. Hoeijmakers (2002). Premature aging in mice deficient in DNA repair and transcription. *Science*, **296**, 1276–1279.
45. M. Ljungman (2000). Dial 9-1-1 for p53: mechanisms of p53 activation by cellular stress. *Neoplasia*, **2**, 208–225.
46. M. Ljungman, H.M. O'Hagan, M.T. Paulsen (2001). Induction of ser15 and lys382 modifications of p53 by blockage of transcription elongation. *Oncogene*, **20**, 5964–5971.
47. M. Ljungman, F. Zhang (1996). Blockage of RNA polymerase as a possible trigger for u.v. light-induced apoptosis. *Oncogene*, **13**, 823–831.
48. D. Yarosh, L.G. Alas, V. Yee, A. Oberyszyn, J.T. Kibitel, D. Mitchell, R. Rosenstein, A. Spinowitz, M. Citron (1992). Pyrimidine dimer removal enhanced by DNA repair liposomes reduces the incidence of UV skin cancer in mice. *Cancer Res.*, **52**, 4227–4231.
49. P. Wolf, H. Maier, R.R. Mullegger, C.A. Chadwick, R. Hofmann-Wellenhof, H.P. Soyer, A. Hofer, J. Smolle, M. Horn, L. Cerroni, D. Yarosh, J. Klein, C. Bucana, K. Dunner Jr., C.S. Potten, H. Honigsmann, H. Kerl, M.L. Kripke (2000). Topical treatment with liposomes containing T4 endonuclease V protects human skin in vivo from ultraviolet-induced upregulation of interleukin-10 and tumor necrosis factor-alpha. *J. Invest. Dermatol.*, **114**, 149–156.
50. H. Stege, L. Roza, A.A. Vink, M. Grewe, T. Ruzicka, S. Grether-Beck, J. Krutmann (2000). Enzyme plus light therapy to repair DNA damage in ultraviolet-B-irradiated human skin. *Proc. Natl. Acad. Sci. USA*, **97**, 1790–1795.

51. D. Kulms, B. Poppelmann, D. Yarosh, T.A. Luger, J. Krutmann, T. Schwarz (1999). Nuclear and cell membrane effects contribute independently to the induction of apoptosis in human cells exposed to UVB radiation. *Proc. Natl. Acad. Sci. USA*, **96**, 7974–7979.
52. W. Schul, J. Jans, Y.M. Rijksen, K.H. Klemann, A.P. Eker, J. de Wit, O. Nikaido, S. Nakajima, A. Yasui, J.H. Hoeijmakers, G.T. van der Horst (2002). Enhanced repair of cyclobutane pyrimidine dimers and improved UV resistance in photolyase transgenic mice. *EMBO J.*, **21**, 4719–4729.
53. K.H. Kraemer (1997). Sunlight and skin cancer: another link revealed. *Proc. Natl. Acad. Sci. USA*, **94**, 11–14.
54. T. Schwarz (1998). UV light affects cell membrane and cytoplasmic targets. *J. Photochem. Photobiol. B*, **44**, 91–96.
55. Y. Aragane, D. Kulms, D. Metze, G. Wilkes, B. Poppelmann, T.A. Luger, T. Schwarz (1998). Ultraviolet light induces apoptosis via direct activation of CD95 (Fas/APO-1) independently of its ligand CD95L. *J. Cell Biol.*, **140**, 171–182.
56. D. Lackinger, B. Kaina (2000). Primary mouse fibroblasts deficient for c-Fos, p53 or for both proteins are hypersensitive to UV light and alkylating agent-induced chromosomal breakage and apoptosis. *Mutat. Res.*, **457**, 113–123.
57. S. Nagata (1997). Apoptosis by death factor. *Cell*, **88**, 355–365.
58. N.N. Danial, S.J. Korsmeyer (2004). Cell death: critical control points. *Cell*, **116**, 205–219.
59. M. Leverkus, M. Yaar, B.A. Gilchrest (1997). Fas/Fas ligand interaction contributes to UV-induced apoptosis in human keratinocytes. *Exp. Cell Res.*, **232**, 255–262.
60. J. Kibitel, V. Hejmadi, L. Alas, A. O'Connor, B.M. Sutherland, D. Yarosh (1998). UV-DNA damage in mouse and human cells induces the expression of tumor necrosis factor alpha. *Photochem. Photobiol.*, **67**, 541–546.
61. M. Bennett, K. Macdonald, S.W. Chan, J.P. Luzio, R. Simari, P. Weissberg (1998). Cell surface trafficking of Fas: a rapid mechanism of p53-mediated apoptosis. *Science*, **282**, 290–293.
62. L. Zhuang, B. Wang, G.A. Shinder, G.M. Shivji, T.W. Mak, D.N. Sauder (1999). TNF receptor p55 plays a pivotal role in murine keratinocyte apoptosis induced by ultraviolet B irradiation. *J. Immunol.*, **162**, 1440–1447.
63. A. Ouhtit, A. Gorny, H.K. Muller, L.L. Hill, L. Owen-Schaub, H.N. Ananthaswamy (2000). Loss of Fas-ligand expression in mouse keratinocytes during UV carcinogenesis. *Am. J. Pathol.*, **157**, 1975–1981.
64. L.L. Hill, V.K. Shreedhar, M.L. Kripke, L.B. Owen-Schaub (1999). A critical role for Fas ligand in the active suppression of systemic immune responses by ultraviolet radiation. *J. Exp. Med.*, **189**, 1285–1294.
65. M.S. Shin, W.S. Park, S.Y. Kim, H.S. Kim, S.J. Kang, K.Y. Song, J.Y. Park, S.M. Dong, J.H. Pi, R.R. Oh, J.Y. Lee, N.J. Yoo, S.H. Lee (1999). Alterations of Fas (Apo-1/CD95) gene in cutaneous malignant melanoma. *Am. J. Pathol.*, **154**, 1785–1791.
66. S.H. Lee, M.S. Shin, H.S. Kim, W.S. Park, S.Y. Kim, J.J. Jang, K.J. Rhim, J. Jang, H.K. Lee, J.Y. Park, R.R. Oh, S.Y. Han, J.H. Lee, J.Y. Lee, N.J. Yoo (2000). Somatic mutations of Fas (Apo-1/CD95) gene in cutaneous squamous cell carcinoma arising from a burn scar. *J. Invest. Dermatol.*, **114**, 122–126.
67. A. Schwarz, R. Bhardwaj, Y. Aragane, K. Mahnke, H. Riemann, D. Metze, T.A. Luger, T. Schwarz (1995). Ultraviolet-B-induced apoptosis of keratinocytes: evidence for partial involvement of tumor necrosis factor-alpha in the formation of sunburn cells. *J. Invest. Dermatol.*, **104**, 922–927.

68. R.J. Moore, D.M. Owens, G. Stamp, C. Arnott, F. Burke, N. East, H. Holdsworth, L. Turner, B. Rollins, M. Pasparakis, G. Kollias, F. Balkwill (1999). Mice deficient in tumor necrosis factor-alpha are resistant to skin carcinogenesis. *Nat. Med.*, **5**, 828–831.
69. J.M. Rivas, S.E. Ullrich (1994). The role of IL-4, IL-10, and TNF-alpha in the immune suppression induced by ultraviolet radiation. *J. Leukoc Biol.*, **56**, 769–775.
70. C. Nishigori, D.B. Yarosh, S.E. Ullrich, A.A. Vink, C.D. Bucana, L. Roza, M.L. Kripke (1996). Evidence that DNA damage triggers interleukin 10 cytokine production in UV-irradiated murine keratinocytes. *Proc. Natl. Acad. Sci. USA*, **93**, 10354–10359.
71. S. Kondo, B. Wang, H. Fujisawa, G.M. Shivji, B. Echtenacher, T.W. Mak, D.N. Sauder (1995). Effect of gene-targeted mutation in TNF receptor (p55) on contact hypersensitivity and ultraviolet B-induced immunosuppression. *J. Immunol.*, **155**, 3801–3805.
72. P. Amerio, P. Toto, C. Feliciani, H. Suzuki, G. Shivji, B. Wang, D.N. Sauder (2001). Rethinking the role of tumour necrosis factor-alpha in ultraviolet (UV) B-induced immunosuppression: altered immune response in UV-irradiated TNFR1R2 gene-targeted mutant mice. *Br. J. Dermatol.*, **144**, 952–957.
73. A. Rehemtulla, C.A. Hamilton, A.M. Chinnaiyan, V.M. Dixit (1997). Ultraviolet radiation-induced apoptosis is mediated by activation of CD-95 (Fas/Apo-1). *J. Biol. Chem.*, **272**, 25783–25786.
74. P.K. Kim, R. Weller, Y. Hua, T.R. Billiar (2003). Ultraviolet irradiation increases FADD protein in apoptotic human keratinocytes. *Biochem. Biophys. Res. Commun.*, **302**, 290–295.
75. V.N. Ivanov, O. Fodstad, Z. Ronai (2001). Expression of ring finger-deleted TRAF2 sensitizes metastatic melanoma cells to apoptosis via up-regulation of p38, TNFalpha and suppression of NF-kappaB activities. *Oncogene*, **20**, 2243–2253.
76. V.N. Ivanov, J.H. Kehrl, Z. Ronai (2000). Role of TRAF2/GCK in melanoma sensitivity to UV-induced apoptosis. *Oncogene*, **19**, 933–942.
77. S. Wu, H.N. Loke, A. Rehemtulla (2002). Ultraviolet radiation-induced apoptosis is mediated by Daxx. *Neoplasia*, **4**, 486–492.
78. L.B. Owen-Schaub, W. Zhang, J.C. Cusack, L.S. Angelo, S.M. Santee, T. Fujiwara, J.A. Roth, A.B. Deisseroth, W.W. Zhang, E. Kruzel, R. Radinsky (1995). Wild-type human p53 and a temperature-sensitive mutant induce Fas/APO-1 expression. *Mol. Cell Biol.*, **15**, 3032–3040.
79. M. Muller, S. Wilder, D. Bannasch, D. Israeli, K. Lehlbach, M. Li-Weber, S.L. Friedman, P.R. Galle, W. Stremmel, M. Oren, P.H. Krammer (1998). p53 activates the CD95 (APO-1/Fas) gene in response to DNA damage by anticancer drugs. *J. Exp. Med.*, **188**, 2033–2045.
80. D. Munsch, R. Watanabe-Fukunaga, J.C. Bourdon, S. Nagata, E. May, E. Yonish-Rouach, P. Reisdorf (2000). Human and mouse Fas (APO-1/CD95) death receptor genes each contain a p53-responsive element that is activated by p53 mutants unable to induce apoptosis. *J. Biol. Chem.*, **275**, 3867–3872.
81. C. Gutierrez-Steil, T. Wrone-Smith, X. Sun, J.G. Krueger, T. Coven, B.J. Nickoloff (1998). Sunlight-induced basal cell carcinoma tumor cells and ultraviolet-B-irradiated psoriatic plaques express Fas ligand (CD95L). *J. Clin. Invest.*, **101**, 33–39.
82. E. Filipowicz, P. Adegboyega, R.L. Sanchez, Z. Gatalica (2002). Expression of CD95 (Fas) in sun-exposed human skin and cutaneous carcinomas. *Cancer*, **94**, 814–819.
83. Z. Xia, M. Dickens, J. Raingeaud, R.J. Davis, M.E. Greenberg (1995). Opposing effects of ERK and JNK-p38 MAP kinases on apoptosis. *Science*, **270**, 1326–1331.
84. C. Tournier, P. Hess, D.D. Yang, J. Xu, T.K. Turner, A. Nimnual, D. Bar-Sagi, S.N. Jones, R.A. Flavell, R.J. Davis (2000). Requirement of JNK for stress-induced activation of the cytochrome c- mediated death pathway. *Science*, **288**, 870–874.

85. K. Lei, R.J. Davis (2003). JNK phosphorylation of Bim-related members of the Bcl2 family induces Bax-dependent apoptosis. *Proc. Natl. Acad. Sci. USA*, **100**, 2432–2437.
86. J. Hildesheim, R.T. Awwad, A.J. Fornace Jr. (2004). p38 Mitogen-activated protein kinase inhibitor protects the epidermis against the acute damaging effects of ultraviolet irradiation by blocking apoptosis and inflammatory responses. *J. Invest. Dermatol.*, **122**, 497–502.
87. W.J. Kozumbo, P.A. Cerutti (1986). Antioxidants as antitumor promoters. *Basic Life Sci.*, **39**, 491–506.
88. P.A. Cerutti (1985). Prooxidant states and tumor promotion. *Science*, **227**, 375–381.
89. S.K. Katiyar, N.J. Korman, H. Mukhtar, R. Agarwal (1997). Protective effects of silymarin against photocarcinogenesis in a mouse skin model. *J. Natl. Cancer Inst.*, **89**, 556–566.
90. D. Kulms, E. Zeise, B. Poppelmann, T. Schwarz (2002). DNA damage, death receptor activation and reactive oxygen species contribute to ultraviolet radiation-induced apoptosis in an essential and independent way. *Oncogene*, **21**, 5844–5851.
91. Y.P. Lu, Y.R. Lou, X.H. Li, J.G. Xie, D. Brash, M.T. Huang, A.H. Conney (2000). Stimulatory effect of oral administration of green tea or caffeine on ultraviolet light-induced increases in epidermal wild-type p53, p21(WAF1/CIP1), and apoptotic sunburn cells in SKH-1 mice. *Cancer Res.*, **60**, 4785–4791.
92. Z. Assefa, M. Garmyn, A. Vantieghem, W. Declercq, P. Vandenabeele, J.R. Vandenheede, P. Agostinis (2003). Ultraviolet B radiation-induced apoptosis in human keratinocytes: cytosolic activation of procaspase-8 and the role of Bcl-2. *FEBS Lett.*, **540**, 125–132.
93. D. Decraene, K. Smaers, D. Gan, T. Mammone, M. Matsui, D. Maes, L. Declercq, M. Garmyn (2004). A synthetic superoxide dismutase/catalase mimetic (EUK-134) inhibits membrane-damage-induced activation of mitogen-activated protein kinase pathways and reduces p53 accumulation in ultraviolet B-exposed primary human keratinocytes. *J. Invest. Dermatol.*, **122**, 484–491.
94. D.E. Brash, J.A. Rudolph, J.A. Simon, A. Lin, G.J. McKenna, H.P. Baden, A.J. Halperin, J. Ponten (1991). A role for sunlight in skin cancer: UV-induced p53 mutations in squamous cell carcinoma. *Proc. Natl. Acad. Sci. USA*, **88**, 10124–10128.
95. A.J. Nataraj, J.C. Trent 2nd, H.N. Ananthaswamy (1995). p53 gene mutations and photocarcinogenesis. *Photochem. Photobiol.*, **62**, 218–230.
96. H.N. Ananthaswamy, A. Ouhtit, R.L. Evans, A. Gorny, P. Khaskina, A.T. Sands, C.J. Conti (1999). Persistence of p53 mutations and resistance of keratinocytes to apoptosis are associated with the increased susceptibility of mice lacking the XPC gene to UV carcinogenesis. *Oncogene*, **18**, 7395–7398.
97. N.D. Lakin, S.P. Jackson (1999). Regulation of p53 in response to DNA damage. *Oncogene*, **18**, 7644–7655.
98. E. Appella, C.W. Anderson (2001). Post-translational modifications and activation of p53 by genotoxic stresses. *Eur. J. Biochem.*, **268**, 2764–2772.
99. Y. Higashimoto, S. Saito, X.H. Tong, A. Hong, K. Sakaguchi, E. Appella, C.W. Anderson (2000). Human p53 is phosphorylated on serines 6 and 9 in response to DNA damage-inducing agents. *J. Biol. Chem.*, **275**, 23199–23203.
100. L.J. Bean, G.R. Stark (2002). Regulation of the accumulation and function of p53 by phosphorylation of two residues within the domain that binds to Mdm2. *J. Biol. Chem.*, **277**, 1864–1871.
101. L.J. Bean, G.R. Stark (2001). Phosphorylation of serines 15 and 37 is necessary for efficient accumulation of p53 following irradiation with UV. *Oncogene*, **20**, 1076–1084.
102. K. Oda, H. Arakawa, T. Tanaka, K. Matsuda, C. Tanikawa, T. Mori, H. Nishimori, K. Tamai, T. Tokino, Y. Nakamura, Y. Taya (2000). p53AIP1, a potential mediator of p53-dependent apoptosis, and its regulation by Ser-46-phosphorylated p53. *Cell*, **102**, 849–862.

103. J. Bartek, J. Lukas (2003). Chk1 and Chk2 kinases in checkpoint control and cancer. *Cancer Cell*, **3**, 421–429.
104. N.D. Lakin, B.C. Hann, S.P. Jackson (1999). The ataxia-telangiectasia related protein ATR mediates DNA-dependent phosphorylation of p53. *Oncogene*, **18**, 3989–3995.
105. R.S. Tibbetts, K.M. Brumbaugh, J.M. Williams, J.N. Sarkaria, W.A. Cliby, S.Y. Shieh, Y. Taya, C. Prives, R.T. Abraham (1999). A role for ATR in the DNA damage-induced phosphorylation of p53. *Genes Dev.*, **13**, 152–157.
106. S.Y. Shieh, J. Ahn, K. Tamai, Y. Taya, C. Prives (2000). The human homologs of checkpoint kinases Chk1 and Cds1 (Chk2) phosphorylate p53 at multiple DNA damage-inducible sites. *Genes Dev.*, **14**, 289–300.
107. K. Unsal-Kacmaz, A.M. Makhov, J.D. Griffith, A. Sancar (2002). Preferential binding of ATR protein to UV-damaged DNA. *Proc. Natl. Acad. Sci. USA*, **99**, 6673–6678.
108. C. Huang, W.Y. Ma, A. Maxiner, Y. Sun, Z. Dong (1999). p38 kinase mediates UV-induced phosphorylation of p53 protein at serine 389. *J. Biol. Chem.*, **274**, 12229–12235.
109. R. Sanchez-Prieto, J.M. Rojas, Y. Taya, J.S. Gutkind (2000). A role for the p38 mitogen-acitvated protein kinase pathway in the transcriptional activation of p53 on genotoxic stress by chemotherapeutic agents. *Cancer Res.*, **60**, 2464–2472.
110. M. Takekawa, M. Adachi, A. Nakahata, I. Nakayama, F. Itoh, H. Tsukuda, Y. Taya, K. Imai (2000). p53-inducible wip1 phosphatase mediates a negative feedback regulation of p38 MAPK-p53 signaling in response to UV radiation. *EMBO J.*, **19**, 6517–6526.
111. Q.B. She, W.Y. Ma, Z. Dong (2002). Role of MAP kinases in UVB-induced phosphorylation of p53 at serine 20. *Oncogene*, **21**, 1580–1589.
112. T. Miyashita, J.C. Reed (1995). Tumor suppressor p53 is a direct transcriptional activator of the human bax gene. *Cell*, **80**, 293–299.
113. P.F. Li, R. Dietz, R. von Harsdorf (1999). p53 regulates mitochondrial membrane potential through reactive oxygen species and induces cytochrome c-independent apoptosis blocked by Bcl-2. *EMBO J.*, **18**, 6027–6036.
114. N.D. Marchenko, A. Zaika, U.M. Moll (2000). Death signal-induced localization of p53 protein to mitochondria. A potential role in apoptotic signaling. *J. Biol. Chem.*, **275**, 16202–16212.
115. F. Bernerd, A. Sarasin, T. Magnaldo (1999). Galectin-7 overexpression is associated with the apoptotic process in UVB-induced sunburn keratinocytes. *Proc. Natl. Acad. Sci. USA*, **96**, 11329–11334.
116. M. Scott, P. Bonnefin, D. Vieyra, F.M. Boisvert, D. Young, D.P. Bazett-Jones, K. Riabowol (2001). UV-induced binding of ING1 to PCNA regulates the induction of apoptosis. *J. Cell Sci.*, **114**, 3455–3462.
117. K.J. Cheung Jr., G. Li (2002). p33(ING1) enhances UVB-induced apoptosis in melanoma cells. *Exp. Cell Res.*, **279**, 291–298.
118. N.M. Wikonkal, E. Remenyik, D. Knezevic, W. Zhang, M. Liu, H. Zhao, T.R. Berton, D.G. Johnson, D.E. Brash (2003). Inactivating E2f1 reverts apoptosis resistance and cancer sensitivity in Trp53-deficient mice. *Nat. Cell Biol.*, **5**, 655–660.
119. M. Schreiber, B. Baumann, M. Cotten, P. Angel, E.F. Wagner (1995). Fos is an essential component of the mammalian UV response. *EMBO J.*, **14**, 5338–5349.
120. S. Haas, B. Kaina (1995). c-Fos is involved in the cellular defence against the genotoxic effect of UV radiation. *Carcinogenesis*, **16**, 985–991.
121. L. Zhou, H. Steller (2003). Distinct pathways mediate UV-induced apoptosis in Drosophila embryos. *Dev. Cell*, **4**, 599–605.

Chapter 16

Opposing roles of UV-induced apoptosis in early skin cancer

Wengeng Zhang and Douglas E. Brash

Table of contents

Abstract

Mutations in the *p53* tumor suppressor gene are well-established as a very early genetic event in ultraviolet (UV) carcinogenesis. Clones of *p53*-mutant keratinocytes are the earliest known step of UV-induced skin cancer. Keratinocyte apoptosis (sunburn cell formation) induced by UV and mediated by wild-type *p53* appears to be a critical pathway for preventing the occurrence of precancerous mutations in the skin.

Recent research in our laboratory provides evidence suggesting that UV-induced apoptosis may also *facilitate* the development of skin cancer, since it preferentially spares cells with *p53* mutations. The situation is analogous to the finding that hypoxia favors the growth of *p53*-mutant cancer cells over normal cells, with hypoxia selectively inducing apoptosis in the latter. Many factors affect UV carcinogenesis, including DNA damage and repair, apoptosis, cell proliferation, the immune response, and aging. Here we focus on events leading to UV-induced apoptosis and the role of this apoptosis in UV-induced skin cancer.

16.1 Photoproducts and precancers

After UVB irradiation, the absorption of photons by DNA causes formation of cyclobutane pyrimidine dimers and pyrimidine (6-4) pyrimidone photoproducts. Both photoproducts can change DNA structure, bending the DNA or creating a facsimile of an abasic site. UV-induced mutations usually occur at sites of adjacent pyrimidines because the photoproducts affect such sites. If these lesions cannot be repaired, they can lead to permanent mutations in the DNA sequence. Two thirds of the mutations are C to T substitutions and about 10% are CC to TT substitutions; together, this pattern is known as the UV mutation signature [1]. Evidence indicates that these photoproducts, especially cyclobutane dimers, are responsible for most of the mutagenic and carcinogenic effects of UV irradiation.

At least two pathways prevent damaged cells containing UV photoproducts from resulting in tumors. One strategy is repair of the damage by the nucleotide excision or recombinational repair mechanism, evolutionarily conserved approaches for repairing different kinds of DNA damage. Another strategy is to eliminate the damaged cells by apoptosis. Both of these can prevent mutations from being passed on to daughter cells and leading to carcinogenesis. P53, the guardian of genome, plays an important role in both repair of UV-damaged DNA and induction of apoptosis. If DNA in actively transcribed genes cannot be repaired, P53 is induced [2,3]. It then arrests cells in G1 phase by inducing inhibitor of cyclin-dependent kinase P21 and prevents cells from entering S phase, thus putatively allowing time to repair DNA and preventing creation of a mutant cell when the damaged DNA is replicated. As discussed later, induced P53 can also lead to apoptosis, perhaps in conjunction with other signaling pathways. If these protective mechanisms are inadequate, a mutant cell results.

One of the most frequently mutated genes is *p53* itself. Results from research on *p53* mutations in human skin cancers, in UV-irradiated mouse skin, and in UV-induced murine skin cancers, have provided new insights into the molecular mechanisms of UV-induced skin cancer. Over 90% of human cutaneous squamous cell cancers and about half of human basal cell cancers contain UV-specific mutations somewhere in the *p53* gene [4–6]. In fact, *p53* gene mutation occurs very early during UV carcinogenesis. Jonason et al. [7] used whole-mount preparations of human skin and immunohistochemical staining to check mutant P53 expression and found that the *p53* mutant keratinocytes appeared as clonal patches. These clonal patches in clinically normal skin arose from the dermal-epidermal junction and hair follicles and contained UV specific mutations. The clones in sun-exposed skin were both more frequent and larger than those in sun-shielded skin. The authors concluded that sunlight acts as both initiator and promoter by favoring the clonal expansion of *p53*-mutant cells. Similar clones have been observed in paraffin sections of human skin [8] and can persist long after sunlight exposure [9].

To investigate the timing of the *p53* alterations in the process of UVB carcinogenesis, Berg et al. [10] exposed mouse skin to UV for either 17 or 30 days, a protocol which would cause skin tumors at 80 or 30 weeks, respectively. Using CM-5 polyclonal antibody, they detected clusters of cells with a strong immunostaining of P53 protein, of which $\approx$70% reacted with the mutant-specific monoclonal antibody Pab240. Although most clusters disappeared within a few weeks after cessation of UV, some were detectable as long as 56 days after 17 days of exposure. This result suggests that *p53* mutation occurs well before skin cancer development. It also suggests that *p53* mutation, and the appearance of keratinocytes as *p53*-mutant clusters, can serve as a very early biological endpoint for research in UV-induced skin cancer.

Indeed, Rebel et al. [11] found that the density of *p53*-mutant foci can serve as a direct measure of tumor risk after UV irradiation. They exposed hairless SKH-1 mice to daily broadband UV and found that with 500 J/m^2 of UV irradiation, the average number of *p53* positive foci in a dorsal skin area of 7.2 cm^2 increased rapidly from 9 at 15 days to 470 at 40 days. At a dose of 250 J/m^2 of UV irradiation, the induction rate of *p53* positive foci was reduced by a factor of 1.49 ± 0.15, a figure which was very similar to their observed 1.54 ± 0.02 fold reduction in squamous cell carcinomas. In a double-log plot of the average number of *p53* foci versus time, the curves for the two different exposure regimens were similar to the curves for the number of tumors versus time. Further, they found that *p53* foci were induced at significantly increased rates in either homozygous X*pa* knockout or heterozygous *p53* knockout mice upon UV irradiation, corresponding to the increased rates of carcinoma formation in these strains.

Therefore the early induction of keratinocyte patches (clusters or clones) expressing mutant P53 is closely, and most likely causally, associated with the eventual tumors. These *p53* patches may thus serve as early indicators of tumor risk [12].

16.2 Evidence that UV-induced apoptosis protects against carcinogenesis

Apoptosis appears to play a crucial role in protecting against UV-induced tumor formation, both at the earliest stage in which *p53* mutant clones form, and at the later stage of malignant carcinoma. First, Ziegler et al. [13] found that *p53* is required for sunburn cell formation. Sunburn cells are a term in dermatology for apoptotic keratinocytes in UV-exposed epidermis. After using varying doses of UVB (0, 500, and 1,000 J/m^2) to irradiate the dorsal skin of wild-type, *p53*$^{+/-}$ heterozygous, and *p53*$^{-/-}$ homozygous mice, sunburn cells were examined 24 hours later by both haematoxylin-eosin staining and fluorescent in situ end-labeling. The highest dose corresponds to a just detectable sunburn. As Figure 1 shows, the number of sunburn cells in all three genotypes increased with the dose of UVB. However, the slope of induction was 7-fold less for *p53*$^{-/-}$ animals. The frequency of apoptosis was intermediate in *p53*$^{+/-}$ heterozygous mice. Cultured mouse keratinocytes derived from homozygous *p53* deficient mice (*p53*$^{-/-}$) were also markedly resistant to UVB-induced apoptosis compared with keratinocytes from wild or heterozygous *p53* deficient mice. Since *p53* mutations are associated with skin cancer, these results suggested that elimination of UV-damaged cells by apoptosis is an epidermal cancer prevention mechanism.

A role for UV-induced apoptosis in preventing skin mutagenesis was supported by a study by Hill et al. [14]. *p53* is a transcriptional target for Fas, the FasL receptor; UV radiation is known to upregulate both Fas and FasL. The authors exposed FasL-deficient mice (C3H/HeJ *gld/gld*) and wild-type C3H/HeJ mice to a

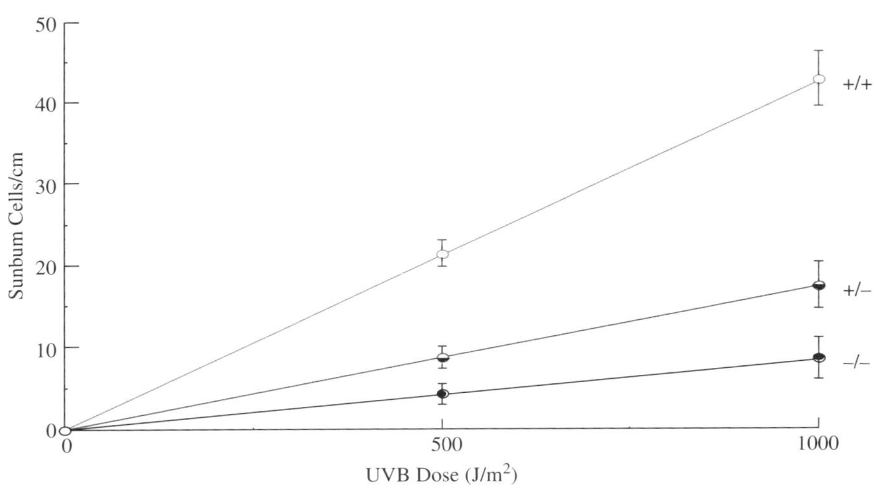

Figure 1. *p53* is necessary for UV-induced apoptosis in mouse skin. Apoptotic cells ("sunburn cells") were defined as having intensively eosinophilic cytoplasm, pyknotic nuclei, and separation from adjacent cells. +/+, wild-type *p53* mouse; +/−, heterozygote; −/−, homozygous *p53* knockout. [Reprinted with permission from A. Ziegler et al. 13.]

single dose of 5,000 J/m^2 UVB radiation and measured keratinocyte apoptosis. FasL-deficient animals demonstrated a 4-fold reduction in UV-induced keratinocyte apoptosis. Further, a single week of UV radiation caused 70% of the FasL-deficient mice, but only 5% of wild-type mice, to accumulate *p53* mutations in the epidermis. This result suggests that normal Fas-FasL induced apoptosis of damaged cells is one of the suppressor pathways for UV-induced skin tumorigenesis. The same group also showed that chronic UV exposure leads to the loss of FasL expression very early, and this loss results in the inhibition of apoptosis, an increase in *p53* mutations, and initiation of skin cancer [15].

Similarly, our laboratory found that 5 weeks of 1,250 J/m^2 UVB irradiation induced about 10 times more *p53* mutant clones in $p53^{+/-}$ mice than in $p53^{+/+}$ mice (unpublished data). Rebel et al. [11] also found many more *p53* mutant clones in $p53^{+/-}$ mice than in $p53^{+/+}$. Similar tumor-prevention effects of apoptosis are observed in other cases. In postnatal lens, for example, inactivating RB protein with a virus results in ectopic cell proliferation and apoptosis. But if P53 protein is also inactivated, apoptosis is reduced and a tumor develops [16]. In embryonic lens and in the choroid plexus, reduced apoptosis leading to tumor development is also observed [17,18]. A caveat is that FasL and P53 affect functions in addition to apoptosis, including DNA repair and cell cycle arrest [19,20].

Studying a later stage in skin cancer development, Lu et al. [21] irradiated female SKH-1 hairless mice with 300 J/m^2 of UVB twice a week for 20 weeks and three weeks later treated the mice topically on the back with acetone, caffeine, or (−)-epigallocatechin gallate (EGCG), once a day, 5 days a week for 18 weeks. 24 hours after the last treatment the mice were sacrificed. Tumors, apoptosis, and cell proliferation were quantitated. Their results showed that topical application of caffeine decreased the number of nonmalignant and malignant skin tumors per mouse by 44% and 72%, respectively. Topical application of EGCG decreased the number of nonmalignant and malignant tumors per mouse by 55% and 66%. Immunohistochemical analysis of the number of caspase 3-positive cells showed that topical application of EGCG or caffeine increased apoptosis by 72% or 87% in nonmalignant skin tumors, respectively, and by 56% or 92% in squamous cell carcinomas. In contrast, there was no effect on apoptosis in nontumor areas of the epidermis. Again, it is possible that non-apoptotic effects of the pharmacological agents played a role.

The X*pc* gene affects global genomic excision repair, which removes most photoproducts from the DNA but not those that induce the UV-apoptosis signal [2,3]. Ananthaswamy et al. [22] found that UV induced *p53* mutations at codon 270 at high frequencies in both $Xpc^{-/-}$ and $Xpc^{+/+}$ mouse skin after 1 week of 2,500 J/m^2 UV irradiation. But after 3~4 weeks of chronic UV irradiation, the mutation was found only in $Xpc^{-/-}$ mouse skin. Accordingly, the authors found that the number of TUNEL-positive cells did not show any difference after 1 or 2 weeks of UV irradiation between $Xpc^{-/-}$ and wild-type mice. But after 3~4 weeks UV irradiation, $Xpc^{-/-}$ mouse had significantly fewer apoptotic cells in the epidermis than did $Xpc^{+/+}$ mice. Using proliferating cell nuclear antigen as a marker of cell proliferation, they also found that $Xpc^{-/-}$ mice were more sensitive to UV-induced epidermal hyperplasia than $Xpc^{+/+}$ mice. The authors concluded that *p53* mutations arise early and persistence of UV-induced *p53* mutations in $Xpc^{-/-}$ mouse skin resulted from decreased apoptosis and increased proliferation of keratinocytes. These events may accelerate the development of UV-induced skin cancer in the $Xpc^{-/-}$ mouse.

16.3 Evidence that apoptosis may promote carcinogenesis

If a DNA photoproduct in the *p53* gene cannot be repaired, and apoptosis does not supervene, a single *p53*-mutant cell results. This single cell may clonally expand into a *p53*-mutant clone and thence to a precancer and cancer. But what drives the single *p53* mutant keratinocyte's clonal expansion?

An indication has come from studies of the cellular and molecular response to stresses found in the tumor microenvironment. Low oxygen tension (hypoxia) is a common feature of solid tumors. In studying the influence of hypoxia on the progression of transformed and tumor cells, Graeber et al. [23] found that hypoxia stimulated P53 protein accumulation and induced apoptosis in c-Myc transformed Rat1 fibroblasts and E1A- and Ha-ras-transformed *p53*-expressing mouse embryonic fibroblasts, but not in E1A- and Ha-ras-transformed $p53^{-/-}$ cells. Using a 1 to 1,000 ratio of E1A- and Ha-ras-transformed $p53^{-/-}$ and $p53^{+/+}$ albina mixed culture system, they found that the infrequent oncogenically transformed $p53^{-/-}$ cells overtook the $p53^{+/+}$ cells after 7 rounds of hypoxia and aerobic recovery treatment, suggesting that hypoxia selected for cells with defects in apoptosis. The researchers also found that highly apoptotic regions strongly correlated with hypoxic regions in transplanted tumors expressing wild-type *p53*. In contrast, little apoptosis was found in hypoxic regions of *p53*-deficient tumors. Therefore, they suggested that hypoxia provides a physiological selection pressure in tumors for the expansion of cells that have lost their apoptotic potential, and in particular for cells that have acquired *p53* mutations.

Does a similar mechanism occur for UVB in the early, precancerous stages of skin cancer? Research in our laboratory provides evidence that UV, like hypoxia in solid tumors, provides a physiological selection pressure for expansion of a *p53* mutant clone [24]. The transition from normal cell to *p53*-mutant clone in normal skin requires a cell to escape both DNA repair and the defense mechanism that leads to apoptosis of UV-damaged cells [13]. While these events can be studied in humans only after the fact, events in mice can be carefully controlled. It is now clear that the mouse is one of the best models for human disease research. More than 90% of the murine and human genomes can be divided into corresponding regions of conserved synteny, reflecting chromosome segments in which the gene order has been conserved in both species [25]. After chronic UVB irradiation, mice develop squamous cell carcinomas that resemble the human disease. Here we will focus on our research performed using the mouse model.

Past studies [26–29] have shown that mouse dorsal skin is organized into hexagonal domains termed "epidermal proliferation units" (EPU). Each EPU contains a single stem cell located in the center, its descendent transient amplifying cells within the basal layer, and the subsequent terminally differentiated cells forming the stratified, suprabasal layers. Thus, an EPU is considered a stem cell compartment. (Additional stem cells are found in the bulge region of hair follicles.) Each EPU contains ≈12~14 nucleated cells and 6~8 overlying differentiating cells. After several weeks of a daily dose of 1,250 J/m^2 UVB irradiation, which is equivalent to a barely perceptible sunburn each day, the epidermal morphology still looked similar to the normal EPU structure of mouse dorsal skin showing polygonal structure of nonoverlapping corneocytes (Figure 2A). Regularly spaced groups of two to three dark-stained overlapping nuclei were located in the center and were surrounded with lightly stained nonoverlapping nuclei, which are consistent with the central column of nuclei reported in the normal EPU (Figure 2B). The *p53*-mutant clones generated by several weeks of UVB irradiation ranged in size from 3 to 2,000 cells. About half of these clones were greater than 20 cells in size and

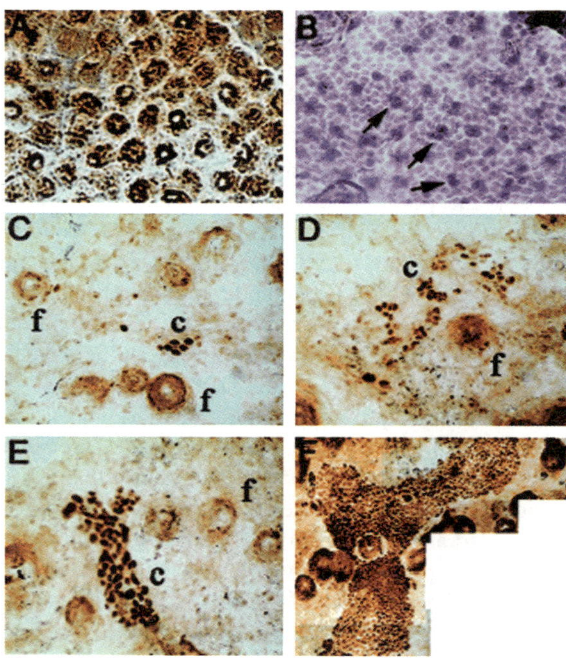

Figure 2. Epidermal domain structure and *p53*-mutant clones in epidermal sheets from UVB-irradiated mouse dorsal skin. (A) After chronic UVB exposure, the polygonal structure of nonoverlapping superficial corneocytes similar to normal epidermal proliferative units still exists; counterstain is diaminobenzamide. (B) A group of 2 or 3 densely-stained overlapping keratinocyte nuclei (arrow) is regularly located in the center of each EPU and is surrounded by lightly-stained nonoverlappling single nucleus of transit amplifying cells; hematoxylin staining. (C–F) Different sizes of *p53*-mutant clones. UVB dose was 1,250 J/m^2 daily for 5 weeks except (F) 11 weeks. Mutant clones (c) and follicles (f) are indicated. Digital photomicrography was ≈ ×400, except (F) ×200. [Reprinted with permission from W. Zhang et al. 24.]

therefore they must have expanded beyond their initial stem cell compartment (Figure 2C–F) [24]. What is the force that drove the clonal expansion?

To check whether the UVB-induced *p53* positive clusters are clonal, we first excised samples of 100~500 cells in size from *p53*-mutant clones, isolated DNA, amplified the whole genomic DNA, PCRed the specific *p53* fragments, and sequenced them. In each case, the mutation was present throughout the whole sample. We then microdissected biopsies of ≥40 cells from opposite end of six clones. Using the same approach we found that each pair carried at least one identical mutation. This result clearly shows the clonal origin of chronic UVB-induced *p53* positive clusters.

To clonally expand, a mutant cell must first proliferate to fill its own EPU and then escape from that EPU to colonize or displace an adjacent EPU. How does this happen? The fact that the normal epidermal structure remained after chronic UVB exposure seems to exclude the possibility of a simple EPU enlargement. By quantitative analysis of the clone sizes, we next found that the clones consisted of multiples of about 12~16 cells (Figure 3). This size is similar to the number of nucleated cells in a mouse EPU. These results indicate that the clonal expansion of *p53* mutant clusters is proceeding by successively colonizing adjacent EPU's, rather than by continuous invasion.

Is the clonal expansion of *p53* mutant clusters driven by a further mutation or by a physiological event such as growth factors or selection pressure? The multiple-hit theory of carcinogenesis states that the first genetic change leads to an initiated cell that will remain at this stage if no more mutations occur. If, however, a further

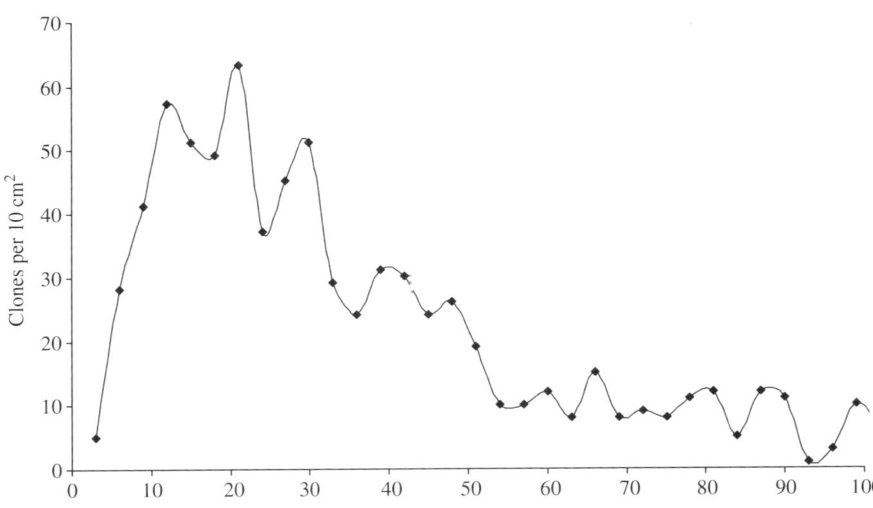

Figure 3. Quantized distribution of sizes of *p53*-mutant clones. When 868 clones were grouped into size bins 3 cells in width, maxima occurred in the clone size distribution at intervals of approximately 12–16 cells/clone. Clones larger than 100 cells are not shown. [Reprinted with permission from W. Zhang et al. 24.]

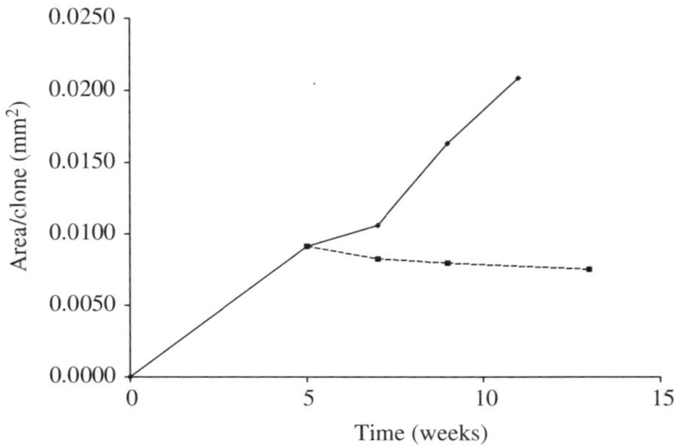

Figure 4. Clonal expansion requires a physiological event rather than a second mutation. Clone area no longer increases after UVB ceases (dashed line). [Reprinted with permission from W. Zhang et al. 24.]

mutation in an initiated cell gives it a proliferative advantage, the cell can expand autonomously without needing the continuous presence of a carcinogen. Therefore, genetic and physiological drivers of clonal expansion can be distinguished by determining whether clonal expansion depends on continuous carcinogen exposure. We found that after chronic UVB exposure ended, the area of *p53* positive clones ceased to increase (Figure 4). If UVB was instead continued, the area of the *p53* positive clones increased radically. The requirement for sustained UVB exposure rules out the possibility that clonal expansion was cell-autonomous, and so it could not have been caused by a mutation. Instead, clonal expansion was forced by the continuing UVB exposure – a physiological event. This physiological event is consistent with a selection pressure due to differential apoptosis susceptibility.

We conclude that: UV mutates the *p53* gene (as well as others such as *PTCH*); a *p53* mutation confers apoptosis-resistance; clonal expansion of apoptosis-resistant cells is driven by a UV-induced physiological event, not a second mutation; and this physiological event may be apoptosis of surrounding normal cells. By affecting the number of cells that are targets for the next UV exposure, these cellular level events may significantly affect the time needed to accumulate alterations in the other genes needed for skin cancer to arise.

References

1. N.M. Wikonkal, D.E. Brash (1999). Ultraviolet radiation induced signature mutations in photocarcinogenesis. *J. Invest. Dermatol. Symp. Proc.*, **4**, 6–10.
2. M. Ljungman, F. Zhang (1996). Blockage of RNA polymerase as a possible trigger for U.V. light-induced apoptosis. *Oncogene*, **13**, 823–831.
3. D.E. Brash, N.M. Wikonkal, E. Remenyik, G.T.J. van der Horst, E.C. Friedberg, D.L. Cheo, H. van Steeg, A. Westerman, H.J. van Kranen (2001). The DNA damage

signal for Mdm2 regulation, Trp53 induction, and sunburn cell formation *in vivo* originates from actively transcribed genes. *J. Invest. Dermatol.*, **117**, 1234–1240.

4. D.E. Brash, J.A. Rudolph, J.A. Simon, A. Lin, G J. McKenna, H.P. Baden, A.J. Halperin, J. Pontén (1991). A role for sunlight in skin cancer: UV-induced p53 mutations in squamous cell carcinoma. *Proc. Natl. Acad. Sci. USA*, **88**, 10124–10128.
5. A. Ziegler, D.J. Leffell, S. Kunala, H.W. Sharma, M. Gailani, J.A. Simon, A.J. Halperin, H.P. Baden, P.E. Shapiro, A.E. Bale, D.E. Brash (1993). Mutation hotspots due to sunlight in the p53 gene of non-melanoma skin cancers. *Proc. Natl. Acad. Sci. USA*, **90**, 4216–4220.
6. P. Rady, F. Scinicariello, R.F. Wagner, S.K. Tyring (1992). p53 mutations in basal cell carcinomas, *Cancer Res.*, **52**, 3804–3806.
7. A.S. Jonason, S.Kunala, G.J. Price, R.J. Restifo, H.M. Spinelli, J.A. Persing, D.J. Leffell, R.E. Tarone, D.E. Brash (1996). Frequent clones of p53-mutated keratinocytes in normal human skin. *Proc. Natl. Acad. Sci. USA*, **93**, 14025–14029.
8. Z.P. Ren, F. Ponten, M. Nister, J. Ponten (1996). Two distinct p53 immunohistochemical patterns in human squamous cell skin cancer, precursors, and normal epidermis. *Int. J. Cancer*, **69**, 174–179.
9. G. Ling, A. Persson, B. Berne, M. Uhlen, J. Lundeberg, F. Ponten (2001). Persistent p53 mutations in single cells from normal human skin. *Am. J. Pathol.*, **159**, 1247–1253.
10. R.J. Berg, H.J. van Kranen, H.G. Rebel, A. de Vries, W.A. van Vloten, C.F. Van Kreijl, J.C. van der Leun, F.R. de Gruijl (1996). Early p53 alterations in mouse skin carcinogenesis by UVB radiation: immunohistochemical detection of mutant p53 protein in clusters of preneoplastic epidermal cells. *Proc. Natl. Acad. Sci. USA*, **93**, 274–278.
11. H. Rebel, L.O. Mosnier, R.J.W. Berg, A. Westeman-de Vires, H. Steeg, H.J. van Kranen, F. de Gruijl (2001). Early p53-positive foci as indicators of tumor risk in ultraviolet-exposed hairless mice: kinetics of induction, effects of DNA repair deficiency, and p53 heterozygosity. *Cancer Res.*, **61**, 977–983.
12. F.R.de Gruijl, J.C. van der Leun, Physical variables in experimental photocarcinogenesis and quantitative relationships between stages of tumor development. *Front. Biosci.*, **7**, d1525–d1530.
13. A. Ziegler, A.S. Jonason, D.J. Leffell, J.A. Simon, H.W. Sharma, J. Kimmelman, L. Remington, T. Jacks, D.E. Brash (1994). Sunburn and p53 in the onset of skin cancer. *Nature*, **372**, 773–776.
14. L.L. Hill, A. Ouhtit, S.M. Loughlin, M.L. Kripke, H.N. Ananthaswamy, L.B. Owen-Schaub (1999). Fas ligand: a sensor for DNA damage critical in skin cancer etiology. *Science*, **285**, 898–900.
15. A. Ouhtit, A. Gorny, H.K. Muller, L.L. Hill, L. Owen-Schaub, H.N. Ananthaswamy (2000). Loss of Fas-Ligand Expression in Mouse Keratinocytes during UV Carcinogenesis. *Am. J. Pathol.*, **157**, 1975–1981.
16. H. Pan, A.E. Griep (1994). Altered cell cycle regulation in the lens of HPV-16 E6 or E7 transgenic mice: implications for tumor suppressor gene function in development. *Genes Devel.*, **8**, 1285–1299.
17. S.D. Morgenbesser, B.O. Williams, T. Jacks, R.A. DePinho (1994). *p53*-dependent apoptosis produced by *Rb*-deficiency in the developing mouse lens. *Nature*, **371**, 72–73.
18. H. Symonds, L. Krall, L.Remington, M. Saenz-Robles, S. Lowe, T. Jacks, T. Van Dyke (1994). p53-dependent apoptosis suppresses tumor growth and progression *in vivo*. *Cell*, **78**, 703–711.
19. G.M. Wahl, A.M. Carr (2001). The evolution of diverse biological responses to DNA damage: insights from yeast and p53. *Nature Cell Biol.*, **3**, E277–E286.

20. L.B. Owen-Schaub (2002). Fas function and tumor progression: Use it and lose it. *Cancer Cell*, **2**, 95–96.
21. Y.P. Lu, Y.R. Lou, J.G. Xie, Q.Y. Peng, J. Liao, C.S. Yang, M.T. Huang, A.H. Conney (2002). Topical applications of caffeine or (−)-epigallocatechin gallate (EGCG) inhibit carcinogenesis and selectively increase apoptosis in UVB-induced skin tumors in mice. *Proc. Natl. Acad. Sci. USA*, **99**, 12455–12460.
22. H.N. Ananthaswamy, A. Ouhtit, R.L. Evans, A. Gorny, P. Khaskina, A.T. Sands, C.J. Conti (1999). Persistence of p53 mutations and resistance of keratinocytes to apoptosis are associated with the increased susceptibility of mice lacking the XPC gene to UV carcinogenesis. *Oncogene*, **18**, 7395–7398.
23. T.G. Graeber, C. Osmanian, T. Jacks, D.E. Houseman, C.J. Koch, S.W. Lowe, A.J. Giaccia (1996). Hypoxia-mediated selection of cells with diminished apoptotic potential in solid tumors. *Nature*, **379**, 88–91.
24. W. Zhang, E. Remenyik, D. Zelterman, D.E. Brash, N.M. Wikonkal (2001). Escaping the stem cell compartment: sustained UVB exposure allows p53-mutant keratinocytes to colonize adjacent epidermal proliferating units without incurring additional mutations. *Proc. Natl. Acad. Sci. USA*, **98**, 13948–13953.
25. Mouse Genome Sequencing Consortium, Initial sequencing and comparative analysis of the mouse genome. *Nature*, **420**, 520–562.
26. C.S. Potten, C. Booth (2002). Keratinocyte stem cells: a commentary. *J. Invest. Dermatol.*, **119**, 888–899.
27. T.M. Kolodka, J.A. Garlick, L.B. Taichman (1998). Evidence for keratinocyte stem cells in vitro: long term engraftment and persistence of transgene expression from retrovirus-transduced keratinocytes. *Proc. Natl. Acad. Sci. USA*, **95**, 4356–4361.
28. I.C. Mackenzie (1998). Retroviral transduction of murine epidermal stem cells demonstrates clonal units of epidermal structure. *J. Invest. Dermatol.*. **109**, 377–383.
29. S.M. Janes, S. Lowell, C. Hutter (2002). Epidermal stem cells. *J. Pathol.*, **197**, 479–491.

Chapter 17

Acquired activation of signalling pathways in skin tumours from DNA repair-deficient xeroderma pigmentosum patients

Jean-Claude Ehrhart, Fabien P. Gosselet, Raphaël M. Culerrier and Alain Sarasin

Table of contents

Abstract

Sunlight is the most important genotoxic agent involved in skin cancers. The UVB component of the solar spectrum induces DNA lesions at bipyrimidine sites which can give rise to mutations in absence of an efficient error-free repair. Xeroderma pigmentosum (XP) is a human syndrome transmitted as an autosomal, recessive trait, characterized by early and numerous cutaneous tumours on sun-exposed sites. These patients are defective in nucleotide excision repair which is the major pathway for removing bulky DNA adducts. Cultured XP cells are hypermutable after UVB irradiation. This increased mutation frequency has been found in key genes, governing signalling pathways implicated in the regulation of cell cycling. A high level of base pair substitutions was detected in the *P53*, *PTCH-1*, *SMO*, $P16^{INK4A}$ or $P14^{ARF}$ genes. Seventy to 80% of XP skin cancers exhibit one or several mutations in these genes, the type and frequency of mutated genes being different between squamous cell (SCCs) and basal cell carcinomas (BCCs). Activation of the corresponding signalling pathways can seriously alter the checkpoints activated after genotoxic damage, therefore allowing a damaged cell to replicate and give rise to mutated daughter cells, then eventually to the development of the carcinogenic process. We have developed in this review the biological consequences of acquired activation of the $P14^{ARF}$, $P16^{INK4A}$ and Sonic Hedgehog pathways on skin tumour induction.

17.1 Introduction

Two main pathways can transform normal skin keratinocytes into tumoral cells after UVB irradiation. The first one corresponds to early mutations inactivating a caretaker gene [1]. In such a case, mutation rates will increase at each cell division and tumoral cells will appear due to the accumulation of genetic modifications. The DNA repair genes are the best candidates for this pathway as exemplified by the human hereditary xeroderma pigmentosum (XP) syndrome [2,3]. Persistence of unrepaired DNA damage produced by exposure to UVB sunlight is associated, in the XP syndrome, with a very high incidence of early and numerous cutaneous carcinomas in sun-exposed body sites.

The second pathway corresponds to early mutations inactivating a gatekeeper gene [1]. In such a case, genetic abnormalities also will strongly increase at each cell division. Oncogenes such as *RAS*, *MYC*, *SMO*, ... and inactivated tumour suppressor genes such as *P53*, $P16^{INK4A}$, $P14^{ARF}$, *PTCH-1*, ... can fulfil this action in skin cancers. An hereditary disease such as the Gorlin's syndrome (or Nevoid Basal Cell Carcinoma Syndrome, NBCCS), where the *PTCH-1* gene shows a germline mutation which is at the origin of numerous BCCs, is also an excellent example of the involvement of gatekeeper genes in skin tumours [4].

17.2 Role of DNA repair in the development of UV-induced skin cancers

UVB radiation is the most harmful component of sunlight for the genome. Direct absorption of UVB photons results in the formation of dimeric photolesions at bipyrimidine sites (cyclobutane pyrimidine dimers, pyrimidine (6-4) pyrimidone photoproducts and related Dewar valence isomers) [5–7]. These bulky DNA base adducts are removed by Nucleotide Excision Repair (NER) which is the most flexible of all DNA repair mechanisms. Two subpathways have been identified in NER: the rapid transcription-coupled repair (TCR) of the transcribed strand of expressed genes, and the slower global genome repair (GGR) of bulk DNA, including the non-transcribed strand of active genes [3].

In the case of inefficient NER of photolesions, damage remaining on DNA can give rise to mutations during replication. This is highly pronounced in XP patients who are defective in NER and is correlated with the very high incidence of skin cancers which arise in these patients at an early age [3]. Compared to the same type of tumours in the normal population, a higher level of oncogene (*RAS*, *MYC*) and tumour suppressor gene (*P53*, $P16^{INK4A}$, $P14^{ARF}$, *PTCH-1*) mutations has been demonstrated in XP tumours. They are targeted at bipyrimidine sites and over 60% are CC to TT tandem transitions compared to 10–15% in sporadic skin tumours from NER-proficient patients. In XP-C patients, whose NER defect is limited to the GGR, all base pair substitutions in active genes such as *P53* are due to translesion synthesis of unrepaired damage remaining only on the non-transcribed strand. This result demonstrates the existence of TCR in humans. Therefore XP is a pertinent model for the understanding of sunlight-associated skin cancer development in the normal population [3].

Increased mutation frequency is found in key gatekeeper genes governing signalling pathways implicated in the regulation of cell cycle and control of cellular proliferation. Acquired activation of these pathways then allows a damaged cell to replicate, giving rise to mutated daughter cells and eventually to the development of the carcinogenic process.

17.3 The $P14^{ARF}$ pathway

Activation of P53 through genotoxic stressors such as UVB leads to the disruption of HDM2 (human MDM2)-mediated P53 proteasomal proteolysis, resulting in stabilization of P53 followed by P53-mediated cell cycle arrest or apoptosis [8]. UVB-induced damage results in phosphorylation of P53 N-terminal residues by the ATR (ataxia telangiectasia and Rad3-related) kinase [9], destabilizing the interaction between P53 and HDM2 (an E3-ubiquitin ligase). The sequences encoding $P14^{ARF}$ are embodied in three exons (designated 1β, 2, and 3), the exon 2 being common to $P14^{ARF}$ and $P16^{INK4A}$ mRNAs but translated according to an alternative open reading frame (ARF) [10]. P14 binds to HDM2 through an N-terminal domain encoded by exon 1β whereas two domains, one of which being encoded by exon 2, are involved in the nucleolar localization of P14 [11]. Thus P14^{ARF} sequesters HDM2 in the nucleolus and enables P53 levels to stabilize and rise in appropriate conditions (Figure 1).

P14^{ARF} protein has not been detected in human epidermis and cultured epidermal keratinocytes before and after UVB irradiation at various doses and irradiation regimens [12]. This is in agreement with reports which have shown that P14^{ARF} is not induced by DNA damaging agents [13].

Mutations uniquely targeting the P14^{ARF}-encoding gene are rare. More than 40% of the INK4A-ARF locus alterations affect both *P16^{INK4A}* and *P14^{ARF}* because they are located in exon 2. P14^{ARF} is functionally impaired in a small number of melanoma-prone kindreds [14]. UV-induced mutations of *P14^{ARF}* have been reported in skin carcinomas from sporadic patients and from XP patients [15]. Of the nine *P16^{INK4A}* mutants reported in sporadic skin carcinomas, seven (78%) had a change in the P14^{ARF} reading frame because it occurred in exon 2. All the XP-associated

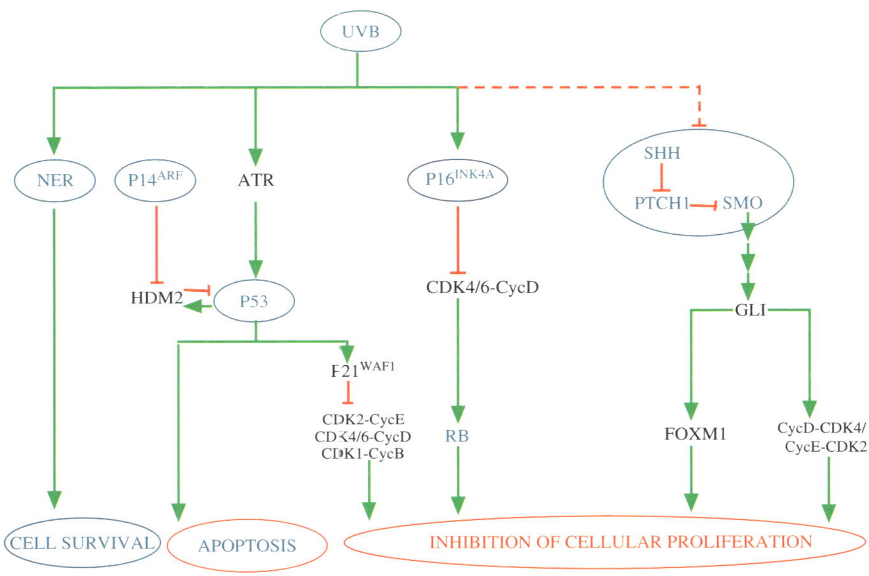

Figure 1. Functional signalling pathways in human epidermal keratinocytes. Several pathways: (*i*) P14^{ARF}; (*ii*) P53; (*iii*) P16^{INK4A}; and (*iv*) SHH/GLI are implicated in the control of proliferation, differentiation and/or apoptosis of keratinocytes. Effects of UVB irradiation on these different signalling pathway activities are depicted by stimulatory or inhibitory arrows. Frequencies of UVB-induced mutations in key gatekeeper genes governing these pathways in human skin: *P14^{ARF}*, *P53, P16^{INK4A}*, and *PTCH-1* are described, and their effects analyzed, in sections 3, 4, and 5, respectively. Abbreviations: ATR: Ataxia Telangiectasia and Rad3-related protein kinase; Cyc: Cyclin; CDK: Cyclin-Dependent Kinase; FOXM1: Forkhead Box M1; GLI: Glioma transcription factor family; HDM2: Human MDM2; NER: Nucleotide Excision Repair; PTCH: PATCHED; RB: Retinoblastoma tumour suppressor protein; SHH: Sonic Hedgehog; SMO: SMOOTHENED.

carcinomas with *P16*^{INK4A} mutations have also *P14*^{ARF} gene mutations. Thus P14 may be relevant in the tumorigenesis of non-melanoma skin cancers.

Compelling evidence shows that dysregulation of the P53 pathway is involved in the pathogenesis of basal cell carcinomas (BCCs) [2,16]. It is believed that point mutations on *TP53* are an early event in skin carcinogenesis since one can find them in pretumoral lesions for skin squamous cell carcinomas (SCCs) such as actinic keratoses [17]. Human sporadic skin cancers from NER-proficient patients are often associated with alterations of the *P53* gene (mutations in *TP53* occur in ~50% of BCCs), with the exception of melanomas which present low incidence (7%) of *TP53* mutations. In BCCs, SCCs and melanomas from XP-C patients, *TP53* is mutated in between 50 and 75% of cases, more than 50% of mutations corresponding to CC to TT tandem transitions [17,18]. The mutation spectrum is not specific to *TP53* because it is similar in *RAS* [19], *P16*^{INK4A} [15] and *P14*^{ARF} [15].

17.4 The P16^{INK4A} pathway

The normal function of this pathway is to mediate arrest of the cell cycle (Figure 1). P16^{INK4A} binds to and inhibits the cyclin D-dependent kinases CDK4 and CDK6, thereby preventing phosphorylation of the Retinoblastoma tumour suppressor protein (RB). Thus RB remains able to repress the transcription of genes whose products are required for DNA synthesis.

P16^{INK4A} expression is induced in UVB-irradiated human keratinocytes [12,20]. P16 accumulation does not take place at the transcriptional level in normal human primary cultured keratinocytes irradiated with different UVB doses and different irradiation regimens [12].

P16^{INK4A} is an important target gene in skin cancers [21]. In up to 20–30% of melanoma-prone families, disruption of the RB pathway occurs through inactivating mutations on the *P16*^{INK4A} gene [22]. In contrast, point mutations on *P16*^{INK4A} occur in only about 10% of sporadic melanomas with about the same proportion showing homozygous deletion of the gene [23].

Loss of heterozygosity (LOH) of *P16*^{INK4} has been reported in sporadic SCCs. UVB radiation-induced mutations of *P16*^{INK4A} have been found in 12% of sporadic skin carcinomas (including 24% of SCCs and 3.5% of BCCs). A higher level (28%) of UVB-induced mutations of *P16*^{INK4A} was found in skin carcinomas from XP-patients. 85% of these mutations were typically UVB-induced, including 54% of CC to TT tandem transitions. This frequency of tandem mutations is much higher than in sporadic skin carcinomas (25%) from NER-proficient patients [15].

17.5 The Sonic Hedgehog pathway

The crucial importance of a disrupted Sonic Hedgehog (SHH) signalling pathway for development of sporadic and hereditary BCCs is widely accepted. Although dysregulation of the P53 pathway is also involved in BCC tumorigenesis, the dominating findings relate to the SHH pathway [24].

Activation of the SHH pathway normally requires antagonism of the PATCHED-1 (PTCH-1) transmembrane protein by the SHH secreted protein ligand, thus releasing the transmembrane protein SMOOTHENED (SMO) for activation of target genes through the glioma (GLI) family of transcription factors (Figure 2). The actual mechanism of suppression of SMO activity by PTCH-1 in absence of SHH remains unclear despite several models [25].

The GLI1 and GLI2β transcription factors have a stimulatory effect on epidermal cell proliferation [26]. No proven human target genes of GLI transcription factors are known. However, it is established that: (*i*) the human *cyclin D2* promoter contains a GLI1 binding consensus element that interacts with GLI1 protein [27]; (*ii*) constitutive activation of *FOXM1* (*FORKHEAD BOX M1*) transcription is observed in GLI1 retrovirally-transduced primary human keratinocytes. FOXM1 is specifically up-regulated in BCCs, in contrast to normal skin and SCCs [28]. *PTCH-1* transcription is also induced by GLI1, thus generating a negative feedback loop [24].

Mutations of *SMO* and *SHH* have been detected in sporadic BCCs, although at a much lower frequency than *PTCH-1*, indicating that modifying one or more of the three proteins can lead to neoplasia. Skin SCCs never present mutations on *PTCH-1*,

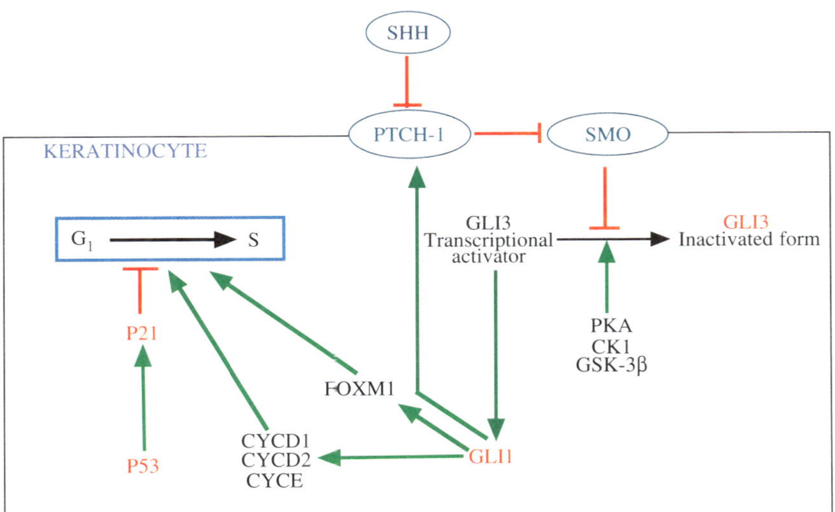

Figure 2. The SHH-PTCH-SMO complex of the Sonic Hedgehog pathway as a basal cell carcinoma gatekeeper. The function of the Sonic Hedgehog (SHH) pathway is to stimulate proliferation of epidermal keratinocytes. Activation of the SHH signalling pathway requires antagonism of PATCHED-1 (PTCH-1) by the secreted SHH ligand, thus releasing SMOOTHENED (SMO) for activation of target genes through the glioma (GLI) family of transcription factors. Among these target genes are *PTCH-1* and, potentially, *Cyclin D2* and *FOXM1*. LOH and mutations of *PTCH-1* are a hallmark of hereditary BCCs (Gorlin's syndrome), and frequently occur in sporadic BCC from NER-proficient and -deficient patients. PKA, CK1 and GSK-3β: cascade of phosphorylations of GLI3 catalyzed by Protein Kinase A, Casein Kinase 1 and Glycogen Synthase Kinase-3β.

suggesting that this gene is not expressed similarly in stem cells responsible for BCC and SCC development [4].

Genetic studies of the Gorlin's syndrome, which predisposes patients to the early development of multiple BCCs, have shown that mutational inactivation of *PTCH-1* is the primary event in the formation of the BCCs [29]. LOH in the *PTCH-1* locus was observed in 89% of hereditary BCCs. The majority (61–71%) of germline *PTCH-1* mutations are rearrangements. Most mutations (<80%) are likely to represent null mutations since they are predicted to result in truncation of PTCH-1 [30].

In sporadic BCCs from NER-proficient patients, the frequency of *PTCH-1* mutations ranges from 12 to 38% and LOH of markers on the region of chromosome 9q22.3 encompassing *PTCH-1* is observed in 50–75% of BCCs. 68% of *PTCH-1* alterations are point mutations and 32% are rearrangements. Thus in sporadic BCCs from normal individuals, the level of deletion/inversion events decreases by a factor of 2 compared to germline mutations [4,30]. The majority of *PTCH-1* mutations are also predicted to result in premature protein termination. In sporadic BCCs from NER-proficient patients, 38% of C to T transitions are found at bipyrimidine sites and 10% of UVB-specific CC to TT tandem transitions are detected. These two characteristic alterations demonstrate a major role for solar UVB in BCC tumorigenesis, but only for about 50% of the mutations [4]. Mutations of *SHH* and *SMO* were also detected in sporadic BCCs from normal individuals. Activating mutations on *SMO* have been found in 10–20% of BCC cases [4]. BCC formation occurs when mutation of regulatory genes leads to constitutive activation of downstream SHH signalling pathway responsive genes. GLI1 and GLI2β mRNAs are highly expressed in sporadic BCCs [26,31]. FOXM1 mRNA and protein are up-regulated in BCCs (n = 7), in contrast to SCCs (n = 4) and normal interfollicular epidermis [28].

A very high mutation frequency (73–90%) of the *PTCH-1* gene was found in BCCs from XP patients compared to NER-proficient patients [30,32,33]. A minority (11%) of these mutations are rearrangements. The majority of *PTCH-1* mutations (79%) are UVB-specific, 63% of which are CC to TT tandem transitions and 16% are C to T transitions at bipyrimidine sites [32,33]. The inability of XP patients to repair UVB-induced *PTCH-1* mutations might significantly contribute to the early and frequent appearance of BCCs observed in these patients. The majority of *PTCH-1* mutations in BCCs from XP patients belonging to the complementation group C (the most frequent in western world) are located on the non-transcribed strand of the gene. This was also the case for *TP53*, confirming the existence of preferential repair (TCR) in XP-C patients. Interestingly, in 50% of XP BCCs analysed, both *PTCH-1* and *P53* genes were mutated [32].

17.6 Conclusions and perspectives

Several target genes of the signalling pathways described herein have now been identified and some of their functions are consistent with control of cell cycle checkpoints, cellular proliferation, differentiation and survival (Figures 1 and 2).

Compelling evidence shows that dysregulation of P53 and SHH pathways are deeply implicated in the pathogenesis of BCCs. A set of members of the SHH signalling pathway have been shown to be directly and specifically involved in BCC development. It remains unclear how mutations in *SHH*, *PTCH-1*, and *SMO*, and consequent changes in SHH signalling target gene expression produce the abnormal proliferation, differentiation and adhesion characteristics of BCC tumour cells. Increased expression of GLI1 and GLI2β have been observed in BCCs and it has long been proposed that overexpression of GLI1 is a necessary step for development of BCCs. The precise roles played by the GLI transcription factors have not thus far been fully defined. The demonstration that *Cyclin D2* and *FOXM1* may be target genes in human keratinocytes requires additional work before a coherent understanding of skin carcinogenic events can be achieved.

Acknowledgements

We thank gratefully Dr. A. Stary for critical reading of the manuscript.

References

1. K.W. Kinzler, B. Vogelstein (1997). Cancer-susceptibility genes. Gatekeepers and caretakers. *Nature*, **386**, 761, 763.
2. A. Sarasin (1999). The molecular pathways of ultraviolet-induced carcinogenesis. *Mutat. Res.*, **428**, 5–10.
3. A. Stary, A. Sarasin (2002). The genetics of the hereditary xeroderma pigmentosum syndrome. *Biochimie*, **84**, 49–60.
4. L. Daya-Grosjean, A. Sarasin (2000). UV-specific mutations of the human patched gene in basal cell carcinomas from normal individuals and xeroderma pigmentosum patients. *Mutat. Res.*, **450**, 193–199.
5. K. Hemminki, G. Xu, L. Kause, L.M. Koulu, C. Zhao, C.T. Jansen (2002). Demonstration of UV-dimers in human skin DNA in situ 3 weeks after exposure. *Carcinogenesis*, **23**, 605–609.
6. D. Perdiz, P. Grof, M. Mezzina, O. Nikaido, E. Moustacchi, E. Sage (2000). Distribution and repair of bipyrimidine photoproducts in solar UV-irradiated mammalian cells. Possible role of Dewar photoproducts in solar mutagenesis. *J. Biol. Chem.*, **275**, 26732–26742.
7. J.L. Ravanat, T. Douki, J. Cadet (2001). Direct and indirect effects of UV radiation on DNA and its components. *J. Photochem. Photobiol. B*, **63**, 88–102.
8. B. Vogelstein, D. Lane, A.J. Levine (2000). Surfing the p53 network. *Nature*, **408**, 307–310.
9. R.T. Abraham (2001). Cell cycle checkpoint signaling through the ATM and ATR kinases. *Genes Dev.*, **15**, 2177–2196.
10. C.J. Sherr (2000). The Pezcoller lecture: cancer cell cycles revisited. *Cancer Res.*, **60**, 3689–3695.
11. P.A. Clark, S. Llanos, G. Peters (2002). Multiple interacting domains contribute to p14ARF mediated inhibition of MDM2. *Oncogene*, **21**, 4498–4507.

12. M. Chazal, C. Marionnet, L. Michel, K. Mollier, J.E. Dazard, V. Della Valle, C.J. Larsen, M.P. Gras, N. Basset-Seguin (2002). P16(INK4A) is implicated in both the immediate and adaptative response of human keratinocytes to UVB irradiation. *Oncogene*, **21**, 2652–2661.
13. F.J. Stott, S. Bates, M.C. James, B.B. McConnell, M. Starborg, S. Brookes, I. Palmero, K. Ryan, E. Hara, K.H. Vousden, G. Peters (1998). The alternative product from the human CDKN2A locus, p14(ARF), participates in a regulatory feedback loop with p53 and MDM2. *EMBO J.*, **17**, 5001–5014.
14. H. Rizos, S. Puig, C. Badenas, J. Malvehy, A.P. Darmanian, L. Jimenez, M. Mila, R.F. Kefford (2001). A melanoma-associated germline mutation in exon 1beta inactivates p14ARF. *Oncogene*, **20**, 5543–5547.
15. N. Soufir, L. Daya-Grosjean, P. de La Salmoniere, J.P. Moles, L. Dubertret, A. Sarasin, N. Basset-Seguin (2000). Association between INK4a-ARF and p53 mutations in skin carcinomas of xeroderma pigmentosum patients. *J. Natl. Cancer Inst.*, **92**, 1841–1847.
16. G. Ling, A. Ahmadian, A. Persson, A.B. Unden, G. Afink, C. Williams, M. Uhlen, R. Toftgard, J. Lundeberg, F. Ponten (2001). PATCHED and p53 gene alterations in sporadic and hereditary basal cell cancer. *Oncogene*, **20**, 7770–7778.
17. G. Giglia, N. Dumaz, C. Drougard, M.F. Avril, L. Daya-Grosjean, A. Sarasin (1998). p53 mutations in skin and internal tumors of xeroderma pigmentosum patients belonging to the complementation group C. *Cancer Res.*, **58**, 4402–4409.
18. A. Spatz, G. Giglia-Mari, S. Benhamou, A. Sarasin (2001). Association between DNA repair-deficiency and high level of p53 mutations in melanoma of Xeroderma pigmentosum. *Cancer Res.*, **61**, 2480–2486.
19. L. Daya-Grosjean, C. Robert, C. Drougard, H. Suarez, A. Sarasin (1993). High mutation frequency in ras genes of skin tumors isolated from DNA repair deficient xeroderma pigmentosum patients. *Cancer Res.*, **53**, 1625–1629.
20. S. Pavey, S. Conroy, T. Russell, B. Gabrielli (1999). Ultraviolet radiation induces p16CDKN2A expression in human skin. *Cancer Res.*, **59**, 4185–4189.
21. B. Bressac-de-Paillerets, M.F. Avril, A. Chompret, F. Demenais (2002). Genetic and environmental factors in cutaneous malignant melanoma. *Biochimie*, **84**, 67–74.
22. N. Soufir, N. Basset-Seguin (2001). [The INK4a-ARF locus: role in the genetic predisposition to familial melanoma and in skin carcinogenesis]. *Bull. Cancer*, **88**, 1061–1067.
23. M. Ruas, G. Peters (1998). The p16INK4a/CDKN2A tumor suppressor and its relatives. *Biochim. Biophys. Acta*, **1378**, F115–177.
24. A. Ruiz i Altaba, P. Sanchez, N. Dahmane (2002). Gli and hedgehog in cancer: tumours, embryos and stem cells. *Nat. Rev. Cancer*, **2**, 361–372.
25. J. Taipale, M.K. Cooper, T. Maiti, P.A. Beachy (2002). Patched acts catalytically to suppress the activity of Smoothened. *Nature*, **418**, 892–896.
26. G. Regl, G.W. Neill, T. Eichberger, M. Kasper, M.S. Ikram, J. Koller, H. Hintner, A.G. Quinn, A.M. Frischauf, F. Aberger (2002). Human GLI2 and GLI1 are part of a positive feedback mechanism in Basal Cell Carcinoma. *Oncogene*, **21**, 5529–5539.
27. J.W. Yoon, Y. Kita, D.J. Frank, R.R. Majewski, B.A. Konicek, M.A. Nobrega, H. Jacob, D. Walterhouse, P. Iannaccone (2002). Gene expression profiling leads to identification of GLI1-binding elements in target genes and a role for multiple downstream pathways in GLI1-induced cell transformation. *J. Biol. Chem.*, **277**, 5548–5555.
28. M.T. Teh, S.T. Wong, G.W. Neill, L.R. Ghali, M.P. Philpott, A.G. Quinn (2002). FOXM1 is a downstream target of Gli1 in basal cell carcinomas. *Cancer Res.*, **62**, 4773–4780.
29. J. Taipale, P.A. Beachy (2001). The Hedgehog and Wnt signalling pathways in cancer. *Nature*, **411**, 349–354.

30. M. Aszterbaum, A. Rothman, R.L. Johnson, M. Fisher, J. Xie, J.M. Bonifas, X. Zhang, M.P. Scott, E.H. Epstein Jr. (1998). Identification of mutations in the human PATCHED gene in sporadic basal cell carcinomas and in patients with the basal cell nevus syndrome. *J. Invest. Dermatol.*, **110**, 885–888.
31. J.M. Bonifas, S. Pennypacker, P.T. Chuang, A.P. McMahon, M. Williams, A. Rosenthal, F.J. De Sauvage, E.H. Epstein Jr. (2001). Activation of expression of hedgehog target genes in basal cell carcinomas. *J. Invest. Dermatol.*, **116**, 739–742.
32. N. Bodak, S. Queille, M.F. Avril, B. Bouadjar, C. Drougard, A. Sarasin, L. Daya-Grosjean (1999). High levels of patched gene mutations in basal-cell carcinomas from patients with xeroderma pigmentosum. *Proc. Natl. Acad. Sci. USA*, **96**, 5117–5122.
33. M. D'Errico, A. Calcagnile, F. Canzona, B. Didona, P. Posteraro, R. Cavalieri, R. Corona, I. Vorechovsky, T. Nardo, M. Stefanini, E. Dogliotti (2000). UV mutation signature in tumor suppressor genes involved in skin carcinogenesis in xeroderma pigmentosum patients. *Oncogene*, **19**, 463–467.

Chapter 18

Chaos theory and self-organized criticality describe the DNA damage signal transduction network

Daniel B. Yarosh

Table of contents

Abstract

Cellular responses to UV induced photoproducts in DNA are among the best understood signal transduction pathways in molecular biology. Damage to mammalian nuclear DNA initiates a cascade of signaling events reflecting the importance of timely and accurate responses to insults to the cell's metabolic and genetic controller. These responses stop cell cycling, activate and focus DNA repair machinery, increase expression of cytokines that communicate with other cells, and ultimately determine if the cell should commit suicide. Despite the vital nature of the signals, the molecular transduction pathways appear to be overly complicated, in defiance of the presumption that critical systems should be selected for simplicity.

The elements of the signal transduction pathways may provide a clue as to their origin and structure. The system is composed of proteins that change their binding specificity or activity based on interaction with other proteins. Chaos theory predicts that at a critical concentration of just such interacting elements, a complex network of pathways may spontaneously appear. The features of such interacting units at this "self-organized criticality" closely resemble the characteristics of DNA damage response signal transduction pathways. The self-organized criticality model of signal transduction helps explain the origin of the network, and suggests which interactions are most important for further study. The model also shifts the paradigm of signal transduction from linear pathways to network interactions.

18.1 Signal transduction after DNA damage

UV-induced DNA damage is among the most common challenges faced by mammalian cells, and mutations that result from its improper repair are responsible for the most common human cancer, skin cancer. The first cellular factor to encounter damage to the DNA template is often RNA polymerase II as it transcribes an active gene, and this leads to a stalled transcription fork [1]. The stalled fork then attracts the sequential loading of protein factors that eventually lead to the DNA damage response, including DNA repair [2]. Among these factors is a complex of nine TFIIH proteins that includes helicase and protein kinase activity [3]. It is the directed recruitment of these factors to the stalled transcription fork that likely explains the preferential repair of actively transcribed genes, and of the transcribed strand of such genes [4]. However, DNA damage is also encountered and repaired during DNA replication, and is detected and repaired in non-transcribed DNA.

The attraction of protein factors to the damaged DNA site is a trigger for signal transduction. At this focal point several proteins, such as p53, are phosphorylated by members of the DNA-dependent protein kinase family, DNA-PK, ATM, ATR and FRAP [5,6], as well as by the TFIIH kinase activity. The events that are triggered subsequent to this key moment of DNA damage recognition are being unraveled first in yeast, with homologs in the human cell [7]. What is clear is that at least four main responses are invoked: [3,8]

1. Cell cycling is paused - this involves the *cdk* family of protein kinases;
2. DNA repair is focused on the transcribed strand – this involves the TFIIH protein complex;
3. Apoptosis is initiated at some threshold of damage – this involves the p53 protein; and
4. Extracellular signaling molecules are released – this involves expression of the TNFα and IL-1 cytokine genes.

Proteins involved in this response network were organized into a molecular interaction map by Kohn [9] (Figure 1). The map was constructed from literature reports and is organized by function; it is therefore externally ordered. Nevertheless, one cannot avoid being astonished (and perhaps intimidated) by the complexity. The basic DNA damage response elements are represented. In segment G1 of Figure 1 is single-stranded DNA (ssDNA), which arises during stalled transcription or replication. This is related by protein interactions to the TFIIH transcription complex in segment C1, and the excision repair complex of proteins in segment H3-J3. These interact with the p53 protein (segment F7-F9). Cell cycling control proteins are also related to p53 by protein interactions (segment E1-F1).

This dauntingly complex pattern of protein-protein interactions has some clear characteristics:

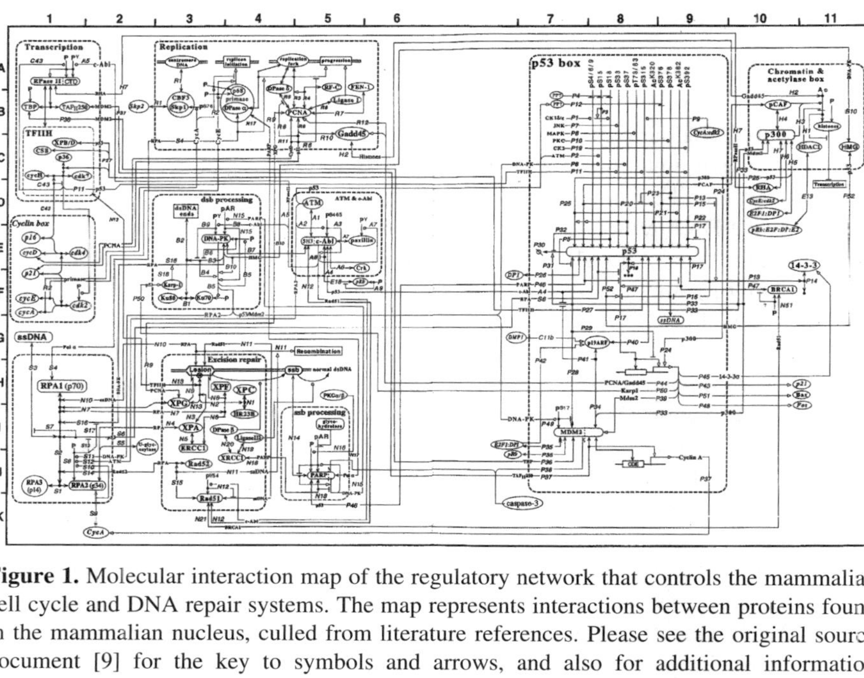

Figure 1. Molecular interaction map of the regulatory network that controls the mammalian cell cycle and DNA repair systems. The map represents interactions between proteins found in the mammalian nucleus, culled from literature references. Please see the original source document [9] for the key to symbols and arrows, and also for additional information contained in color-coding not visible in this reproduction. Reproduced with permission.

1. Proteins are allosterically controlled, i.e. they change their binding specificity or enzymatic activity when they themselves are bound or modified;
2. Proteins interact through long pathways of sequential protein-protein interactions;
3. Proteins interact through multiple, parallel pathways composed of different protein-protein interactions;
4. Pathways converge on some "node" proteins, such as p53, which have many more interactions than other proteins.

18.2 The problem of complexity

Our reductionist method of studying molecular biology is not well suited to systems with this complexity of interactions. For example, if a target protein that phosphorylates p53 is knocked-out by genetic manipulation and no change is observed in signal transduction, then can we conclude that the target protein does not participate in signal transduction? Or is it possible that a parallel pathway may compensate? This map makes it clear that we need a new way to test hypotheses in networks of pathways.

More troubling questions are: "How did this system of interacting proteins develop? Why does it have to be so complicated?" These concerns echo a long-standing criticism of Darwinism that intermediates in the evolution from simplicity to complexity are of no value. Of what adaptive value is half an eye? Molecular biologists have treated signal transduction as a descriptive science, and the problem of the origin of the networks remains unaddressed.

18.3 Chaos theory and self-organized criticality

Chaos theory is a field of mathematics in which the systems it describes are capriciously sensitive to their initial conditions. We are most interested in a portion of that theory that describes self-organization at critical concentrations in a non-equilibrium system – self-organized criticality. This self-organization occurs at a critical concentration about the phase transition. Self-organized criticality has been well described by Kaufman [10] and Bak [11].

Consider a system of reactants as in Figure 2 (see also ref. [10], Figure 3.4). When the concentration of reactants is low, the frequency of reactions is low and greatly dependent on the chaotic meeting of the reactants. At high concentrations of reactants, each possible reaction occurs immediately and all the time, which is represented by the crystal state. However, as the concentration of reactants increases to a critical concentration a phase transition occurs. Systems that operate with this critical concentration range may self-organize into pathways that are necessarily complex. What is amazing is that the self-organization and complexity are based on very simple reactions.

The study of complex systems arising from simple reactions has been described by Wolfram [12]. Very simple rules controlling the reaction of arbitrary units,

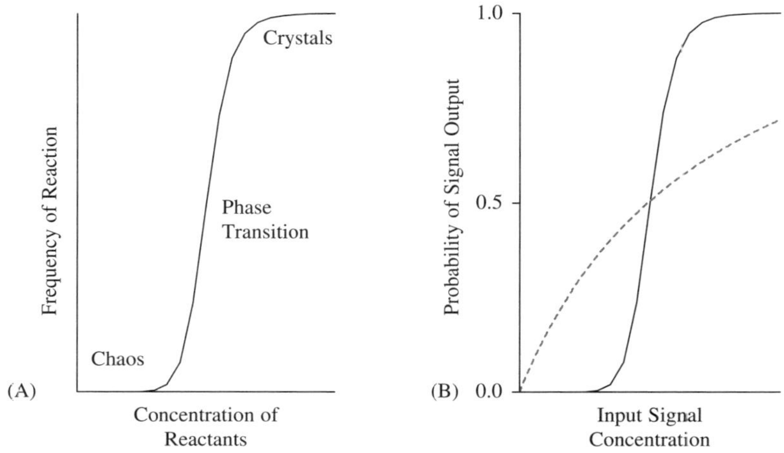

Figure 2. (A) Diagram of the phase transition state between chaos and crystals. A plot of the frequency of reactions as the concentration of reactants increases. The states are described as chaos at low concentrations, crystals at high concentration, and a phase transition state at a critical concentration in between. (B) Model of signal transduction by node proteins. The probability of signal output is plotted against the concentration of signal input. Solid black line: the output signal requires 11 input signals; Dotted gray line: the output signal requires 1 input signal. Both reactions follow the equation $N(s) = (s^{nH})/(K_m+s^{nH})$, where N(s) is the probability of signal output, s is the concentration of signal inputs, K_m is the Michaelis constant for the reaction, and nH is the number of input signals required to trigger an output signal (from [22]).

called cellular automata, can lead to very complex patterns. A rule can be as simple as "if a certain cell reacts one way, and another cell reacts a second way, then a third cell reacts the second way; otherwise the third cell remains the same." Complex patterns arise spontaneously from some of these simple rules, even though no one unit controls the pattern and no one has designed the complexity. Some of these simple rules produce a stable, complex pattern after a few rounds of rule application, even if the starting conditions are random. This is extremely important, because it means that a system following such rules will spontaneously reach a complex state, and if it is perturbed it will automatically return to its stable, complex state.

These units might be considered as reactants in a signal transduction cascade. The reactants act like cellular automata because they are modified by other reactants and then react with further reactants, all based on biochemical rules. For example, "if protein A is phosphorylated at one site, but unphosphorylated at a second site, then protein A will phosphorylate protein B; otherwise protein A will do nothing." Signal transduction reactions, such as phosphorylation, acetylation and protein-protein binding, are well-known biochemical reactions that fit this model of reactant activity. Allosteric control of enzyme activity is also a well-known phenomenon, where binding or modification of an enzyme changes its activity.

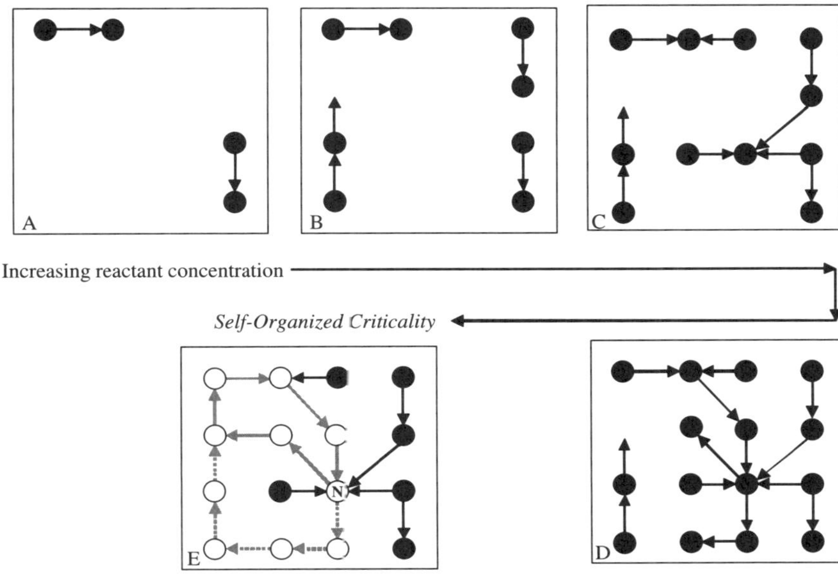

Figure 3. Schematic representation of the origin of self-organization from increasing reactant concentration. As the concentration of reactants increases (Panel A to D), the frequency of interactions increases, until the system reaches self-organized criticality (Panel E). At this concentration there are long pathways (solid gray lines), parallel pathways (solid and dotted lines) and signal transduction nodes (N) with high numbers of interactions with other proteins.

Allosteric control is precisely the type of "rule-making" that leads to complexity based on very simple rules.

The self-organization of complex pathways at a critical concentration of biomolecules is illustrated in Figure 3 (see also ref. [10], Figure 3.3). At low concentration of reactants, the numbers of reactions that occur are much less than the number of reactants, and those that do are randomly selected. However, as the concentration of reactants increases, the frequency of reactions increases. At the phase transition, where the number of reactants is about twice the number of reactions, chains of reactions spontaneously arise, and long chains of self-organizing reactions spontaneously arise.

This protein interaction network is a complex of protein activity following rules set by the allosteric interactions, and has the following properties:

1. Two proteins interact through a long chain of intermediates in a signal transduction pathway (solid gray line);
2. Parallel pathways may connect the same two proteins (solid and dotted gray line);
3. Some proteins are at the convergence of multiple pathways.

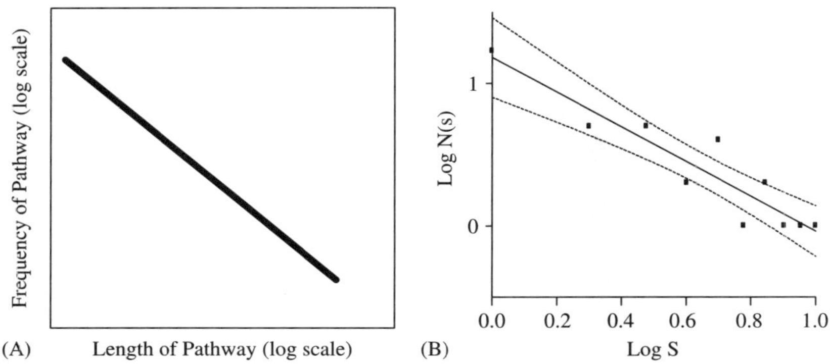

Figure 4. The Power law in signal transduction networks. (A) At self-organized criticality, the length of the reaction pathway is linearly related to its frequency, on a log-log plot, according to the equation $N(s) = s^{-\tau}$. (B) Application of the Power law to the protein-protein interaction network connected by SH3 domains in yeast. The frequency of the number of connections [N(s)] as a function of the number of connections (S), on a log-log plot. The solid line is a linear regression plot ($\tau = 1.2$), and the dotted lines are the 95% confidence interval (goodness of fit: $r^2 = 0.87$). The data is taken from Figure 3B of ref. [20].

A fourth important property is that the distribution of protein-protein interactions follows the power law, $N(s) = s^{-\tau}$, where 'N(s)' is the frequency of an event and 's' is the size of the event. At self-organized criticality, the log-log plot of the size of the longest reaction pathway is linearly related to its frequency, with a slope of τ (Figure 4). Simply put, long reaction chains are much less likely than short reactions. It also follows that increasing the length of reaction pathways among a limited number of reactants means that reactants must participate in increasing number of pathways. Therefore, the frequency of reactions per reactant also follows the power law, and this is a diagnostic feature of systems at self-organized criticality.

18.4 Signal transducers

The reactions in the DNA damage signal transduction network follow simple rules, consistent with cellular automata rules for generating complexity. They are generally protein-protein binding interactions, in which a protein is bound and modified, for example, by phosphorylation. Protein kinases are the largest single enzyme family in the human genome, with 2,000 genes, and their targets are the approximately 30% of all human proteins that covalently bind phosphate [13]. This modification then alters the enzymatic or binding activity of the target protein, so that the signal is propagated, such as in a protein kinase cascade. The self-organized network hypothesis requires that proteins participate in both input and output reactions, and in fact, on average a cellular protein has between 2 and 10 interactions with other proteins [14], although not every interaction is available to a protein at any one instant.

These reaction rules are further simplified because the recognition between reactants is controlled by short, defined peptide motifs that appear repeatedly within the amino acid sequence of signaling molecules [15]. The protein recognition domain SH2, for example, binds to protein sites containing a phosphorylated tyrosine. The SH3 domain binds to proline-rich sequences, and the FHA (fork-head assembly) domain recognizes phosphothreonine-containing sites. The family of protein kinases was likely built up by duplication of an enzymatic kinase domain and recombination with a variety of protein recognition domains [15]. Thus many signaling proteins follow the simple rule that if it becomes phosphorylated in its recognition domain, the kinase domain is activated and phosphorylates the recognition domain in its target protein. Phosphatases, which also are governed by these protein recognition motifs, then stop the signaling by removing the phosphate.

These simple rules are at work in the complex DNA damage response pathway. For example, UV-induced cyclobutane pyrimidine dimers activate DNA-protein kinases from the family of phosphoinisitol-kinase related kinases, which include DNA-PK, ATM, ATR and FRAP (Mec1 and Tel1 in yeast). These kinases then phosphorylate p53, and the rad9 protein in yeast, at threonines within the target sequence for FHA binding [16]. The Chk2 protein in humans (yeast Rad53 protein) has an FHA domain that then binds these target proteins and activates its kinase activity, which controls cell cycle checkpoint progression [17]. Rad53 itself becomes phosphorylated at a threonine moiety within the FHA-target sequence, and then binds the FHA domain of the yeast protein Dun-1. This binding by the Dun-1 FHA domain is essential to the transcriptional induction of DNA repair genes in yeast [18].

A second example involves the phosphorylation by the DNA-protein kinase family of serine residues in target proteins. Following DNA damage, these kinases phosphorylate target proteins, including p53 and the phosphatase cdc25, at serines within key recognition sequences. This creates a binding site for the 14–3-3 scaffolding protein, which then sequesters the phosphatase activity in the cytoplasm, leaving cdc2 phosphorylated and blocking cell cycle progression [19].

Many other proteins, such as the scaffolding protein BRCA-1, become phosphorylated after UV-induced DNA damage but the role of recognition domains in these interactions has not been worked out. Some recognition domains, such as BARD and BRCT-1, occur repeatedly in DNA damage response proteins, but their function is not completely understood. Finally, the phosphorylation potential of most DNA repair proteins is not known. All of this is a fruitful area for research.

Simple rules for protein interactions build up complex networks. Tong et al. constructed a network map in yeast of protein-protein interactions governed only by SH3 domains, and confirmed by both phage-display and two-hybrid analysis [20]. The network contains 39 proteins with between 1 and 10 interactions, and follows the Power law (Figure 4B). This is consistent with the hypothesis that a signaling network controlled by protein recognition domains and following cellular automata rules arose by self-organization.

18.5 Nodes

A characteristic of signaling networks both predicted by self-organized criticality and observed in cellular pathways is the occasional node protein that interacts with many other proteins. An example of such a protein is the p53 protein, which has 11 sites for phosphorylation and acetylation, and in turn binds to at least 15 other proteins and also DNA sequences to alter transcription and translation.

After DNA damage and release from MDM-2, the p53 population swells and samples the inputs from the incoming kinase and acetylation pathways. Considering all the possible combinations, the p53 protein can exist in a cell in any one of about 2,000 conformations. The p53 pool may therefore act as an integrated information processor, calculating the precise state of damage in the cell [9].

Information processing by a node protein is important because it serves to validate incoming signals and reduce erroneous transduction that may be catastrophic for a cell, such as transduction of a signal for apoptosis. For example, if a linear apoptotic pathway has 5 steps, and each step has an error frequency of 1%, then the whole pathway has a 5% chance of spontaneously triggering cell death. However, if a node protein requires input from 5 pathways before transducing the signal, and each pathway has a 5% error rate, then the chance of a wrongful cell death is now reduced to 0.00003%. Another example is a cell cycle checkpoint protein that halts cell cycling until it receives signals that adequate nutritional, genetic and environmental conditions are met [21].

The processing of cellular information by a node protein may explain how a cell knows when the UV damage it has sustained is too great for DNA repair and that apoptosis is necessary. The switch may occur as p53 molecules accumulate modifications from input phosphorylation and acetylation pathways toward a critical number of modifications required to trigger a response. Figure 2B shows that the p53 output reaction becomes a trigger when eleven modifications are required, compared to a continuous output when a single modification is required. Nash et al. [22] have described a similar system of 6 phosphorylation sites in the degradation of Sic1, a cell cycle inhibitor. Since these curves show the sharp inflection with as few as 5 required inputs, experimental elimination of any single modification site or a single upstream kinase is not likely to reveal much about the overall signaling process of p53. A node protein converts many simple modifications into an information computation that switches on signal transduction at a threshold of protein modification.

Networks that have multiple pathways connecting nodes have a surprising stability and resistance to local failures, but such systems are vulnerable to severe disruption by malfunctioning of one of a few key nodes at the extreme of connectivity [23]. The prediction follows that in the DNA damage network, the most connected proteins will be tumor suppressor genes, where a loss of function leads to massive disruption of DNA repair, leading to genome instability. Two of the node proteins identified so far, p53 and BRCA-1, are in fact products of tumor suppressor genes, and we expect that as more are identified they will be highly connected proteins, and that proteins which are found to be highly connected to others are very likely products of tumor suppressor genes.

18.6 DNA damage response network

Recently, Ho et al. [24] constructed protein interaction maps in *S. cerevisiae*, based on the mass spectrometric identification of proteins immunoprecipitated with "bait" proteins. One diagram (Figure 5) shows interactions among proteins binding to 86 proteins implicated in DNA damage responses that bear striking resemblances to the characteristics of a self-organized network. Proteins interact through long, multistep pathways, and parallel pathways connect some proteins. Pathways converge on key node proteins, which are the focal proteins in key DNA damage responses such as nucleotide excision repair (e.g., Rad1), TFIIH transcription/repair controls (e.g., Rad3), and cell cycle checkpoint protein kinases (e.g., Dun1). The authors note that the connectivity distribution among the data set of 1,578 different proteins follows the power law [24].

These protein interaction approaches, as well as techniques such as gene chip array, capture a snapshot of dynamic cell functions and may miss key responses that unfold during response to DNA damage. However, the data to date are consistent

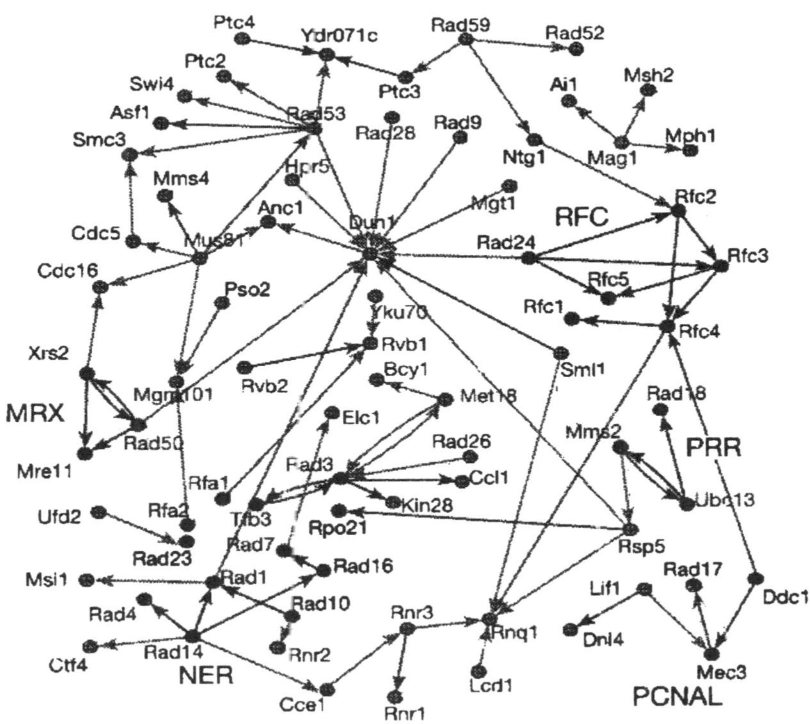

Figure 5. DNA damage response network in *S. cerevisiae*. Interactions of proteins related to DNA repair, transcription and cell cycling, detected by high-throughput mass spectrometric protein complex identification, from ref. [24] (*n.b.*: not all detected interactions with all components are shown, and not all interactions are direct). [Reprinted by permission from Nature, 10 January 2002, pg. 181, Figure 2A, copyright 2002 Macmillan Publishers Ltd.]

with self-organized pathways as described here, and no doubt the principles of self-organized criticality help us to understand and evaluate additional data in signal transduction as it is gathered. In an analogous cellular network, that of enzymes that perform core metabolic functions, the enzymes are the nodes, the substrates are the connecting signals, and the connectivity distribution analyzed among 43 organisms follows the power law [25]. It may be that the self-organization of DNA damage signal transduction is part of a larger pattern of the structure of complex networks in cell biology [26]. Here we seek to identify the larger pattern, and then correlate its theoretical units with the actual proteins participating in DNA repair signal transduction, for the purpose of uncovering and understanding new proteins and protein interactions.

18.7 Summary

DNA damage responses are characterized by long, serial and parallel signal transduction pathways that converge on key node proteins to modulate and direct the overall reaction. The components of these signal transduction pathways, however, are simple protein-protein interactions controlled by minor modifications such as phosphorylation and protein binding. The origin of signal transduction complexity from these simple reactions is modeled by the theory of self-organized criticality. One prediction of the model is that the connectivity of the networks follows the power law, and there is experimental data supporting this prediction. A second prediction is that highly connected nodes will be products of tumor suppressor genes, and that many tumor suppressor genes will code for highly connected nodes, and the examples of p53 and BRCA-1 also support this. It is likely that self-organization is a characteristic of all signal transduction pathways, including signal transduction mediated by membrane receptor binding, and may well be a principle of many biological networks.

This model implies that signal transduction arose *de novo* as complex pathways and did not evolve from more simple pathways. A third prediction is that it is unlikely that we will find evolutionary intermediates of these networks in any organisms. As our understanding increases we will find that we cannot ultimately dissect signal transduction into linear reactions. We need to develop new ways of designing experiments to test network responses.

References

1. D. Tantin (1998). RNA polymerase II elongation complexes containing the Cockayne syndrome group B protein interact with a molecular complex containing the transcription factor IIH components xeroderma pigmentosum B and p62. *J. Biol. Chem.*, **273**, 27794–27799.
2. M. Volker, M.J. Moné, P. Karmakar, A. van Hoffen, W. Schul, W. Vermeulen, J.H.J. Hoeijmakers, R. van Driel, A.A. van Zeeland, L.H.F. Mullenders (2001). Sequential assembly of the nucleotide excision repair factors *in vivo*. *Mol. Cell*, **8**, 213–224.

3. P. Frit, E. Bergmann, J.M. Egly (1999). Transcription factor IIH: a key player in the cellular response to DNA damage. *Biochimie*, **81**, 27–38.
4. V. Bohr (1991). Gene specific repair. *Carcinogenesis*, **12**, 1983–1992.
5. C.A. Hall-Jackson, D.A.E. Cross, N. Morrice, C. Smythe (1999). ATR is a caffeine-sensitive, DNA-activated protein kinase with a substrate specificity distinct from DNA-PK. *Oncogene*, **18**, 6707–6713.
6. D. Yarosh, P. Cruz, I. Dougherty, N. Bizios, J. Kibitel, K. Goodtzova, D. Both, S. Goldfarb, B. Green, D. Brown (2000). FRAP DNA-dependent protein kinase mediates a late signal transduced from ultraviolet-induced DNA damage. *J. Invest. Dermatol.*, **114**, 1005–1010.
7. J. Rouse, S.P. Jackson (2002). Interfaces between the detection, signaling and repair of DNA damage. *Science*, **297**, 547–551.
8. J. Kibitel, V. Hejmadi, L. Alas, A. O'Connor, B.M. Sutherland, D. Yarosh (1998). UV-DNA damage in mouse and human cells induces the expression of tumor necrosis factor α. *Photochem. Photobiol.*, **67**, 541–546.
9. K. Kohn (1999). Molecular interaction map of the mammalian cell cycle control and DNA repair systems. *Molec. Biol. Cell*, **10**, 2703–2734.
10. S. Kaufman (1995). *At Home in the Universe*. Oxford University Press Inc., New York, NY.
11. P. Bak (1996). *How Nature Works*. Springer-Verlag, New York Inc., New York, NY.
12. S. Wolfram (2002). *A New Kind of Science*. Wolfram Media, Inc., Champaign, IL.
13. S.B. Ficarro, M.L. McCleland, P.T. Stukenberg, D.J. Burke, M.M. Ross, J. Shabanowitz, D.F. Hunt, F.M. White (2002) Phosphoproteome analysis by mass spectrometry and its application to Saccharomyces cerevisiae. *Nature Biotech.*, **20**, 301–305.
14. G. Bader, I. Donaldson, C. Wolting, B.F.F. Ouellette, T. Pawson, C.W.F. Hogue (2001). BIND – The bimolecular interaction network database. *Nucl. Acids Res.*, **29**, 242–245.
15. T. Pawson, G. Gish, P. Nash (2001). SH2 domains, interaction modules and cellular wiring. *TRENDS Cell Biol.*, **11**, 504–511.
16. D. Durocher, J. Henckel, A R. Fersht, S.P. Jackson (1999). The FHA domain is a modular phosphopeptide recognition motif. *Molec. Cell*, **4**, 387–394.
17. J. Falck, C. Lukas, M. Protopopova, J. Lukas, G. Selivanova, J. Bartek (2001). Functional impact of concomitant versus alternative defects in the Chk2-p53 tumor suppressor pathway. *Oncogene*, **20**, 5503–5510.
18. A. Hammet, B.L. Pike, K.I. Mitchellhill, T. Teh, B. Kobe, C.M. House, B.E. Kemp, J. Heierhorst (2000). FHA domain boundaries of the dun1p and rad53p cell cycle checkpoint kinases. *FEBS Lett.*, **14**, 141–146.
19. T. Pawson, J.D. Scott (1997). Signaling through scaffold, anchoring, and adaptor proteins. *Science*, **278**, 2075–2080.
20. A.H.Y. Tong, B. Drees, G. Nardelli, G.D. Bader, B. Brannetti, L. Castagnoli, M. Evangelista, S. Ferracuti, B. Nelson, S. Paoluzi, M. Quondam, A. Zucconi, C.W.V. Hogue, S. Fields, C. Boone, G. Cesareni (2002). A combined experimental and computational strategy to define protein interaction networks for peptide recognition modules. *Science*, **295**, 321–324.
21. H. McAdams, A. Arkin (1999). It's a noisy business! Genetic regulation at the nanomolar scale. *Trends Genet.*, **15**, 65–69.
22. P. Nash, X. Tankg, S. Orlicky, Q. Chen, F. Gertier, M. Mendenhall, F. Sicheri, T. Pawson, M. Tyers (2001). Multisite phosphorylation of a CDK inhibitor sets a threshold for the onset of DNA replication. *Nature*, **414**, 514–521.
23. R. Albert, H. Jeong, A.-L. Barabasi (2000). Error and attack tolerance of complex networks. *Nature*, **406**, 378–382.

24. Y. Ho, A. Gruhler, A. Heilbut, G. Bader, L. Moore, S.L. Adams, A. Millar, P. Taylor, K. Bennett, K. Boutilier, L. Yang, C. Wolting, I. Donaldson, S. Schandroff, J. Shewnarane, M. Vo, J. Taggart, M. Goudreault, B. Muskat, C. Alfarano, D. Dewar, Z. Lin, K. Michalickova, A. Willems, H. Sassi, P. Nielson, K. Rasmussen, J. Andersen, L. Johansen, L. Hansen, H. Jespersen, A. Podtelejnikov, E. Nielson, J. Crawford, V. Poulsen, B. Sorensen, J. Matthiesen, R. Hendrickson, F. Gleeson, T. Pawson, M. Moran, D. Durocher, M. Mann, C. Hogue, D. Figeys, M. Tyers (2002). Systematic identification of protein complexes in Saccharomyces cerevisiae by mass spectrometry. *Nature*, **415**, 180–183.
25. H. Jeong, B. Tombor, R. Albert, Z. Ottvai, A.-L. Barbasi (2000). The large-scale organization of metabolic networks. *Nature*, **407**, 651–654.
26. S. Maslov, K. Sneppen (2002). Specificity and stability in topology of protein networks. *Science*, **296**, 910–913..

Subject Index